Technological Forecasting for Decision Making

Second Edition

Technological Forecasting for Decision Making

Second Edition

Joseph P. Martino
University of Dayton Research Institute, Ohio

North-Holland
New York • Amsterdam • Oxford

Elsevier Science Publishing Co., Inc.
52 Vanderbilt Avenue, New York, New York 10017

Sole distributors outside the United States and Canada:

Elsevier Science Publishers, B.V.
P.O. Box 211, 1000 AE Amsterdam, The Netherlands

Library of Congress Cataloging in Publication Data

Martino, Joseph Paul, 1931–
Technological forecasting for decision making.

Includes bibliographical references and index.
1. Technological forecasting. 2. Decision–making. I. Title.
T174.M38 1983 658.4′0355 82-12449
ISBN 0-444-00722-9

Manufactured in the United States of America

This book is dedicated to
my wife, Mary,
whose constant encouragement
and support made it possible.

Contents

Preface to the Second Edition xiii
Preface to the First Edition xv

Chapter 1. Introduction **1**

1. What Is Technological Forecasting? 1
2. Why Forecast Technology? 4
3. Alternatives to Forecasting 5
4. Will It Come True? 8
5. Stages of Innovation 9
6. Remainder of the Book 11

References 12
For Further Reference 12
Problems 12

Chapter 2. Delphi **14**

1. Introduction 14
2. Advantages of Committees 15
3. Disadvantages of Committees 15
4. The Delphi Procedure 16
5. Conducting a Delphi Sequence 17
6. Variations on Delphi 20
7. Delphi as a Group Process 22
8. The Precision of Delphi 24
9. The Reliability of Delphi 25
10. Selecting Delphi Panel Members 26
11. Guidelines for Conducting a Delphi Sequence 29
12. Constructing Delphi Event Statements 33
13. Summary 35

References 36
For Further Reference 36
Problems 36

Chapter 3. Forecasting by Analogy **39**

1. Introduction 39
2. Problems of Analogies 39
3. Dimensions of Analogies 40
4. Deviations from a Formal Analogy 49
5. Summary 50

References 51
Problems 51

Chapter 4. Growth Curves **53**

1. Introduction 53
2. Substitution Curves 54
3. The Pearl Curve 57
4. The Gompertz Curve 58
5. Choosing the Proper Growth Curve 58
6. The Base 10 Pearl Curve 59
7. The Fisher–Pry Curve 60
8. Estimating the Upper Limit to a Growth Curve 61
9. Selecting Variables for Substitution Curves 62
10. An Example of a Forecast 63

References 66
For Further Reference 67
Problems 67

Chapter 5. Trend Extrapolation **69**

1. Introduction 69
2. Exponential Trends 70
3. An Example of a Forecast 73
4. A Model for Exponential Growth 73
5. Nonexponential Growth 79
6. Qualitative Trends 81
7. A Behavioral Technology 81
8. Summary 84

References 84
For Further Reference 85
Problems 85

Chapter 6. Measures of Techology **86**

1. Introduction 86
2. Scoring Models 88
3. Constrained Scoring Models 93

4. Planar Tradeoff Surfaces 94
References 97
Problem 97

Chapter 7. Correlation Methods **98**

1. Introduction 98
2. Lead–Lag Correlation 98
3. Technological Progress Function 103
4. Maximum Installation Size 104
5. Correlation with Economic Factors 106
References 108
Problems 109

Chapter 8. Causal Models **110**

1. Introduction 110
2. Technology-Only Models 111
3. A Technoeconomic Model 117
4. A Simulation Model 121
5. Summary 126
References 127
For Further Reference 127
Problems 127

Chapter 9. Forecasting Breakthroughs **129**

1. Introduction 129
2. Examples of Breakthroughs 130
A. Atomic Energy 130
B. The Transistor 133
C. Penicillin 133
3. Monitoring for Breakthroughs 134
4. Where to Look for Signals 136
5. An Example 138
6. Summary 142
References 142
For Further Reference 143
Problems 143

Chapter 10. Combining Forecasts **145**

1. Introduction 145
2. Trend and Growth Curves 145
3. Trend and Analogy 146
4. Components and Aggregates 147
5. Scenarios 148
6. Cross Impact Models 150
7. Summary 155

Reference 155
For Further Reference 155
Problems 156

Chapter 11. Normative Methods **159**

1. Introduction 159
2. Relevance Trees 159
3. Morphological Models 163
4. Mission Flow Diagrams 165
5. Summary 167

Reference 168
For Further Reference 168
Problems 169

Chapter 12. Planning and Decision Making **171**

1. Introduction 171
2. The Purpose of Planning 172
3. The Role of the Forecast 174
4. Decision Making 175
5. Summary 177

Problems 177

Chapter 13. Technological Forecasting for Research and Development Planning **178**

1. Introduction 178
2. Research 179
3. Technology Advancement 182
4. Product Development 184
5. Testing and Evaluation 187
6. Summary 189

For Further Reference 189
Problems 190

Chapter 14. Technological Forecasting in Business Decision **192**

1. Introduction 192
2. What Business Are We In? 193
3. Business Planning for Technological Change 195
4. Secondary Impacts 199
5. Summary 200

For Further Reference 200
Problems 201

Chapter 15. Technological Forecasting in Government Planning 202

1. Introduction 202
2. Internal Operations of Government 202
3. Regulatory Agencies 204
4. Public Goods 207
5. Changes in the Form of Government 210
6. Summary 212

For Further Reference 212
Problems 213

Chapter 16. Technology Assessment 215

1. Introduction 215
2. Some Historical Cases 216
3. The Role of Technological Forecasting 221
4. An Example 222
5. Summary 224

For Further Reference 225
Problems 225

Chapter 17. Some Common Forecasting Mistakes 226

1. Introduction 226
2. Environmental Factors That Affect Forecasts 229
3. Personal Factors That Affect Forecasts 235
4. Core Assumptions 240
5. Summary 246

References 247
Problems 247

Chapter 18. Evaluating Forecasts as Decision Information 250

1. Introduction 250
2. The Interrogation Model 251
3. Summary 259

Problems 259

Chapter 19. Presenting the Forecast 261

1. Introduction 261
2. Making the Forecast Useful 262
3. Making the Forecast Credible 271
4. A Checklist for Forecasts 278
5. Summary 281

References 282
For Further Reference 282
Problems 282

Appendix 1. Regression Analysis of Time Series **285**
1. Introduction 285
2. Statistical Analysis 286
3. Simple Regression 299
4. Parabolic Regression 312
5. Multiple Linear Regression 318
6. Summary 326
References 326
Problems 327

Appendix 2. Statistical Tables **329**

Appendix 3. Historical Data Tables **333**

Appendix 4. Computer Programs **373**
1. Introduction 373
2. KSIM 373
3. Growth 376
4. Regress 379

Index **383**

Preface to the Second Edition

Since the first edition of this book there have been significant changes in the state of the art of technological forecasting. These include refinements and improvements on older techniques, as well as some completely new techniques. In addition, there has been a change in emphasis among techniques; for instance, a decade ago computer models were hardly used, whereas their use is now widespread. This new edition brings together the new techniques and the changed emphases and integrates them with the older techniques.

Another important change since the first edition is the increased use of technological forecasting for a variety of applications. It has become widely accepted in industry, government, and universities, and the chapters on applications have been updated to reflect this.

Finally, this second edition has benefited from considerable feedback from users, both individual readers and those who have used it as a text in formal classes. Some of the background material and historical illustrations have been shortened or eliminated to sharpen the focus on the more important points. In addition, lengthy derivations have been omitted where they did not contribute to an understanding of the techniques presented. Readers can refer to the original literature to find these derivations, and those interested only in applications of techniques will find them presented more compactly here.

Joseph P. Martino

Preface to the First Edition

This book is intended to provide an advanced treatment of technological forecasting. While no prior acquaintance with the subject is assumed, the treatment does assume a fairly strong background in one of the sciences or one of the engineering disciplines, and its presentation some prior knowledge of calculus. Some knowledge of statistics is also desirable, but a statistical appendix is provided for those with no prior acquaintance with the discipline.

This is a "how to" book: how to do technological forecasting, how to apply it in specific decision situations, and how to avoid some of the more common errors and difficulties in the preparation of individual forecasts. It is written primarily for those—graduates and undergraduates—who wish to become technological forecasters. It is also intended for professional technological forecasters who wish to know more about the subject. In addition, the book is intended for those practicing scientists and engineers who wish to know more about technological forecasting, for administrators, in the scientific and engineering fields, and for other decision makers whose work is heavily influenced by technological change.

For the student of forecasting as well as the professional forecaster, the book presents a detailed discussion of all the important techniques currently in use. For each technique, the discussion includes the rationale behind it, a description of the methodology including mathematical derivations where these are appropriate, and one or more historical examples. In addition, the proper conditions for application of each technique are described, and the strengths and weaknesses of each technique discussed. Each chapter concludes with several problems intended to give the user an opportunity to practice the methods covered therein.

For the practicing scientist or engineer, the book provides information he will need to understand technological forecasting. There are three reasons why he will need such understanding. First, more and more scientists and engineers are going to be called upon to participate in the making of forecasts in the areas in which they are well informed. They should understand the methods that will be used by the professional forecasters with whom they will be required to collaborate. Second, technological forecasting will be increasingly used in the planning of scientific and technological activity. It is advantageous for them, therefore, to understand the bases upon which decisions will be made about their work. Third, technological forecasting can be of value in expanding the application of their work. Since most scientists and engineers are interested in seeing their work applied, technological forecasts can be of direct benefit to them. Regardless of which of these reasons is of most importance in a specific case, the practicing scientist or engineer will find the information in this book to be of value in both making and using forecasts.

Administrators in science and engineering and other decision makers whose work will be influenced by technological change will not usually be making detailed technological forecasts. Forecasts will be prepared for them by technical specialists and professional forecasters. Nevertheless, it is important for these decision makers to be aware of the methods commonly used, to know when these methods are appropriate and when they are not, and particularly to know the strengths and weaknesses of the various methods. This book can provide them with adequate background to make fully effective use of the forecasts they will receive.

The material in the book is divided into three parts. The first part contains a discussion of forecasting methods; the second describes applications of technological forecasting to various types of decisions; the third provides instruction and guidance for the preparation and presentation of specific forecasts to be utilized in specific decisions. The student of forecasting and the professional forecaster will find the first and third parts most useful; they will draw on the second part as the occasion demands. Other users will probably make most frequent use of the second part; they will perhaps consult the first and third parts for specific information about forecasting methods or for evaluating forecasts they have been given.

This book differs from most others on the subject in two respects. First, it maintains continuity and orderly growth of the subject; the material is presented in sequence of increasing difficulty and increasing rigor. Second, it stresses the viewpoint that technological forecasts are not made for their own sake but as inputs to decisions. The book emphasizes that technological forecasting is an important element in the decision making process. However, the distinction between the role of the forecaster who *advises* a decision maker and that of the decision maker himself is repeatedly emphasized.

Joseph P. Martino

Technological Forecasting for Decision Making

Second Edition

Chapter 1

Introduction

1. What Is Technological Forecasting?

This book is about technological forecasting: How to do it, how to apply it in specific situations, and how to prepare forecasts so that they can be useful from the decision-making standpoint.

Before discussing technological forecasting we must first define *technology*. *Webster's Seventh Collegiate Dictionary* defines it as "the totality of the means employed to provide objects necessary for human sustenance and comfort." This definition is adequate for our purposes, provided that we understand *objects* to include not only goods but services. Thus technology here will mean the tools, techniques, and procedures used to accomplish some desired human purpose; that is, technology is not restricted to hardware only but may include "know-how" and "software."

The term *technology* is frequently used in the narrower sense of applied science or the use of science-based knowledge. In actuality, however, most of the technology in the world is not science-based; it is largely empirical in nature. Before the growth of science in the 17th century virtually all technology was empirical rather than based on scientific theories.

There is no intent here to disparage science nor its use in the advancement of technology; instead, the point is simply to recognize that much of technology is still based on practical experience and not on scientific theory. Even this purely empirical technology, however, is the province of the technological forecaster. Thus we are concerned not solely with science-based technology but with all means used to provide the tools, techniques, and procedures necessary for human sustenance and comfort.

Webster defines *forecast* as "to calculate or predict (some future event or condition) usually as a result of rational study and analysis of available pertinent data." The idea here is to state what is going to occur.

Combining the ideas of technology and forecasting, then, we define a technological forecast as "a prediction of the future characteristics of useful machines, procedures, or techniques."

There are two important points contained in this definition. First, a technological forecast deals with characteristics, such as levels of performance (e.g., speed, power, temperature). It does not have to state how these characteristics will be achieved. That is, the forecaster is not required to invent the technology being forecast. Even though the forecaster may predict characteristics that exceed the limitations of current technical approaches, the forecast need not state how these will be achieved. The forecaster's obligation is fulfilled by warning that these limitations will be surpassed.

Second, technological forecasting deals with useful machines, procedures, or techniques. In particular, this is intended to exclude from the domain of technological forecasting those items intended for luxury or amusement, which depend more on popular tastes than on technological capability. It does not seem possible to predict these rationally; however, the forecaster might be concerned with the means by which popular tastes will be formed, such as advertising or propaganda.

Having defined technological forecasting in a general sense, we now want to characterize it more precisely. A technological forecast (just as any good forecast) has four elements: the time of the forecast, the technology being forecast, a statement of the characteristics of the technology, and a statement of the probability associated with the forecast. What are these individual elements?

The time of the forecast is the future date when the forecast is to be realized. This may be a single point in time or a time span. In either case the time of the forecast should be stated clearly.

The technology being forecast may be stated in either of two ways, depending upon the intent and nature of the forecast; but before specifying this further, we need to define two terms.

The first term is *technical approach*. This means a specific technical means of solving a problem or performing a particular function. For instance, piston engines and jet engines are two different technical approaches to the general function of powering aircraft; incandescent lamps and fluorescent lamps are two different technical approaches to the function of providing illumination. Sometimes a technical approach can be further subdivided. Jet engines can be divided into turbojets and turbofans. For some purposes the forecaster will consider jet engines as a single technical approach; in other cases he will consider turbojets and turbofans as alternative technical approaches to the function of powering an aircraft.

The second term is *technology*. By this we mean a family or series of technical approaches that have some major characteristic in common, or which perform the same function. For instance, when the forecaster wishes to distinguish between turbojets and turbofans as technical approaches, the entire class of jet engines is then a technology, to be distinguished from the technology of piston engines. When the forecaster wishes to distinguish between incandescent and fluorescent lamps, the entire class of electric lights is a technology, to be distinguished from other technologies such as gas lights. On the basis of context, this use of the word *technology* will be easily distinguished from the broader use defined above.

Armed with these definitions, we can be precise about the technology element of a technological forecast: The forecast must state whether it is for a single technical approach or for a more general technology. If it is for a technical approach, the forecast must be clear about how that approach is to be distinguished from other approaches in the same general technology; if it is for a technology, the forecast must be clear about how that technology is to be distinguished from others used for the same function.

The third element of the forecast, the characteristics of the technology, are given in terms of "functional capability." Technology is intended to perform some function, and functional capability is a quantitative measure of its ability to carry out this function. This definition is broader than simply the technical performance of a machine. For instance, a function to be performed is the transportation of people. One measure of the functional capability available is speed. (Clearly this is not the only such measure, and for some purposes not even the best.) Hence a means of transportation may have its functional capability measured in miles per hour. However, a specific device used to carry out this function may also have one or more technical measures of performance, such as miles per gallon of fuel or percentage of efficiency. These may be of interest to the technological forecaster only indirectly, as they bear on the functional capability, or they may be of direct concern (e.g., a forecast used in a R&D program to reduce the fuel consumption of an engine). The level of functional capability is a numerical measure of the functional capability available at any time; thus at any time the level of functional capability in "people moving" may be the actual speed in miles per hour of the devices used. The third element of a technological forecast, therefore, is a specification of the functional capability being forecast, and a numerical measure of its level.

The fourth element of the forecast is a probability, which may be stated in many ways: The forecast may give the probability of achieving a given level of functional capability at all, it may state the probability of achieving a given level by a certain time, or it may state the probability distribution over the levels that might be achieved by a specific time. When the probability is not stated, it is assumed to be 100%.

The amount of information available in a forecast is limited; however, it may be distributed among the four elements of the forecast in different ways. If the forecast is precise about the date, it must be less precise about the level of functional capability or the probability distribution or something else; if the forecast is precise about the level of functional capability, it may have to be less precise about the date or the probability. Increased precision in one element of the forecast always means less precision in the other elements, given the fixed amount of information available to the forecaster.

How should the information be distributed among the elements of the forecast? Which should be given high precision and which lower precision? This depends upon the use to which the forecast is put. In all cases the forecast should be tailored to the needs of the user, providing greater precision where required, at the cost of lesser precision in other elements of the forecast.

2. Why Forecast Technology?

The question, Why forecast technology? has a false implication; it implies that it is possible *not* to forecast technology. But forecasting technology is no more avoidable than is forecasting the weather. All people implicitly forecast the weather by their choice of whether or not to wear a raincoat, carry an umbrella, and so on. Any individual, organization, or nation that can be affected by technological change inevitably engages in forecasting technology with each and every decision which allocates resources to particular purposes. A change in technology may completely invalidate a particular decision about allocating resources. Every decision, then, carries within it the forecast that technology either will not change at all or will change in such a way as to make the decision a good one.

Given that technological forecasting is unavoidable, however, there is still the issue of what specific reasons people have for making technological forecasts. In actuality, people make technological forecasts for the same reasons they make other forecasts:

1. To maximize gain from events external to an organization.
2. To maximize gain from events that are the result of actions taken by an organization.
3. To minimize loss associated with uncontrollable events external to an organization.
4. To offset the actions of competitive or hostile organizations.
5. To forecast demand for production and/or inventory control.
6. To forecast demand for facilities and capital planning.
7. To forecast demand to assure adequate staffing.

8. To develop administrative plans and policy internal to an organization (i.e., personnel and budget).
9. To develop policies that apply to people who are not part of an organization.

Most of the items listed here boil down to the idea of maximizing gain or minimizing loss from future conditions. Each item could be a reason for technological forecasting, as well as for economic, business, political, or weather forecasting.

Throughout this book the emphasis will be on forecasting for decision making; that is, the entire justification for producing forecasts will be their use in making decisions. The implication of this is that using forecasts helps make better decisions. In particular, the forecasts play specific roles in improving the quality of decision making. Ralph Lenz, pioneer in technological forecasting, has presented the following list of specific roles that forecasts can play in improving the quality of decisions.

1. The forecast identifies limits beyond which it is not possible to go.
2. It establishes feasible rates of progress, so that the plan can be made to take full advantage of such rates; it does not demand an impossible rate of progress.
3. It describes the alternatives that are open for choice.
4. It indicates possibilities that might be achieved if desired.
5. It provides a reference standard for the plan. The plan can thus be compared with the forecast at any point in time to determine whether it can still be fulfilled or whether, because of changes in the forecast, it has to be changed.
6. It furnishes warning signals, which can alert the decision maker that it will not be possible to continue present activities.

In playing these roles, the forecast provides specific pieces of information needed by the decision maker. We have so far tacitly assumed that the improved quality of the decision more than offsets the cost of the forecast. However, this is not at all certain. Sometimes forecasts cost more than they are worth. This is a topic we will take up in a later chapter. For the moment we simply note that forecasts do provide specific information that can improve the quality of decisions.

3. Alternatives to Forecasting

The definition of forecasting given above incorporated notions of rationality and analysis of data. However, there are many "alternatives" to rational and analytic forecasting. Many of these alternatives are widely used for the purposes given for forecasting in the previous section, and therefore it is worthwhile to briefly examine them.

No Forecast. This alternative means facing the future blindfolded. If taken literally, it means that no attempt is made to determine what the future will be like and that decisions are made with no regard whatsoever to their future consequences, be they favorable or unfavorable. It should be clear that any organization operating on this basis will not survive. Even if the environment is unchanging, most decisions will be wrong, since they will not even take into account a forecast of the constancy of the environment. If the environment is changing rapidly, disaster may come even more quickly, since a decision that is right for a short time may be rendered inappropriate in the longer run. In most cases, however, the concept of "no forecast" is not meant literally but is really intended to mean a forecast of constancy or negligible rate of change. Thus when a decision maker claims not to believe in or use technological forecasts, what is really meant is that the assumption of an unchanging technology is made; decisions are made on the basis of a forecast that the technology in existence at the time during which these decisions have their impact will be the same as the technology of today. Thus this is really not an alternative to forecasting but a very specific, though implicit, forecast.

Anything Can Happen. This represents the attitude that the future is a complete gamble, that nothing can be done to influence it in a desired direction, and that there is no point, therefore, in attempting to anticipate it. It is doubtful if there is any decision maker who runs his or her personal life this way. Even if decision makers claim to have this attitude, they are still likely to take a raincoat on cloudy days. Therefore, if they pretend to adopt this attitude toward their professional decisions, it really amounts to a cover for something else—perhaps an attempt to avoid the effort of thinking through the implications of a forecast. Obviously, any decision makers who really act on the basis of this attitude are headed for trouble; in particular they may find their organizations unable to withstand the competition from other organizations that do make an attempt to anticipate the future through rational means. An organization run on this basis can only be short-lived.

The Glorious Past. This represents an attitude that looks to the past and ignores the future. Many organizations can point to significant achievements at some time or the other in the past. Their very survival over an extended period indicates that they must have done the right things. Unfortunately, when conditions change, it is very unlikely that the policies and decisions that led to success in the past will continue to be suitable. Stubbornly clinging to visions of the glorious past, under the assumption that the glorious past guarantees a glorious future, is a certain road to disaster. In short, an organization that concentrates on its past instead of the future can end up only in becoming a museum piece.

Window-Blind Forecasting. This involves the attitude that technology moves on a fixed track, like an old-fashioned roller window blind, and that the only direction is up. This attitude is encapsulated in expressions such as "higher, faster, and farther," or "bigger and better"; it assumes that the future will be like the past, only more so. While this attitude does at least recognize that changes do take place and is therefore somewhat better than the other alternatives, it fails to recognize that there are other directions besides up. A particular technical approach, for instance, may come to a halt or move sideways if another technical approach supersedes it. An organization that depends on window-blind forecasting will sooner or later be taken by surprise as some unanticipated technological change brings an end to the track it was following.

Crisis Action. This can best be described as "pushing the panic button." It consists of waiting until the problem or crisis has arrived and then taking some immediate action to attempt to alleviate the impact of the crisis. Over the long term, crisis action means that the organization is not making any net progress toward its goals. As a result of expedient responses to crises, it may only be zig-zagging instead of proceeding directly toward an objective. Furthermore, this alternative is based on the assumption that there will be time to respond effectively after a crisis has arrived. If this assumption is belied by a specific crisis, the organization fails to survive. Finally, this alternative ignores the fact that had a proper forecast been used, the crisis might have been avoided completely. Hence, although sometimes used, it is not really an acceptable alternative to proper forecasting.

Genius Forecasting. This is really not an alternative to forecasting, since it does involve the preparation of a forecast. However, it is an alternative to the use of rational and explicit methods for obtaining forecasts. This method consists in finding a "genius" and asking him or her for an intuitive forecast. It must be recognized that many "genius forecasts" made in the past have been successful. Unfortunately, there have also been many so wide of the mark as to be useless. Ralph Lenz has described the shortcomings of genius forecasting as follows: It is impossible to teach, expensive to learn, and allows no opportunity for review by others. Obviously, there is no sure-fire way of obtaining a genius, and even if a genius has been located, his or her forecast cannot be checked by anyone else, even by another genius. It must be taken on faith. It may perhaps be that in some cases there is no alternative to a genius forecast;

however, it should be clear that where rational and explicit methods are available, they are much to be preferred. In fact, rational and explicit methods relieve the decision maker of the burden of ferreting out geniuses.

The whole purpose of this recitation of alternatives, of course, is to show that there really is no alternative to forecasting. If a decision maker has several alternatives open, he or she will choose the one which provides the most desirable outcome. Thus every decision is inevitably based on a forecast. Hence the decision maker does not have a choice as to whether or not to make or use a forecast. The only choice is whether the forecast is obtained by rational and explicit methods, or by intuitive means from the depths of someone's subconscious.

The virtue of the use of rational methods is that they are teachable and learnable; they can be described and explained; they provide a procedure that can be followed by anyone who has been able to absorb the necessary training. In some cases, in fact, the methods are even guaranteed to produce the same forecast, regardless of who uses them.

Another virtue of the use of explicit methods is that they can be reviewed by others; in particular, the forecast can be reviewed by several people prior to its acceptance by the decision maker. It can be checked to see if any mistakes have been made in the application of the method, in the calculations, or in the manipulation of the data. Furthermore, the forecast can be reviewed at any subsequent time to see if it is still acceptable. If conditions have changed enough to invalidate the forecast, the plans that were based on it can be altered accordingly. If the forecast is not rational and explicit, it cannot be subjected to subsequent review to assure that it is still acceptable, with the risk that the plans based on it would be left unchanged despite the fact that they are no longer appropriate.

Even though decision makers may agree that they have no alternative to forecasting, and even though they may agree to the virtues of rational and explicit methods, there is yet another issue with which the decision maker may confront the technological forecaster, and this is what we will now take up.

4. Will It Come True?

It might appear that a good forecast is one that comes true; after all, what good is a forecast which is wrong? A person seeking a weather forecast, for instance, wants to know what the weather will be in order to prepare for it. A forecast that turns out to be wrong is totally useless. The user may have made unnecessary preparations for bad weather when the weather actually turns out to be good, or may have failed to make routine preparations for bad weather because the forecast predicted good weather. Clearly, a weather forecast has to be correct if it is to be useful.

Many people apply the same criterion to technological forecasts (as well as to economic and political forecasts); however, there are two things wrong with this criterion: The first is that it cannot be applied before the fact, and the second is that it fails to take into account self-altering forecasts.

A self-altering forecast is one that, by virtue of having been made, alters the outcome of the situation. Suppose someone forecasts an undesirable situation. Then suppose a decision maker accepts the forecast and acts to prevent the undesirable situation. Clearly the forecast did not come true. Was it then a bad forecast? On the contrary, it was highly useful. Because of the forecast, someone obtained a better outcome. The same argument can be made for a forecast of a favorable outcome, that comes true only because the forecast is accepted and acted upon. Clearly, the forecast was useful, because it led someone to take an action that made things better; but the value of the forecast was in its usefulness, not in its coming true.

It is important that the forecaster not get trapped into evaluating forecasts by whether or not they came true. It is even more important that forecasters educate forecast users to the idea that the goodness of a forecast lies in its utility for making better decisions and not in whether it eventually comes true.

5. Stages of Innovation

Finally, we come to a concept that will be extremely useful in our later discussions of technological change. This is "stages of innovation." No technological device springs directly from the fertile mind of an inventor to immediate widespread use; it passes through a number of stages along the way, with successive stages representing greater degrees of practicality or use.

Various writers have given lists of the stages through which an innovation passes. The following list is an expanded version of one originated by James R. Bright:

1. Scientific findings.
2. Laboratory feasibility.
3. Operating prototype.
4. Commercial introduction or operational use.
5. Widespread adoption.
6. Diffusion to other areas.
7. Social and economic impact.

Each of these stages is discussed below.

Scientific Findings. At this stage the innovation exists in the form of the scientific understanding of some phenomenon, the properties of some material, the behavior of some force or substance, and so on. It is not in any sense capable of being utilized to solve a problem or carry out a function. Scientific findings, in general, merely represent a knowledge base from which solutions to specific problems can be drawn.

Laboratory Feasibility. At this stage a specific solution to a problem has been identified and a laboratory model has been put together. It is clear that no natural or physical laws are violated by the device and that it is capable of performing the desired function or solving the problem of concern, but only under laboratory conditions. The innovation at this stage may be described by such terms as *breadboard model* or *brassboard model*. It certainly would not function well outside the laboratory nor without the constant attention of a skilled technician.

Operating Prototype. At this stage a device has been built that is intended to function satisfactorily in the operational environment; that is, the device is expected to be rugged enough, reliable enough, and easy enough to operate and maintain that it can perform its intended function in the hands of a typical user. It should be noted that frequently the purpose of building operating prototypes is to verify that the design will in fact function as intended in the intended situation or environment. An operating prototype of a consumer appliance might be use-tested by a panel of "typical" buyers, an operating prototype of a military device might be tested in maneuvers with "typical" troops, and so on. At this stage, the innovation is called upon to demonstrate the adequacy of its design.

Commercial Introduction or Operational Use. This stage represents not only technical and design adequacy, but economic feasibility. The innovation is supposed to demonstrate that people will want it badly enough to give up other things in order to obtain it. The "first production model" is often chosen as representing the point in time at which an innovation has reached this stage.

Widespread Adoption. The innovation now has demonstrated that it is technically and economically superior to whatever else was used in the past to carry out the same function, and bids fair to replacing these prior devices, procedures, and so on on a wide scale. There is no precise definition of when this stage is reached; it might be expressed in terms of a percentage of the total potential use, a percentage of the Gross National Product, a percentage of the sales of all the items performing the same function, a percentage of all the items in use for the same purpose, and so forth. While the forecaster has some freedom to define precisely what is meant by "widespread adoption" in a particular case, once this is done, the definition must be followed consistently.

Diffusion to Other Areas. This is the stage when the innovation has not only supplanted or replaced the devices or techniques formerly used to carry out some function, but has also been adopted for performing functions that the previous devices or techniques were never used for. The transistor, for instance, has reached this stage: Not only has it virtually replaced the vacuum tube in much of conventional electronics, but it has begun to appear in applications for which vacuum tubes were never used, such as transistorized ignitions in automobiles and automatic exposure control in cameras with built-in light meters.

Social and Economic Impact. When an innovation reaches this stage, it has in some way changed the behavior of society or has reached the point where a significant portion of the economy is somehow involved with it. Television, for instance, has certainly had a major impact on American society. The automobile has not only had a major impact on American society, but has reached the point where more than 10% of the U.S. Gross National Product is involved in its manufacture, sale, operation, and maintenance.

Not every innovation goes through all these stages. Some reach a certain stage and go no farther, others may cover the entire list but combine two or more stages, and some, especially empirically based innovations, may skip the early stages. Nevertheless, this list is useful for avoiding confusion about the degree of development, or extent of use, of some device or technique. The concept of stages of innovation will be useful in subsequent chapters.

One particular use will be mentioned here, however. It is often noted that innovations tend to appear in clusters, and many scholars have claimed that these clusters appear at the end of depressions or are in other ways associated with the long cycles of the economy ("Kondratieff waves") (e.g., see Graham and Senge, 1980, and Mensh, 1979). The point is that these innovations do not appear from "nowhere." In many cases they are innovations that were languishing at an appropriate stage of innovation when the external economic conditions made it possible for them to develop rapidly into the succeeding stages.

6. Remainder of the Book

The emphasis of this entire book is on the *use* of technological forecasts for making decisions. The forecast is viewed solely as an imput to the decision-making process. The book is divided into three major sections, each devoted to a particular aspect of the preparation and use of technological forecasts.

The first section, consisting of Chapters 2 to 11, is devoted to *methods* of technological forecasting. Various methods are described and illustrated by examples. The strengths and weaknesses of each method are described.

The second section, consisting of Chapters 12 to 16, is devoted to *applications* of technological forecasting. Planning and decision making are discussed in general at first, and then in detail for each of several major application areas. These applications are illustrated by practical and historical examples.

The third section, consisting of Chapters 17 to 19, is devoted to a discussion of how the forecaster can most effectively present his work so that it will be useful for making decisions. The various chapters discuss good practices, bad practices, and pitfalls.

References

Graham, Alan K., and Peter M. Senge, "A Long-Wave Hypothesis of Innovation," *Technological Forecasting and Social Change* (4), 283–311.

Mensch, Gerhard (1979). *Stalmate in Technology* (Cambridge, MA: Ballinger).

For Further Reference

Balachandra, B. (1980). "Technological Forecasting: Who Does It and How Useful Is It? *Technological Forecasting and Social Change* 16 (1), 75–86.

Problems

1. Which of the following items or activities represent or exemplify technology as it is defined in the text?

Farming	A legislature
Money	Fixed-wing aircraft
Cattle raising	The limited-liability corporation
Roads	Smelting metals from ores
Bridges	The violin

2. Which of the items in Problem 1 was, from its inception, based on a scientific understanding of the principles underlying its operation or activity?

3. Trace the history of radio communication, identifying the dates at which it passed through the stages of innovation given in the text.

4. Consider a technology for which there is a steady demand and that receives broad-based support from a wide variety of sources, each seeking competitive

advantage by achieving an advance in the level of functional capability. Why would one of the groups involved in advancing this technology need a forecast of its probable future progress?

5. Consider a technology for which the demand fluctuates about some norm or regular pattern and that is advanced by a small number of corporations or other groups, each of which is interested in seeing that its products or capabilities do not fall behind those of the other corporations or groups. Why would one of these groups need a forecast of the probable future course of this technology?

6. Consider a technology that is utilized by only one group, whose advance is supported by that group alone, and whose rate of advance is geared to the needs of the group (an example would be instrumentation on a military missile-testing range). Why would this group need a forecast of the probable future course of this monopolistically controlled technology?

Chapter 2

Delphi

1. Introduction

Formal forecasting methods are intended to replacement subjective opinion with objective data and replicable methods. However, there are three types of circumstances under which expert opinion will always be needed. (Note, however, that the selection of a particular "objective" method may involve some subjectivity and implicit assumptions. The importance of these assumptions will be discussed in Chapter 17.)

The first type is when no historical data exist. In technological forecasting this usually involves new technologies. Despite the absence of historical data, a forecast must often be prepared. Expert opinion is then the only possible source of a forecast.

The second type is when the impact of external factors is more important than the factors that governed the previous development of the technology. These external factors may include decisions of sponsors and opponents of the technology, or changes in public opinion. In such a case data about the past may be irrelevant. Expert opinion may be the only possible source of a forecast.

The third type is when ethical or moral considerations may dominate the economic and technical considerations that usually govern the development of technology. These issues are inherently subjective, and expert opinion may be the only possible source of a forecast.

Given that expert opinion is needed, how is it to be obtained? The problems of expert opinion may be overcome to some extent by using several experts—two heads are better than one. In a group of experts individual biases may be canceled out, and the knowledge of one member may compensate for another's lack of it.

On the other hand, "a camel is a horse designed by a committee." A forecast designed by a committee might be equally grotesque. What is needed is some way to obtain the benefits of a committee while minimizing the disadvantages.

2. Advantages of Committees

The first major advantage of a committee is that the sum of the information available to a group is at least as great as that available to any individual member. Adding members to a group does not destroy information. Even if one member knows more than the rest put together, this does not reduce the total information available to the group; the others may still make useful contributions. If the group has been chosen to contain only people who are experts in the subject, the total information available to the group is probably many times that possessed by any single member.

The second major advantage is that the number of factors that can be considered by a group is at least as great as the number which can be considered by a single member. This point is at least as important as the first. Studies of forecasts that have gone wrong show that one very common cause of failure is neglecting to take into account important factors outside the technology being forecast, which in the long run turned out to be more significant than those internal to the technology. This advantage of a group is therefore very important.

3. Disadvantages of Committees

The first major disadvantage of a group is that there is at least as much misinformation available to the group as there is to any single member. One reason for using a group is the hope that the misinformation held by one member may be canceled out by valid information held by another. However, there is no guarantee that this will take place.

The second major disadvantage is the social pressure a group places on its members—pressure to agree with the majority even when the individual feels that the majority is wrong. This is especially true in the production of group forecasts. One member may well give up presenting certain relevant factors if the remainder of the group persists in taking a contrary view.

The third major disadvantage is that a group often takes on a life of its own. Reaching agreement becomes a goal in itself, of greater importance than producing a well thought out and useful forecast. Group forecasts may thus be only a watered-down least common denominator that offends no one, even though no one agrees strongly either.

A fourth major disadvantage is the influence that the repetition of arguments can have. Experiments with small groups show that often it is not the validity but the number of comments for or against a position that carries the day. A strong vocal minority may overwhelm the majority by pushing its views vigorously, even though the arguments may have little objective merit.

A fifth major disadvantage of groups is their vulnerability to the influence of dominant individuals. One individual, by active participation in debate, by putting ideas forward with a great deal of vigor, or through a persuasive personality, may have an undue influence on the group's deliberations. Such an individual may get his or her way simply by wearing down the opposition with persistent argument.

A sixth disadvantage of groups is that members of a group may come to have vested interests in certain points of view, especially if they have presented them strongly at the outset. Their objective becomes one of winning the remainder of the group over, rather than reaching a more valid conclusion. Such members may be impervious to the facts and logic of the remainder of the group. They will concentrate only on winning the argument.

A seventh disadvantage of groups is that the entire group may share a common bias. This often arises from a common culture shared by the members—especially a subculture peculiar to the technology in which the members are experts. The presence of a common bias nullifies the advantage of a group in canceling biases.

4. The Delphi Procedure

Delphi is intended to gain the advantages of groups while overcoming the disadvantages. It was originally developed at the Rand Corporation as a means of extracting opinion from a group of experts. Its first public presentation in a Rand report dealt with a series of technological forecasts, which led to the misunderstanding that Delphi was primarily a technological forecasting method. It is not. It can be used for any purpose for which a committee can be used. While the emphasis here is on technological forecasting, other uses of Delphi are discussed in Linstone and Turoff (1975).

Delphi has three characteristics that distinguish it from conventional face-to-face group interaction: (1) anonymity, (2) iteration with controlled feedback, and (3) statistical group response.

Anonymity. During a Delphi sequence the group members usually do not know who else is in the group. The interaction of the group members is handled in a completely anonymous manner through the use of questionnaires. This avoids the possibility of identifying a specific opinion with a particular person. The originator can therefore change his mind without publicly admitting he has done so. In addition, each idea can be considered on its merits, regardless of whether the group members have high or low opinions of the originator.

Iteration with Controlled Feedback. Group interaction is carried out through answers to questionnaires. The group moderator extracts from the questionnaires only those pieces of information that are relevant to the issue and presents these to the group. Each group member is informed only of the current status of the group's collective opinion and the arguments for and against each point of view. Group members are not subjected to a harangue or an endless restatement of the same arguments. Any viewpoint can be presented to the group, but not in such a manner as to overwhelm the opposition by sheer repetition. The primary effect of this controlled feedback is to prevent the group from taking a life of its own. It permits the group to concentrate on its original objectives rather than self-chosen goals such as winning the argument or reaching agreement for its own sake.

Statistical Group Response. Typically a group will produce a forecast that contains only a majority opinion; it will represent simply that viewpoint on which a majority of the group could agree. At most there may be a minority report. There is unlikely to be any indication of the degree of difference of opinion that existed within the group. Delphi presents instead a statistical response that includes the opinions of the entire group. On a single item, for instance, group responses are presented in statistics that describe both the "center" of the group opinion and the degree of spread about that center.

5. Conducting a Delphi Sequence

The following description is of "classical" Delphi as originated at Rand. Variations from this base line will be taken up in a later section.

Before describing Delphi, some definitions are required. A Delphi sequence is carried out by interrogating a group of experts with a series of questionnaires. Each successive questionnaire is a "round." The term *questionnaire* may be misleading, however. The questionnaires not only ask questions, but provide information to the group members about the degree of group consensus and the arguments presented by the group members for and against various positions. The questionnaire is the medium for group interaction. The set of experts taking part in the Delphi is usually referred to as a "panel." In large Delphis there may be subgroups devoted to specific specialties. These subgroups may be identified by subject, such as the "electronics panel." Either the entire set of experts or a subgroup may be referred to as a panel. Context usually provides a guide as to which use is meant. The person responsible for collecting the panel responses and preparing the questionnaires is called the "moderator."

Delphi will be described in terms of rounds. Each round calls for somewhat different activities on the part of the panelists and moderator. Before the first round there must be preliminary activities such as clarifying the subject and explaining the methods. After these preliminaries the first round can begin.

Round One. The first questionnaire is completely unstructured. The panelists are asked to forecast events or trends in the area for which the panel was assembled. This has some disadvantages, which will be discussed later, but it also has some significant advantages. The panelists have been selected because of their expertise in the area to be forecast. They should know much more than the moderator does about that area. If the first questionnaire were too structured, it might prevent the panelists from forecasting some important events of which the moderator might not be aware.

The questionnaires are returned to the moderator, who then consolidates the forecasts into a single set. Similar items must be combined; items of lesser importance must be dropped to keep the list at a reasonable length; and each event must be stated as clearly as possible. The list of events then becomes the questionnaire for the second round.

Round Two. The panelists receive the consolidated list of events and are asked to estimate the time of occurrence for each event. The estimate may be a date; it may be "never," if they think that the event is impossible; or it may be "later" if some time horizon has been specified for forecasts and they believe that it will occur later than that horizon.

The moderator collects the forecasts from the panel and prepares a statistical summary of the forecasts for each event. This usually consists of the median date and the upper and lower quartile dates for each event (for definitions see Appendix 1). The third questionnaire consists of the set of events and the statistical summary of the forecasts.

Round Three. The panelists receive the questionnaire with events, medians, and quartiles. They are asked to prepare new forecasts for each event, either sticking with their previous forecast or making a new one. If their forecasts fall in either the upper or lower quartile (that is, if they are "outliers"), they must present reasons why they believe they are correct and the majority of the panel incorrect. Their reasons may include references to specific factors the other panelists may be overlooking, facts the other panelists may not be considering, and so on. The panelists are just as free to advance arguments and objections as they would be in a face-to-face group; the only difference is that their arguments are written and anonymous.

When the moderator receives the third-round responses, he or she prepares a statistical summary of the forecasts, as well as a consolidated summary of the panel's reasons for advancing or delaying the forecasts. Similar arguments are combined and lengthy arguments summarized. (Fortunately the need to write the arguments often forces the panelists to be concise.) The questionnaire for the fourth round consists of the list of events, the medians and quartiles for the third round, and the summary of arguments for changing the forecasts of each event.

Round Four. The panelists receive the events and dates and the reasons given for changing their estimates. They are asked to take the reasons into account and make new forecasts for each event. Depending upon the needs of the moderator, they may be asked to justify their position if their forecasts fall in the upper or lower quartile. In addition, the moderator may invite comments from all panelists on the arguments given during the third round.

Upon receiving the forecasts from the panelists, the moderator again computes medians and quartiles and, if comments were requested, consolidates and summarizes them. (If the moderator does not plan to analyze the arguments, there is no point in asking for them.) In some cases, when the panel has not been able to reach a consensus, the moderator may well be interested in the arguments on both sides. In such cases the moderator should ask for comments and be prepared to analyze them.

The date forecast for each event is the median date on the fourth round. In addition, the moderator can determine the amount of disagreement in the panel on the final round from the difference between the quartile dates. The comments on each event provide a summary of those factors the panelists believe are important and that may affect the forecast. The output of a Delphi thus contains a great deal more information than is usually obtained from a committee. In addition, the nature of Delphi focuses this information on the topics of interest to the moderator and organizes it in a readily understandable manner.

Ordinary committees are judged as successes if they reach agreement or consensus. Indeed, committee action is designed to achieve consensus and may force a false one. Delphi is intended to display disagreement where it exists and search for the causes. Delphi sequences are judged as successes when they reach stability, that is, no further change of opinion, with the reasons for divergence clearly displayed. Thus if a particular item reaches stability before the fourth round, it may be dropped. In some cases, however, it may be necessary to restate an event, to split an event into distinct subevents, or to combine separate events in order to obtain agreement on what is really at issue, and thereby reach stability.

General experience is that there is convergence of the panel estimates during the sequence of rounds. The panel members will usually have widely varying estimates on each event on the second round. However, as the panelists offer their reasons for shifting the estimates, the subsequent estimates tend to cluster near preferred dates. This convergence results from actual transfer of information and interaction among the panel members.

Panel members do not always shift their opinions under the influence of the arguments of other panelists. Delphi panelists have just as much opportunity to stick with their original views as do members of a face-to-face group. The advantage of Delphi is that panel members can shift position without losing face when they see convincing reasons from other panel members for a shift of their estimates.

6. Variations on Delphi

Since Delphi was first publicly announced there have been numerous variations on the basic procedure. Some of these are described briefly below.

Providing an Initial List of Event. Classical Delphi has been described as "starting with a blank sheet of paper." While this has advantages, it also seems to bother some panelists, who find themselves confused by the unstructured situation. Some users of Delphi have started with an initial list of events generated by some process before the start of the Delphi. The panelists may be asked to make forecasts for these, effectively going immediately to round two; alternatively, they may be asked to suggest additional events. The augmented event set then becomes round two.

Beginning with a Context. The exact course of the development of a technology will depend upon external political and economic conditions. When these are important, the forecasts will depend upon the assumptions made about these external conditions. If the panel is composed of technology experts, they should not be expected to forecast these economic and political conditions as well. Hence it may be desirable to obtain a political and economic forecast and present this to the panelists prior to round one. This provides the panelists with a common context for their forecasts of technology. If the economic and political forecasts are in error, the resulting technological forecast will also be in error. However, this problem cannot be avoided by failing to provide a context. Doing so simply means that the technology experts will make their own political and economic forecasts. Providing a context can be especially useful in industrial applications of Delphi when a panel of experts has been chosen from the company's technical staff. A context provided by the company's sales, marketing, and top management personnel can provide a helpful guide to the technical specialists on the panel.

Number of Rounds. Classical Delphi includes four rounds. Some Delphis have taken as many as five rounds. Experience indicates that four rounds is usually sufficient. Round four can be deleted if the moderator sees no need to obtain rebuttals to the arguments presented in round three. Round one can be omitted if the panel is started off with a list of events. Thus in some cases two rounds may be sufficient. Since Delphi provides advantages over face-to-face groups, it should be used if at all possible, even when a full four rounds cannot be used. Even two rounds may be better than the use of a single expert or a face-to-face panel.

Multiple Dates. In the classical Delphi each panelist provides one forecast for the date of an event. In some cases this is specified as the date by which the event is 50% likely. In other applications of Delphi, however, panelists may be asked to provide three dates: In addition to the 50% date, they may be asked to provide "barely possible" and "virtually certain" dates. These may be quantified as 10, 50, and 90% probability estimates or some other suitably chosen probabilities. The statistical group response is then obtained by taking the median date for the 50% estimates. The degree of disagreement in the panel is represented by the spread between the median dates for the low-likelihood and high-likelihood dates.

Computerization. Computerized analyses of Delphi results are quite common, especially for Delphis with many people and several panels. However, computerization can go well beyond processing the Delphi responses. In some Delphi sequences panelists have used remote computer terminals to participate. The terminals are connected to a central computer that keeps track of the current status of each event and the last estimate made by each panelist. A panelist participates by "logging on" to the computer via a terminal; the computer displays the median and quartiles of the current estimates of the panelists, reminds the panelist of his or her last estimate and asks whether this estimate should be changed. This approach does away with the round structure. Panelists may log on as often as they choose. Some will do so more frequently than others; some panelists will change their estimates frequently, while others will permit theirs to stand for a longer time. This "real-time, on-line Delphi" can allow participants to achieve a consensus much more rapidly than via written questionnaires sent through the mail. Application of this approach is currently limited by the availability of computer terminals. As terminals become more widely available, this approach to Delphi is likely to become much more widespread.

Delphi with Partial Anonymity. Delphi is sometimes used in face-to-face groups: Arguments are thus made publicly, while estimates are still made anonymously through secret voting. The panelists discuss an event and then make their forecasts. This may go through several rounds as panelists offer reasons why the others should change their forecasts. Paper ballots are often used; however, an electronic device known as a "Consensor" is sometimes employed instead. The Consensor consists of a small computer, a TV-like display screen, and a dozen or so control units connected to the computer by cables. The control units consist of a numbered scale and a knob that can be rotated to one of the numbers. Each participant can "vote" by setting the knob on his control to the number representing his estimate. When all participants have voted, the computer prepares a statistical analysis of the estimates and displays it on the screen. The display may be a bar graph or some other suitable picture that shows the "center" of the estimates and the dispersion about that center. With the Consensor votes may be taken quickly at any point in the discussion. Participants can quickly see how much consensus has been reached; they can decide whether further discussion is worthwhile. The discussion is public, but the cables to the control units can be scrambled so that the voting is untraceable, and the control units can be concealed from the rest of the participants to maintain the anonymity of individual estimates.

7. Delphi as a Group Process

People frequently ask about the accuracy of Delphi; however, this question is misdirected. Delphi is based solely on expert opinion. The accuracy of the forecasts is only as good as the opinions that go into the forecasts. Since Delphi is used when expert opinion is the best forecast available, the proper issue is whether Delphi is a better method for extracting opinion from a group of experts than is any other method.

Much of the work on the accuracy of Delphi, as a group process, goes back to Dalkey's experiments with "almanac" questions. Dalkey asked his subjects questions to which there was a known numerical answer; moreover, the questions were ones to which the subjects were unlikely to know the answer but about which they could make informed judgments. A typical question was, How many telephones were there in Africa in 1965? In Dalkey's experiments each participant made an estimate for each question. Then the participants either received anonymous feedback, as in Delphi, or participated in a face-to-face discussion. The feedback, whether anonymous or face to face, was then followed by another set of individual estimates. The findings were that, more often than not, the anonymous feedback made the median of the second round better. The face-to-face discussion, more often than not, made the median of the second round worse.

Another view of Delphi as a group process comes from an experiment reported by Salancik (1973). The panelists took part in a Delphi to forecast applications for computers. For instance, the panelists were asked to forecast the date by which it was 50% likely that one-half of all physicians would be using computers in a particular application. In addition to his forecast date, each panelist was asked to give reasons for his estimate.

The reasons given by the panelists were categorized as dealing with benefits, costs, or feasibility. Whether a response gave a positive or negative view of the benefits, for instance, it was categorized as a statement of benefits. The net number of benefit statements (number stating positive benefit minus the number stating negative benefit) was computed for each application. The same was done to obtain a net number of statements regarding feasibility and (low or acceptable) cost. The median date forecast for each application on the second round was then regressed on the net number of statements in each of the three categories. The regression equation explained 85% of the variance in median dates. This analysis showed that Delphi panels do assimilate the comments from panel members into their aggregate estimates. Delphi is not simply a repeated poll; group interaction does take place.

Another important finding is that first-round estimates have a log-normal distribution; this is true for both almanac-type questions and actual forecasts. A typical result is shown in Figure 2.1, which is based on a total of 19,000 separate forecasts. For the first-round responses the mean and standard deviation were computed for each event. The response to each event was then standardized by first subtracting the mean for that event and dividing by the standard deviation. The standardized responses were then pooled and stratified. The cumulative frequency was plotted against the response on log-normal paper. The implication of this finding is that people tend to think in ratios; so an estimate half the true value is considered to be of the same size error as an estimate twice the true value. Thus the logarithms of the ratios of estimates to mean values are normally distributed.

The log-normality of first-round estimates shows that estimation is a "lawful" behavior governed by rules that produce regularity in the estimates. The relationship between net numbers of arguments and the median date of the forecast shows that the initial estimates are subjected to a genuine group interaction. Dalkey's findings show that the group process in Delphi does an efficient job of extracting information from a

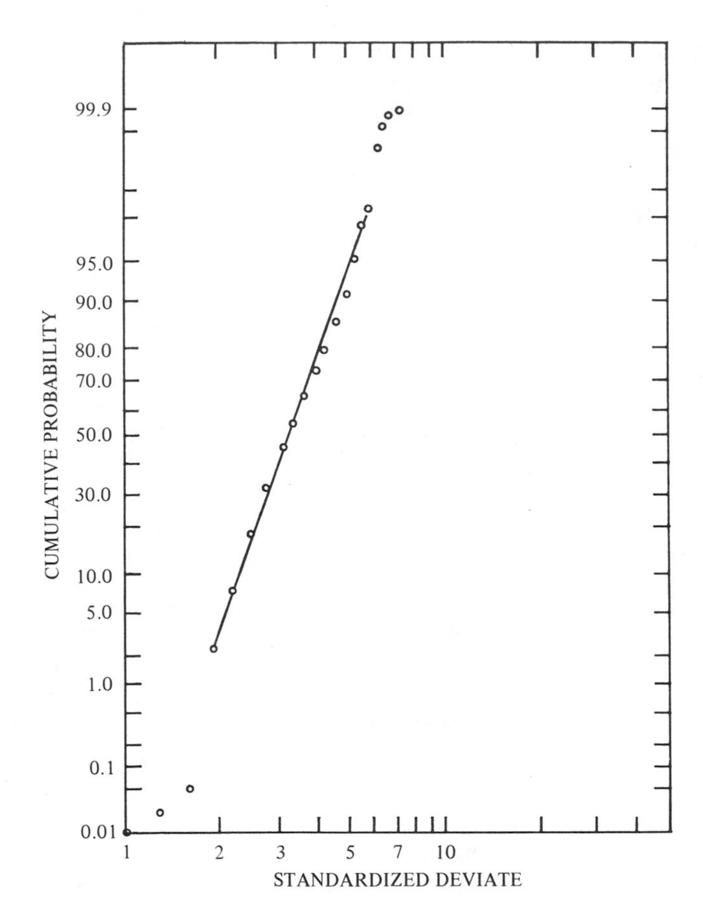

Figure 2.1. Cumulative probability distribution of standardized deviates for 50% likelihood estimates. A value of 3.7 was added to each deviate to make all the results positive.

panel as compared with face-to-face interaction. While none of these findings can "validate" a Delphi forecast, taken together they indicate that when it is necessary to use expert opinion, Delphi is a good way of getting it.

8. The Precision of Delphi

The uncertainty in a Delphi forecast is measured by the interquartile range of the panel's responses or, in some cases, the difference between the dates forecast for very low and very high likelihoods of occurrence. In general, the precision of Delphi estimates varies with the length of the forecast. A typical result is shown in Figure 2.2, which plots the time

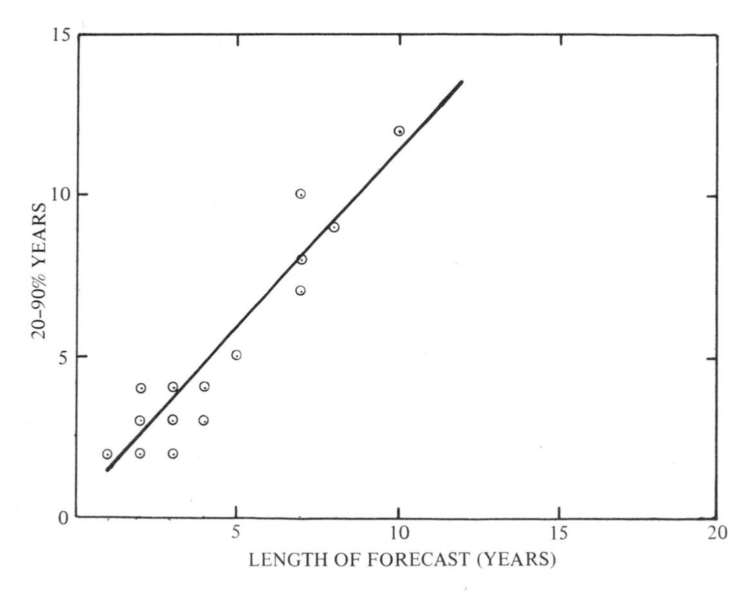

Figure 2.2. Spread of estimates versus length of forecast.

between the dates for the 20 and 90% likelihoods of occurrence against the length of time from the present to the 50% likelihood of occurrence for several forecasts by the same panel. The farther away the event, the greater the uncertainty of the panel, and the less the implied precision of the forecast. The linear growth of uncertainty is one more indication of the "lawful" behavior of Delphi estimates.

9. The Reliability of Delphi

How likely is it that two equally expert Delphi panels will give significantly different forecasts for the same event? Since experts do not always agree, this is a possibility; but if it happened often, Delphi would be a useless method of forecasting.

Dalkey investigated this in his work with almanac-type questions. He took first-round responses and treated them as a population from which he drew samples of various sizes. For each sample he obtained the median and for each sample size he obtained the correlation between the median and the true answer. The results are shown in Figure 2.3, which shows the mean correlation coefficients, over all questions, for several sample sizes. The mean correlation between the median and the true answer increases with increasing sample size. For panels of as few as 11 members the correlation exceeds 0.7. These results indicate that a panel of 15 members, if truly representative of the "expert community" on some

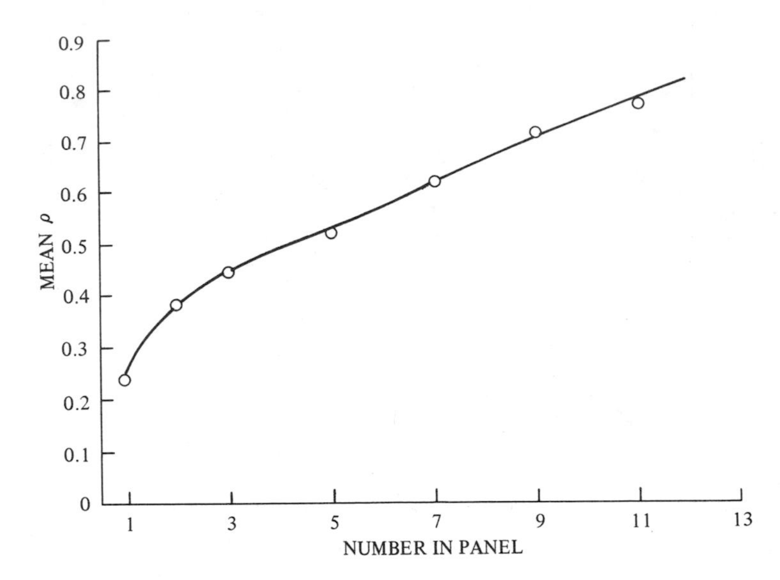

Figure 2.3. Panel reliability versus panel size.

topic, is unlikely to produce forecasts that differ markedly from those of another equally expert panel of the same size.

10. Selecting Delphi Panel Members

Some people today challenge the notion that expert opinion is needed when objective data is lacking. This has been debated in our society for years. One side insists that society is so complex that it can be understood only by the "experts," who should be given control of it; the other side denigrates expertise, claiming that the experts do not know any more about society than the rest of us. Following this latter argument, many people reject Delphi as one more attempt to put the experts in charge.

Both sides in the debate are making the same mistake: They think that there is some small group of people who are experts while everyone else is a nonexpert. This is completely wrong. An expert is someone who has special knowledge about a specific subject. Each one of us is an expert on *something,* and all of us are nonexperts on most things. There is no subset of society that can be called experts in contrast with everyone else.

This point is particularly relevant in selecting experts for Delphi panels. The panelists should be experts in the sense that they know more about the topic to be forecast than do most people. On all other topics, however, the panelists may know even less than most people. A panel member should be selected for expertise *with regard to the topic to be forecast.* Expertise in other areas is irrelevant and is certainly not implied by selection for the panel.

How can the forecaster identify an expert? Here we will focus on identifying experts in technology. Delphis run for other purposes may require other methods for identifying experts in their subject matter.

There are two aspects to selecting experts for a Delphi panel: First, how does one identify an expert? Second, of the experts identified, which should be selected for the panel? A related issue is whether to select experts from inside the organization or from outside.

The question of whether to use inside or outside experts depends primarily on the type of forecast needed and, in some cases, the uses to be made of the results. If the preparation of the forecast requires intimate knowledge of the organization, its history, policies, and so on, then there is little alternative to the use of experts from within the organization. If, however, the forecast does not depend on knowledge of the organization but more on familiarity with some area of technology, then it is probably better to obtain the best people available, and, in general, these will come from outside the organization. Except for organizations like large universities, in general no organization can afford to have on its own staff more than one or two people of the caliber desired for this type of Delphi panel.

If the forecast is intended to be used in some manner that requires that it remain secret to be effective, then again there is little choice but to use experts from within the organization. The federal government, when obtaining a forecast in an area touching on national security, probably would have little difficulty in maintaining the desired degree of secrecy, even if outsiders were employed to help prepare the forecast. A business firm, however, that hopes to gain an advantage over its competitors through the effective use of a forecast, probably cannot count too heavily on maintaining a proprietary status for the forecast if really high-caliber outsiders are to serve on the Delphi panel. Some of the desired people may not be willing to serve if the results are to be maintained in a proprietary status. In such cases the firm is most likely better off using its own people; the employees of the firm may well make up for their lack of expertise, as compared with the best experts available anywhere, with their knowledge of the firm's interests, strengths, and weaknesses.

If the decision is made to use experts from within the organization, the identification of such experts is very much simplified. This is especially true if part of the required expertise is knowledge of the organization. The panel director will look for people in responsible technical or managerial positions who have been with the organization long enough to have acquired the desired knowledge of its special or unusual features. Evaluation of the level of technical expertise can usually be obtained from supervisors, records of merit promotions and pay increases, and so on. In some cases the organization chart will be a sufficient guide.

Once the experts within the organization have been identified, there remains the problem of selecting among them. The biggest problem in this regard is that experts are busy people. This will be more true the higher they are placed within the management structure. This means that they may not have time to give the Delphi questionnaires adequate attention. In practice, a tradeoff must usually be made between getting panelists whose organizational position gives them a sufficiently broad view, and getting panelists who will be able to spend adequate time filling out the questionnaires. There is always a temptation for the panelist to make an estimate coincide with the panel median, simply to avoid the problem of justifying a different viewpoint. If the panelist is a busy executive, trying to fill out the questionnaire in his or her spare time, the temptation may be overwhelming, despite a sincere desire to provide a responsive and useful answer. The hasty opinion of a vice-president is probably not worth as much as the considered opinion of someone two or three levels lower in the organization.

If the decision is made to use outside experts, then the problem of identification is much more difficult. Peer judgment is usually the best criterion for identifying an expert. If the organization has on its own staff one or more specialists in the desired field, they can be asked to nominate outside experts; the outside experts can themselves be asked to nominate others. A good rule of thumb is to select those who have been nominated by at least two other people. In addition to these nominations, there are other selection criteria that have at least the appearance of being objective and which are in any case useful aids to judgment: honors by professional societies, the number of papers published, the number and importance of patents held, citation rates of published papers, and other signs of professional eminence such as holding office in a professional society.

With outside experts, assuring that they will have adequate time to answer questionnaires is not a serious problem. Outside experts are usually chosen from among university faculty members, private consultants, and others who have a significant degree of control over their own time. Their agreement to serve on a Delphi panel can be construed as a commitment to devote adequate time to preparing the forecast. The most serious problem is finding a panel who will not only agree to serve but also be available for the full sequence of questionnaires. University faculty members, for instance, tend to do a great deal of traveling during the summer; thus if the panel is to be staffed mainly with university faculty, the sequence should be timed so that it can be completed during the academic year.

Given that a set of experts has been identified, which of them should be asked to serve on the panel? Or, viewing it from a more practical standpoint, which should be asked first in the hope that they will agree to serve and that it will not be necessary to contact others? How can the panel moderator establish a hierarchy among the potential panelists? Degree of expertness, as determined during the initial search, is probably the most important single consideration. The forecast should represent the best opinion available; hence the panel should be composed of the most knowledgeable experts available. After that, considerations such as likely availability and probable willingness to serve can be taken into account.

There is another factor that must be given consideration during the selection of the panel. As pointed out earlier, one of the difficulties with any forecast prepared by a group is the problem of common or cultural bias. If the members of the panel share some set of biases, these will almost inevitably show up in the forecast. The panelists themselves are unlikely to be aware of them. There is no absolute guarantee that this problem can be eliminated; it can only be minimized by selecting representatives of every major school of thought in the subject area. If there are people within the organization who are sufficiently familiar with the field, they may be asked to identify the major schools of thought and to indicate which experts belong to which schools. The panel moderator can also make use of the various *Who's Who* publications, rosters of professional societies, and so on, to determine the background of each expert. Facts such as previous employers, schools attended, identity of thesis advisor(s), and so on can be used to help assure that a panel is not inadvertently chosen which has a one-sided outlook. If this kind of information is not readily available, then the panel should be chosen to include members with widely varying ages and representing a variety of institutions with as wide a geographical spread as possible.

It cannot be emphasized too strongly that choosing the panel is the most important decision the panel moderator will make, and considerable effort in making a good selection is fully justified.

11. Guidelines for Conducting a Delphi Sequence

It is a mistake to say, "We don't have time to get a forecast by any other methods, so let's do a Delphi." Probably more people have had bad experiences with Delphi for this reason than for any other. Delphi cannot be done "on the cheap"; Delphi takes as much time, effort, and expense as does preparing an equivalent forecast by other means.

Even though Delphi is neither cheap nor easy, it can be done with reasonable cost and effort if the more common mistakes are avoided. The following guidelines can help the user avoid such mistakes. They should not be taken as an indication that Delphi is either cheap or easy.

Obtain an Agreement to Serve on the Panel. If questionnaires are simply sent out to a list of names, without making sure that these people are willing to serve on the panel, the moderator runs the risk of not getting enough answers to be meaningful, especially if the list of names is a very short one. A few attempts to run Delphi sequences have begun by sending the first questionnaire to 200 or 300 names. Response rates typically run to 50% or less, and six to eight weeks are sometimes required to get even that many responses. In addition to the delay involved, there is no assurance that the same people will respond to every round. The moderator may well be putting in a lot of effort and not gaining any of the advantages of Delphi, in fact simply running a poll by mail, and of a very poorly selected group at that. As emphasized in the section on panel selection, choosing the panelists is the most important decision the moderator makes during the course of a Delphi sequence. The moderator must not only select the right people, but also make certain that they will in fact serve. The panel selected should also be slightly larger than the moderator thinks will be necessary (panelists have been known to die during the course of a Delphi sequence). In addition, if the panel includes the best people available, the moderator must expect that from time to time some of the panelists will have to miss a round because of higher-priority demands on their time. If the original panel is just big enough, any losses such as these may seriously reduce the effectiveness of the resulting forecast.

Explain the Delphi Procedure Completely. Delphi is not yet so well known that the moderator can be confident that the experts selected are familiar with or have even heard of it. Even if they are aware of it, they may have only a distorted picture of what is involved and what will be expected of them. It is especially important that they understand the iterative nature of the sequence. Several Delphi sequences have run into problems because some of the panelists did not understand the purpose of the successive questionnaires.

Make the Questionnaire Easy. The format of the questionnaire should be designed to help, not hinder, the panelist, who should be spending his or her time thinking about the forecast, not wrestling with a complicated or confusing questionnaire. A good way of doing this is to make use of "check the block" of "fill in the blank" questions. This is not always possible, especially in the case of events surrounded by considerable debate as to whether they will occur at all. However, it should be done whenever possible. In addition, the arguments for and against each event should be summarized and presented in a compact form that makes it easy for the panelists to follow the arguments and connect them with the question. Finally, there should be ample space on the questionnaire for the panelists to write in their own comments and arguments. In short, the questionnaire should be designed for the convenience of the panelists and not that of the moderator. Efforts in making the questionnaire easier to answer will directly improve the quality of the responses.

The Number of Questions. There is a practical upper limit to the number of questions to which a panelist can give adequate consideration. This number will vary with the type of question. If each question is fairly simple, requiring only a single number in response to a simple event statement, the limit will be higher. If, on the other hand, each question requires considerable thought, with the weighing of conflicting arguments and the balancing of opposing trends, the limit will be lower. As a rule of thumb, 25 questions should be considered a practical upper limit. In special circumstances the number of questions may be higher; however, if the number of questions rises to 50, the moderator should examine them carefully to be sure they are focusing on the points of real interest and not diluting the efforts of the panel on minor matters.

Contradictory Forecasts. When the set of questions is generated by the panelists during the first round, it is entirely possible that contradictory forecasts will appear. These might be, for instance, pairs of events that are both possible but mutually exclusive. In principle, there is no reason why both such events should not be included in the questionnaire, especially if the outcome is of considerable interest to the moderator. However, it should be made clear to the panelists that both events are included because of the responses to the first round. The panelists should not be left with the feeling that the moderator is including contradictory events for the purpose of trapping them in an inconsistency.

Injection of the Moderator's Opinions. From time to time during a Delphi sequence it will appear to the moderator that the two sides in a debate on some event are not effectively meeting each other's arguments or that there is some obvious (to the moderator) argument or fact which both sides are overlooking. Under these circumstances the moderator may be tempted to include his or her own opinions in the feedback on the next round. This temptation must be resisted without fail. Under no circumstances should a moderator inject personal opinions into the feedback. This advice may seem harsh, but there is no alternative. Once the moderator has violated this rule, there is no recognizable place to draw the line. If a little bit of meddling is permissible, why not a little more? And this can continue until the entire forecast is distorted to conform to the views of the moderator. If the moderator's own opinions are injected into the feedback, there is a risk of converting the Delphi sequence into an elaborate and expensive means of fooling the moderator (or worse yet, fooling the clients, who may be impressed by the names of the panelists).

The moderator has gone through considerable trouble picking a panel of experts, people who presumably know a lot more about the subject than anyone else. Their deliberations should not be meddled with. If a moderator becomes convinced that the panelists are overlooking some significant elements of the problem, it should be recognized that somehow the panel selected is unqualified, and the only solution is to discard the forecast produced and repeat the work with another panel. This advice is particularly important, since considerable research (e.g., see Bradley, 1978) has shown that Delphi results can be manipulated by the injection of false or distorted information into the genuine feedback of the participants.

Payment to Panelists. Originally most Delphi forecasts were prepared by unpaid panelists; it was considered almost an honor to be asked to participate. However, those days are over. The moderator of a Delphi panel is asking for time and expert advice from the panelists and should be prepared to pay for these valuable commodities at market rates. The forecast is presumed to be valuable to the organization asking for it; a bad forecast may cost much more than the cost of preparing it. Thus the panelists should be paid at customary consulting rates.

Professional societies and charitable institutions may still be able to obtain unpaid Delphi panelists. Experts may be as willing to lend their time and knowledge to these organizations as they are to donate money or other kinds of effort. Nevertheless, moderators for such Delphis should remember that they are asking for something valuable and are depending on the good will of the panelists, which should not be abused.

Workload Inolved in a Delphi Sequence. During the Delphi the main task of the moderator is to receive and analyze responses from the panelists and prepare the questions for the next round. Experience shows that this will require about two person-hours per panelist per round. The clerical workload in preparing the questionnaire is about the same, but the timing is different.

For large panels, computerizing the analysis is almost essential. Even for panels of 50 the manual-processing workload is so heavy that there is no time for adequate analysis, and the turnaround time becomes excessive. Even for small panels, computerizing the computation of medians and quartiles is often worth the effort.

Turnaround Time Between Questionnaires. Delphis run using the mail usually take about a month between successive questionnaires. When Delphis are carried out within organizations located in a small area (plant, laboratory, university campus, etc.), turnaround can be much shorter. For panels of 10 to 15 members, using interoffice mail or couriers, two weeks has often been sufficient to carry out four full rounds. However, the panelists must be motivated to respond promptly, or the advantages of internal communication can be lost.

12. Constructing Delphi Event Statements

A Delphi questionnaire is neither a public opinion poll nor a psychological test. Many critics have failed to understand this and have complained that Delphis do not follow the rules developed for questionnaires in these fields; of course, there is no reason why Delphis should. However, there are some rules that must be followed if Delphi questionnaires are to obtain the information the moderator wants. Some of the most important rules are the following.

Avoid Compound Events. If the event statement contains one part with which a panelist agrees and another part with which he or she disagrees, there can be no meaningful response. Consider the following event: A commercial nuclear fusion plant for generating electricity using deuterium from sea water will begin operation in the year ______. The panelist who thinks that nuclear fusion will be based on the use of tritium cannot respond to this event: If he or she believes that fusion power will be available commercially at a certain date and gives this date, the response may be interpreted as supporting the use of deuterium from sea water. If he or she responds "never," it may be interpreted as doubting that fusion power will ever become commercial. In general, it is best to avoid event statements of the form, "Capability A will be achieved by method B in the year ______."

The moderator can never be certain to have eliminated all compound events. Despite one's best efforts, some panelists may find two distinct parts to what was intended to be a single event. In such a case the feedback between rounds can help the moderator improve the question. Clarifying an event statement on the basis of feedback may be as important as the forecast itself if it uncovers alternatives that were not apparent at first.

Avoid the Ambiguous Statement of Events. Ambiguity can arise from the use of technical jargon or from terms that "everyone knows." Most ambiguity comes from the use of terms that are not well defined. Consider the following event: By the year ______, remote-access computer terminals will be common in private homes. How common is "common?" Ten percent of all homes? Fifty percent? Ninety percent? If 70% of all homes with incomes over $20,000 have terminals, but only 10% of homes where the income is less than that figure do, is this "common"? Descriptive terms such as *common, widely used, normal, in general use, will become a reality, a significant segment of,* and so on are ambiguous and should not be used.

Ambiguity can often be eliminated by using quantitative statements of events. However, consider the statement, "By 19____, the per capita electric power consumption in Africa will be 25% of the U.S. per capita power consumption." Does this mean 25% of today's U.S. consumption or 25% of the U.S. consumption in the same year? Even though the statement is quantitative, it is not clear. Consider the statement, "By 19____, a majority of all foods sold in supermarkets will be radiation sterilized and will not require refrigeration." Does this mean over 50% of each kind of food? Or does it mean over 50% of the total, but some foods not at all? If the latter, is it 50% by weight, volume, or dollar sales?

Avoid Too Little or Too Much Information in Event Statements. Some research by Salancik et al. (1971) shows that it is just as bad for a statement to have too much information as too little.

The researchers related the degree of consensus in the forecast to the complexity of the statement. They measured complexity by the number of words, which is a crude but objective measure of complexity. They also used a measure of consensus more sophisticated than the interquartile range: They borrowed a concept from Information Theory, where the information content of a message is measured in "bits." One bit is the information contained in a single "yes" or "no" when both are equally likely; that is, one bit of information is just enough to answer a binary question. So the degree of consensus was measured in bits by comparing the actual distribution of forecasts with a condition of complete uncertainty, that is, a uniform distribution of forecasts over the entire possible range of years to the time horizon.

Complete consensus would have provided 2.58 bits; however, the actual degree of consensus provided only an average of 0.6 bits per event, about one quarter of the maximum possible. On the average, the greatest consensus was achieved for event statements about 25 words long, which provided about 0.85 bits. The degree of consensus declined for either longer or shorter statements. The number of bits provided was only about 0.45 for both 10- and 35-word statements.

Some of the event statements dealt with technology, while others dealt with applications. For the technology events the shorter the event statement, the higher the degree of consensus. For application statements the reverse was true. Salancik et al. tested the possibility that the panelists were more familiar with the technology than with applications, which would mean that fewer words were needed to adequately define technology statements. They divided the applications statements into three categories on the basis of degree of use, from common to unusual. For the more common applications the most consensus was reached for the shortest statements. For the unusual applications the most consensus was reached for the longest statements.

The panelists were asked to rate their expertise on each question. For the nonexperts the longer a statement, the greater the consensus reached. For the experts the most consensus was reached at intermediate-length statements, with consensus declining for both longer and shorter statements.

The conclusion is that if an event is unfamiliar to the panelists, the more description given, the greater the degree of understanding and the greater the degree of consensus. If an event is familiar, the more description given, the more confusing the statement appears and the less the degree of consensus. Event statements should therefore be chosen to provide neither too much nor too little information. If the panel has trouble reaching consensus, the problem may be an event statement that provides either too much or too little information. The moderator should attempt to clarify this and reword the statement.

13. Summary

In the years since Gordon and Helmer brought the Delphi procedure to public notice, hundreds of Delphi sequences have been run by a variety of organizations and groups, for a variety of purposes. Descriptions of many of these sequences have been published in report form, as well as in the form of articles in journals devoted to management, planning, and forecasting. On the basis of these studies some conclusions can be drawn that should be of value to those considering the use of Delphi.

Delphi does permit an effective interaction between members of the panel, even though this interaction is highly filtered by the summarization of arguments made by the moderator. Several experiments in which the panelists were asked to give reasons as to why they changed their estimates showed that the panelists were, in fact, reacting to the views of their fellow "experts." However, this cannot be viewed as weakness of will. (In one such experiment, on the contrary, one of the panelists claimed that it made him even more "stubborn" to know that "only I had the right answer.") Panelists do shift their estimates when the arguments of their fellow panelists are convincing; otherwise they will hold tenaciously on to their differing opinions.

At the same time, however, there is ample evidence from a number of experiments that if the panelists feel that the questionnaire is an imposition on them, or if they feel rushed and do not have time to give adequate thought to the questions, they will agree with the majority simply to avoid having to explain their differences. In this respect, therefore, the Delphi procedure is not an absolute guarantee against the degrading influences of the "bandwagon effect" and fatigue. However, in a Delphi these problems are to some extent under the control of the moderator, whereas they are virtually uncontrollable in a face-to-face committee or problem-solving group.

The Delphi procedure is thus a feasible and effective method of obtaining the benefits of group participation in the preparation of a forecast while at the same time minimizing or eliminating most of the problems of committee action. It can take longer to complete than a face-to-face committee, especially if the deliberations are carried out by mail. Since it is unlikely that a long-range forecast would be prepared in a hurry, this delay need not be a disadvantage. Even if a forecast must be obtained by a certain deadline, sufficiently advanced planning can usually make the use of Delphi possible. Thus, whenever adequate time is available, Delphi should be considered as a practical approach to obtaining the required forecast.

References

Linstone, Harold A., and Murray Turoff (1975). *The Delphi Method* (Reading, MA: Addison-Wesley).

Nelson, Bradley W. (1978). "Statistical Manipulation of Delphi Statements: Its Success and Effects on Convergence and Stability," *Technological Forecasting and Social Change* 12 (1), 41–60.

Salancik, J. R. (1973). "Assimilation of Aggregated Inputs into Delphi Forecasts: A Regression Analysis," *Technological Forecasting and Social Change* 5, 243–247.

Salancik, J. R., William Wenger, and Ellen Helfer (1971). "The Construction of Delphi Event Statements," *Technological Forecasting and Social Change* 3, 65–73.

For Further Reference

Dajani, Jarir S., and Michael Z. Sincoff (1979). "Stability and Agreement Criteria for the Termination of Delphi Studies," *Technological Forecasting and Social Change* 13 (1), 83–90.

Kendall, John W. (1978). "Variations of Delphi," *Technological Forecasting and Social Change* 11 (1), 75–86.

Linstone, Harold A., and Murray Turoff (1975). *The Delphi Method: Techniques and Applications* (Reading, MA: Addison-Wesley).

Technological Forecasting and Social Change 7 (2) (1975). Entire issue devoted to Delphi.

Problems

1. Which of the following items are likely to require expert judgment to determine, and which would be better obtained by some objective means?
   a. The likelihood of the Supreme Court upholding a patent on a technological innovation.

b. The profitability of a new device, as compared with the device it will replace.
c. The likelihood of new technology being rejected on moral or ethical grounds.
d. The willingness of the public to accept a specific alternative to the automobile for personal transportation.
e. The Federal Government's probable response to a new technological advance.

2. Assume that laboratory feasibility of a radically new technological device has just been demonstrated. This device is based on new principles and is largely the work of one man, who therefore knows much more about it than anyone else in the world. You want to obtain a forecast of its future level of functional capability and degree of acceptance by potential users. Is it worth supplementing the judgment of the inventor by organizing a committee, with him as member, to prepare the forecast? If your answer is yes, what characteristics would you look for in selecting other members of the committee?

3. If a forecast is being prepared by a committee, would you insist that the committee forecast only those things on which a majority of the members agree? Would you even insist on unanimous agreement? Is insistence on agreement likely to produce a better forecast?

4. What are the relative advantages and disadvantages of asking the panel to suggest events whose times of occurrence are to be forecast in subsequent rounds?
5. When would it be desirable to provide a panel with economic, political, and so on, context for the forecast it is to produce?

6. Your company wants a forecast of technological advances that may supplant its current products or provide it with new ones. It is decided to obtain the forecast using a Delphi panel. What are the relative advantages and disadvantages of the following panel types?
a. A panel of experts from outside the company.
b. A panel of experts from within the company.
c. A panel combining both company experts and outside experts.
d. Two panels, one of company experts and one of outside experts.

7. You are an official of a charitable organization that has in the past supplied funds for a great deal of medical research on a particular class of diseases. Cures or satisfactory preventives for these diseases are expected to be available within the next few years. You need to determine the avenues of medical research toward which your organization should shift its support. What kind of a panel (or panels) would you select to provide forecasts useful in this situation?

8. Your company manufactures a type of device that, traditionally, has been

Problems *(continued)*

bought and installed by the Federal Government for widespread public use. Technological progress in the field has been rapid, with successive devices being rendered obsolete by improvements within a few years. As a guide to your company's long-range planning, you wish to obtain a forecast of likely technological progress in the field over the next 20 years. What type of members would you include on a Delphi panel?

9. Correct the following questions so that they will not cause confusion if used in a Delphi questionnaire.

a. Computer-controlled education for self-teaching will be available in the home by the year _____.

b. The teaching-machine market will be a significant part of the total market for educational materials and equipment by the year _____.

c. Power from nuclear fusion will be a reality by the year _____.

d. Electric automobiles will be in common use as "second cars" by the year _____.

e. A majority of office clerical operations now handled manually will be done by computer by the year _____.

Chapter 3

Forecasting by Analogy

1. Introduction

New technological projects are often compared to older projects in terms such as, "This is as big as the Manhattan Project was for its time." The idea is to convey the relative difficulty of the project with respect to the conditions of the time. This is an analogy.

The use of analogies in forecasting simply builds on this notion. It involves a systematic comparison of the technology to be forecast with some earlier technology that is believed to have been similar in all or most important respects.

But what does it mean to be "similar"? And which respects are "important"? Answering these questions is the whole point behind the idea of systematic comparison. This chapter presents a method for identifying those respects that are important and estimating their degree of similarity.

2. Problems of Analogies

The use of analogies is subject to several problems. These must be understood before a suitable method can be devised to overcome them.

The first problem is the lack of inherent necessity in the outcome of historical situations. A forecaster may discover a "model" historical situation, which is then compared with the situation to be forecast. If the two are sufficiently similar, the forecast would be that the current situation will turn out as the model situation did. However, the current situation will not necessarily follow the pattern of the model situation; only in Greek tragedy is the outcome inevitable. Moreover, a study of historical records will show that the actual outcome of most historical situations turned out to be a surprise to at least some of the participants, who at the time expected something else. Some historical outcomes appear implausible even in retrospect. Even though we know what happened, we find it hard to believe that people acted as they did. Analogies are based on the assumption that there is a "normal" way for people to behave and that given similar situations, they will act in similar ways. However, there is no guarantee that people today will act as people did in the model situation. Hence the forecast is at most probable, never certain.

The second problem is historical uniqueness: No two historical situations are ever alike in all respects. Thus it is important to be able to say which respects are important and which can be ignored. An analogy will be strengthened if there are several historical cases with parallel outcomes that can be compared to the present case to be forecast. However, since each of these cases is unique in itself, it is important to be able to determine whether they are "similar enough" to each other so as to be considered analogous; hence a systematic means for comparing model situations with each other and with the current situation is essential.

The third problem is historically conditioned awareness. Even though a historical situation may be judged to be sufficiently similar to the present situation to be called analogous, people may be aware of the previous outcome. If they do not like the way the previous situation turned out, they may deliberately act differently this time in order to secure a more preferred outcome. Historically conditioned awareness, then, violates the assumption that there is a "normal" way for people to behave and that they always behave that way. Thus, despite a forecaster's best efforts to check for analogous situations, the forecast may be invalidated by people's awareness of the prior outcome.

Despite these problems, analogies can be a very useful method for forecasting technological change. The problems cannot be solved completely, but they can be minimized by using a systematic method for establishing analogies.

3. Dimensions of Analogies

We wish to compare a historical or model situation with a current situation in order to develop an analogy between the two. Since we are primarily concerned with technological change, it is important to compare the two situations on the basis of factors that affect technological change: the invention of some device or procedure, adoption of the invention, and widespread diffusion of the invention. The following list of dimensions is based on studies of factors that have affected technological change, and it therefore provides a suitable basis for comparing situations for a possible analogy:

1. Technological.
2. Economic.
3. Managerial.
4. Political.
5. Social.
6. Cultural.
7. Intellectual.
8. Religious–Ethical.
9. Ecological.

We will discuss briefly how each of these dimensions has affected technological change in the past and why, therefore, it is important to compare two possibly analogous situations with respect to these dimensions.

The Technological Dimension. Technologies do not exist in isolation; every technology exists to perform some function desired by people. There are usually alternate ways of performing the same function that compete with the technology in question. The technology must also draw on supporting technologies. Finally, the technology must mesh with complementary technologies that perform other functions. A comparison of two situations with respect to the technological dimension requires that each of these three aspects be compared.

The technology used as the "model" had to be superior to its competing technologies or else it would not have succeeded. The forecaster must identify the competing technologies in the model situation and in the current situation to be forecast and compare these. What are the weaknesses of the competing technologies then and now? Are they comparable? What are the strengths of the model technology and the technology to be forecast? Are they comparable? Does the technology to be forecast provide the same degree of advantage over competing technologies as did the model technology?

A technology must draw on other technologies for support in such things as production, maintenance, energy supply, and transportation. For instance, the Newcomen steam engine, invented in 1705, was limited in its power output by the size of the cylinder. Machine tool technology was not good enough to allow machinists to bore large cylinders that were round enough to have tight-fitting pistons. In 1744 Wilkinson built the first precision boring lathe, which was used to bore cylinders up to 50 in. in diameter for Watt's steam engine, and do this with unprecedented accuracy. This supporting technology was critical to the success of Watt's steam engine.

A technology must be compatible with complementary technology, usually in terms of input and output. This compatibility may be as simple as a common truck bed height or railroad gauge, or it may be as sophisticated as the pitch and shape of screw threads or the frequency of an alternating-current power supply. The steam engine provides another example of this need for complementarity. During the first half of the 18th century European industry used machines such as grinding and crushing mills driven by wind or water power, which required rotary motion in their motive power. They could not have been driven by a steam engine, because all steam engines of the time produced reciprocating motion (which was, however, ideal for driving pumps). Only after Watt invented the sun-and-planet gear to convert reciprocating into rotary motion was the steam engine used for driving industrial machinery.

In examining the technological dimension, one must look beyond the physical hardware to the theories behind both the model technology and the technology to be forecast. The forecaster must ask whether the predictive and explanatory powers of the theories and the level of understanding of the relevant phenomena are the same in the two cases. The question is not whether one theory is more advanced than the other, since the more recent theory is likely to be more advanced; the question is whether the theories in both cases were equally capable of meeting the demands placed on them.

The Economic Dimension. Technology is intended to perform some function. It will be used for that function only if people are willing to pay for it, that is, give up alternate uses for the resources needed to deploy the technology. Therefore the forecaster must look at the ability and willingness of the relevant people, in both the model and present cases, to pay for the technology.

The relevant cost is not the unit cost to the final consumer but the cost of deploying the entire system. For instance, we would be less concerned with the cost of an airline ticket than with the cost of building a new generation of transport aircraft. Relevant costs may include research and development, capital investment, manufacturing costs, and the costs of support and maintenance. Even the costs of training operators and technicians may have to be included. However, the proper comparison is not between absolute costs but between relative costs. The range of costs of the engineering projects shown in Chapter 5 covers six orders of magnitude. If we were to develop an analogy between two of them, however, we would be concerned not with the dollar cost of each but with the cost of each as a fraction of the total resources available to those supporting the projects.

In addition to comparing the relative costs of two possibly analogous projects, we must look at the financial mechanisms available for mobilizing resources. Can money be raised on the scale necessary? Are risk-spreading mechanisms available so that many segments of society can share in the funding of the project? The essential feature in an analogy is not the similarity of financial mechanisms in the two cases but their respective abilities to raise the necessary resources.

It is also necessary to compare the "market" in the two cases, that is, compare the demand for the technology in terms of market size. History is full of cases of two essentially similar technologies, one of which succeeded because it had an adequate market and the other of which failed because a market was lacking. For instance, McCormick is generally considered to be the inventor of the reaper, yet his machine differed little from any invented in the decades prior to 1841. The reason why McCormick's reaper succeeded where previous ones failed is that McCormick introduced it just when the rolling farmlands of Ohio, Indiana, and Illinois were being opened to wheat farming. The rocky and uneven soil of the older states had been unsuitable for mechanized reaping, and machines invented earlier lacked the market McCormick had.

In addition to the size of resources, it is necessary to compare economic climates. If the historical technology was introduced at a time of economic expansion whereas the present is a time of contraction, an otherwise good analogy may be invalidated.

Finally, it is necessary to compare the two situations in terms of economic theories. Differences in economic theory may spoil an analogy despite agreement on other elements of the economic dimension. As an example, consider the economic theories of the Soviet Union. Marx held that the owner of capital performed no service and therefore deserved no payment for its use. Hence in the Soviet Union industrial enterprises were not charged interest on the money they borrowed to finance expansion; thus capital was costless, but labor had to be paid wages. Soviet factory managers therefore emphasized automated and highly mechanized factories, which led to a great deal of "hidden unemployment." Under similar conditions, a manager who had to pay for both capital and labor would have employed more workers and less automation. Here is a clear-cut case where economic theory had an impact on technological change.

The Managerial Dimension. The introduction of technological change does not just happen; it must be managed. Therefore it is necessary to compare the levels of managerial capability in the model and current situations relative to the size and complexity of the task.

Size depends upon the number of people to be managed and their geographic spread. How big was the model task relative to tasks already successfully managed? Was it comparable or even smaller? Was it bigger? Then, how big is the current task relative to tasks already managed successfully? Is it comparable? If it is larger, is the increase in size no greater than in the model task?

Complexity is measured by the number of different types of activities involved and the number of locations at which these must be carried out. Was the complexity of the model task greater or lesser than the complexity of tasks previously managed successfully? Is the complexity of the current task within the demonstrated capabilities of project management? If it is greater, how does the increase in complexity compare with that in the model project?

What of the pool of available managerial talent? It is not a question of absolute size but of relative size. Does the current project demand a bigger share of the pool of talent than did the model project? If so, the analogy would be invalid.

Finally, it is necessary to compare managerial techniques. The ability to manage "larger" projects comes from better managerial techniques and procedures—managerial "technology." The comparison then demands that the managerial technology in both the model and current cases be equally adequate to the task. A difference in the absolute level of managerial technique is to be expected; however, a difference in the relative adequacy in the two cases would invalidate the analogy.

The Political Dimension. Webster's Seventh Collegiate Dictionary defines politics as "competition between competing interest groups or individuals for power and leadership in a government or other group." Innovation represents a change that will affect the relative power of different groups. The basic questions about the interaction of politics and technological change, therefore, are, Who benefits? and, Who gets hurt?

Those who benefit from a technological change will try to encourage it; those who suffer from it will try to stop it. Hence it is necessary to compare the relative political power of the people who benefit and of those who get hurt in both the model and current situations.

Those who may suffer from the introduction of a new technology include the producers, suppliers, and operators of the old technology. Some other "losers" are the people involved in related institutions, such as craft unions, or government regulatory bodies. Conversely, "winners" include those who will produce, supply, and operate the new technology, as well as their related institutions. Of course, the consumers are also winners, since if the new technology does not benefit them they will not pay for it anyway. However, consumers are often unorganized and unable to take part in political debates about how some service will be supplied to them. They are often the unfortunate victims in political struggles in which they cannot take part. A good example is cable television, which consumers clearly wanted. The television networks feared it, and the Federal Communications Commission hampered its growth for over a decade in order to protect the networks.

In addition to direct winners and losers, there may be people who have nothing against the technology itself but who oppose it because it hinders political goals they support. The use of nuclear explosives for excavating canals and harbors has been halted, not because the technology itself is such a problem, but because growth of that technology might hamper the political goal of nuclear disarmament. Therefore those opposed to nuclear weapons have also objected to "peaceful uses" of nuclear explosives.

In addition to comparing the relative power of winners and losers in the model technology and the current case, it is necessary to compare the political theories in the two cases. What are the rights and duties and the privileges and obligations of different groups in the two cases? If a group wishes to oppose or support a technological change, can it do so within the prevailing political theories and laws? In addition to comparing the model and current situations in terms of political power, the forecaster must compare them in terms of the avenues of support or opposition "legitimately" open to the various groups.

The Social Dimension. Webster's Seventh Collegiate Dictionary defines society as "a community, nation or broad grouping of people having common traditions, institutions and collective activities and interests." Every technological change occurs within a society and both acts on and is acted on by that society. Hence the forecaster must compare a model situation with the present situation in terms of the society into which the innovation was or will be introduced. The comparison must include the people in the society, their institutions, and their traditions and customs.

The people making up a society can be characterized in terms of total population, age distribution, geographic distribution, income distribution, urban/rural distribution, and so on. In comparing two possibly analogous situations the important consideration is not the absolute numbers but the relative sizes. Was the model innovation introduced into a mostly rural society, while the present innovation is being introduced into a mostly urban society? This difference may invalidate the analogy. However, if both the model and current technologies are agricultural technologies, the urban/rural mix may be unimportant by comparison with the age and income distributions of the rural people. The comparison must be made on the basis of the relevant portions of society.

Institutions include the family, schools, business, government, and churches. Each of these can have a favorable or unfavorable impact on technological innovation. Conversely, an innovation may be opposed or supported because it has favorable or unfavorable impacts on one of these institutions. A comparison between two possibly analogous situations must include the effects of these institutions on the technologies being compared; it must also include an analysis of the effects of the technologies on these institutions and possible reactions of the supporters of these institutions.

The traditions and customs of a society bind it together and reflect its self-image. A technological change may be supported if it seems to fit in with this image, or enhance or extend it. Conversely, a technological change may be opposed if it seems to undermine tradition or leads a society away from the ideal image its people hold of it. Thus, in comparing two technological changes for a possible analogy, the forecaster must take into account how each fits in with the traditions and customs of the society into which it was or will be introduced.

The Cultural Dimension. This dimension deals with the values, attitudes and goals of the society into which the innovation is to be introduced. Rescher (1969) gives the following definition: "A value represents a slogan capable of providing for the rationalization of action by encapsulating a positive attitude toward a purportedly beneficial state of affairs." By *slogan*, Rescher means that the value can be named with a catchword; by *rationalization,* Rescher means that it relates to the justification, critique, defense, recommendation, and so forth, of some course of action. People have needs and desires and are capable of using reason. In particular, they are capable of using reason to achieve goals, rather than simply responding to stimuli. Values arise from this fact, since "to have a value is to be able to give reasons for motivating goal-oriented behavior in terms of benefits and costs, bringing to bear explicitly a *conception* of what is in a man's interests and what goes against his interest (Rescher, 1969)."

To make this idea concrete, Rescher lists some values: health, comfort, physical security, economic security, productiveness, honesty, fairness, charitableness, courtesy, freedom, justice, beauty, clearness of conscience, intelligence, and professional recognition. Different societies and people at different times and places have ranked these in differing orders of importance.

Since values provide a rationalization for action, they will affect technological change. This may come about in two ways: The values subscribed to by a society may favor or inhibit change; conversely, a proposed technological change may threaten or enhance an important value. Thus a society may be favorably or unfavorably disposed toward change in general, or a society may oppose or support a particular change because of the technology's impact on values subscribed to by the society. A comparison between two technologies, for a possible analogy, must take social values into account. Two societies must be compared on the basis of the general effect of their values on innovation and on the basis of specific reactions to particular innovations that may affect important values.

The Intellectual Dimension. The preceding two dimensions dealt with society at large. This dimension deals with the intellectual leaders of society, including decision makers for public and private organizations, people who speak on behalf of prestigious institutions or causes, and opinion leaders such as editors, poets, and writers.

The intellectual leaders of a society may support or oppose a technological change on the basis of whether it enhances or diminishes their values and goals. The effect of their opposition or support will, however, depend upon the extent to which their leadership is actually followed by others. A comparison between a model case and a current case for a possible analogy must then examine two things. It must first examine the extent to which intellectual leaders have a following. If the intellectual leadership has a significantly different degree of effectiveness in the two cases, the analogy is invalidated. If the intellectual leadership has a significant impact in both cases, then it is necessary to look at the way in which the two technologies affect and are affected by the values and goals of the intellectual leadership. If the interaction is the same in both cases, then the analogy is supported in this dimension. If intellectual leadership has no impact in either case, then this dimension can be omitted from consideration.

The Religious–Ethical Dimension. Most people judge actions and events on the basis of some standard of right or wrong, and these judgments affect technological change, just as they affect virtually everything else. This dimension has two components: (1) the beliefs that guide ethical judgments and (2) the institutions which promulgate or formulate ethical beliefs.

Beliefs may affect technological change in two ways. First, people may oppose or favor technological change on general grounds. Some religions, for instance, require their members to refrain from using modern technology such as mechanized farm equipment. Second, people may oppose or favor specific technological changes because they impact unfavorably or favorably on religious doctrine. When two cases are compared for a possible analogy, it is necessary to compare both the general beliefs regarding technology and the specific beliefs regarding the two technologies considered for analogy.

Religious institutions may act in an organized manner to oppose or favor technological change on much the same grounds as their individual members. They may have generally favorable or unfavorable attitudes toward technological change. In addition, they may have favorable or unfavorable attitudes toward specific technological changes, depending on how these affect their beliefs, how they impact on their members, or how they impact on the institutions as institutions. Religious institutions may also favor certain technological changes (the printing press, the Radio Gospel, the Electronic Church) because these changes make it easier for them to expand the teaching of their doctrines. When two cases are being compared for an analogy, it is necessary to consider not only the attitudes of religious institutions but also the number of members in each institution and the degree of control the institution has over its members. For a valid analogy the situations should be similar in both cases.

Another type of institution may have an impact on technological change through the ethical dimension. This is the professional society which establishes standards of ethical practice for its members. Codes of ethics established by medical societies and codes of engineering ethics established by engineering societies are examples of institutional measures that may affect technological change, either hastening it or retarding it. In comparing two cases for an analogy, it is important to identify any applicable codes of ethics that affect the practitioners of the relevant technologies. It is necessary to consider not only the existence of the codes but also the membership size of the professional societies and the degree to which the society members actually adhere to the published codes. For two situations to be analogous the applicable codes of ethics should have similar effects (retarding or enhancing) and the degree of membership support for these codes should be comparable.

The Ecological Dimension. In the past, ecological considerations have not had significant effects on technological change. However, now there is a great deal of concern about the ecological impact of new technologies, and it is therefore important to take this dimension into account. The fact that an ecologically damaging technology was accepted in the past cannot be taken as an indication that one will be accepted today.

From the standpoint of forecasting, the most important aspect of this dimension may be the level of ecological damage that is acceptable to the people involved. If a new technology appears to exceed this level, it may not be adopted; if an old technology exceeds this level, it may be replaced by a more benign technology.

It is important that this dimension not be viewed solely in terms of pollution. Other effects besides pollution are important and must be considered.

It is also important to recognize that not all changes to the natural environment are harmful; man improves the ecology as well as damages it. Technologies that promise such an improvement may be adopted more readily than those which do not. Thus analogies should take into account the possibility that the technologies to be compared represent improvements in the ecology, as well as the possibility of their representing threats.

This completes the description of the dimensions of an analogy. Using these dimensions, it is possible to make a systematic comparison between two situations to determine whether a valid analogy exists. Not all the dimensions will be important in each case; however, the forecaster should always check each dimension to ascertain whether or not it is important. The forecaster should not fall into the practice of ignoring certain dimensions simply because they have been found to be unimportant in specific cases. Conversely, the validity of an analogy in one dimension should not be allowed to outweigh the forecaster's judgment that the dimension is not important, particularly if it causes one to overlook the failure of the analogy in other dimensions.

4. Deviations from a Formal Analogy

Suppose that two situations were compared for a formal analogy, and the requirements were found to be violated in several dimensions, but always in the same direction? That is, suppose all the deviations were such as to make innovation more likely? Even though a formal analogy, then, does not exist, can the forecaster still predict that the technology of interest will follow the same pattern as in the model case, only more rapidly because conditions are more favorable? For instance, could one not forecast that space technology likely to reduce environmental degradation will be adopted for nonspace use even more rapidly than railroad technology was adopted for nonrailroad use because of a greater interest in preserving the environment at the present time?

The forecaster cannot do so on the basis of analogy, since a formal analogy does not exist. If a forecast is attempted in this manner, one is actually developing a crude version of an analytical model, a topic we shall take up in Chapter 8. The difficulty of this situation is that there is neither the precedent of an analogy nor the rigor of a true analytical model to support the forecast. The forecaster thus has no basis whatsoever for making a forecast with any precision. The analogy is not based on any assumption of necessity or inevitability, but only on the assumption that under a certain set of circumstances there is a "normal" way in which people behave and that given a repetition of the circumstances, there is likely to be a repetition of the behavior. With the circumstances changed, this support disappears. The analytical model is based on certain assumptions about the inner workings of the system in question and is generally based on a notion of the "necessity" of a certain type of behavior. As we shall see later, this assumption may not be valid. However, in the case of deviations from an analogy, we do not even have this much to provide support for any conclusions.

What is the forecaster to do, then, when there are no strictly analogous cases but one or more pairs of situations where the deviations from a formal analogy all tend in the same direction? Presumably an attempt was made to use analogies because in the case in question the other methods available either could not be applied or were unsatisfactory. As is repeatedly pointed out in this book, one cannot refuse to make a forecast.

Since these pairs of situations are the only material available, the forecaster must make the best of them. The essential point to be remembered is that in this situation one must inevitably depend on a great deal of judgment, one's own and possibly that of as many experts in the subject matter as possible. The utility of the not-quite-analogy, not-quite-model developed is to provide a framework in which judgment can be exercised. The judgments the forecaster uses will be estimates of the impact, one way or another, of the deviations from strict analogy. With this framework, and with the estimates of the impact of deviations, a more or less precise forecast of the course of development of the technology of interest can be prepared. Even though it may not be as satisfactory as complete development of one or more formal analogies, it is better than simply giving up in despair.

5. Summary

In this chapter we have discussed the use of historical analogies for forecasting purposes. In cases where the forecaster wants to make use of something more rigorous than judgment alone, historical analogy may be used. If some previous situation can be found that is similar to the one to be forecast, the forecaster may then predict that the results in the situation of interest will be analogous with the outcome of the historical situation.

The use of analogies is a very common way of thinking and the basis of inductive inference. However, there is nothing inevitable about the outcome of a situation. There is no guarantee that if the circumstances are repeated in detail, the outcome will be repeated also. Nevertheless, by assuming that there is a normal or natural way for people to behave in a particular situation, and that the historical model situation illustrates this natural way of behavior, the forecast can be based on the historical model situation.

Since the use of analogies depends on comparing two situations in detail, it is helpful to have some systematic way of making these comparisons. For this purpose we have developed a set of dimensions in which situations can be compared. Because the primary interest here is in forecasting the progress of technology, the dimensions are based on factors that have historically demonstrated a connection with a tendency either to innovate or to resist innovation. In particular cases, however, other sets of dimensions might be more useful. Thus, while this set has proven to be useful in some technological forecasting situations, the forecaster should be alert to the possibility that for the case in question the set of dimensions may need to be modified to suit the problem.

Finally, in those cases where the forecaster cannot find any cases that are formally analogous to the one to be forecast, but can find some in which the deviations from formal analogy are all in the same direction, a forecast is still possible. This forecast has neither the support of historical precedent nor the rigor of an analytical model; it will be based on judgments about the impact of deviations from strict analogy. However, the use of the framework provided by the incomplete analogy will allow a forecast that should be better than one obtained through the use of completely unstructured judgment.

References

Rescher, N. (1969). *Introduction to Value Theory* (Englewood Cliffs, NJ: Prentice-Hall).

Problems

1. Explain the advantages and disadvantages of forecasting by the use of analogies.

2. What is "historically conditioned awareness," and what effect may it have on analogical forecasts?

3. Develop the technological dimension of a possible analogy between the construction of the Suez Canal and of the Panama Canal. Are they analogous in this dimension?

4. Develop the economic dimension of a possible analogy between the Manhattan Project (U.S. Atomic Bomb program, 1942–1945) and the U.S. Air Force Ballistic Missile Program (1954–1957). Are they analogous in this dimension?

5. Develop the managerial dimension of a possible analogy between the Manhattan Project and the Ballistic Missile Program. Are they analogous in this dimension?

6. Develop the political dimension of a possible analogy between the introduction of aircraft and of ballistic missiles into the American Armed Forces. Are they analogous in this dimension?

7. Develop the social dimension of an analogy between the introduction of railroads into France and into England during the period 1820–1850. Are they analogous in this dimension?

8. Develop the cultural dimension of a possible analogy between the introduction of television in the United States and in the Soviet Union. Are they analogous in this dimension?

Problems *(continued)*

9. Develop the intellectual dimension of a possible analogy between the introduction of motion pictures and of television in the United States. Are they analogous in this dimension?

10. Develop the religious–ethical dimension of a possible analogy between the Methodists' and the Amish Mennonites' attitudes toward technological changes. Are they analogous in this dimension?

11. Develop the ecological dimension of a possible analogy between the introduction of fossil-fueled and of nuclear electric power plants in the United States. Are they analogous in this dimension?

Chapter 4

Growth Curves

1. Introduction

The term *growth curve* represents a loose analogy between the growth in performance of a technology and the growth of a living organism. Despite some attempts to make the analogy rigorous, it is at best a casual one. Nevertheless, the term *growth curve* is descriptive and evokes an image of a pattern of technological change.

Figure 4.1 shows the growth in height and weight of an individual. The growth in height, especially, presents a pattern of rapid initial increase, followed by slower increase, and finally the approach to a limit. Figure 4.2 illustrates the same behavior for two different technical approaches to illumination technology, the incandescent lamp and the fluorescent lamp. The similarity between the curves for organic growth and technological change is the basis for the term *growth curve*.

Growth curves are used to forecast the performance of individual technical approaches to solving some problem. A particular technical approach, such as incandescent lighting, will be embodied in a succession of devices all using the same principles and often much the same materials and design. Any single technical approach is limited in its ultimate performance by chemical and physical laws that establish the maximum performance which can be obtained using a given principle of operation. Adoption of a device using a different principle of operation means a transfer to a new growth curve.

Growth curves are therefore used to forecast how and when a given technical approach will reach its upper limit. History seems to show that when a technical approach is new, growth is slow owing to initial difficulties. Once these are overcome, growth in performance is rapid. However, as the ultimate limit is approached, additional increments in per-

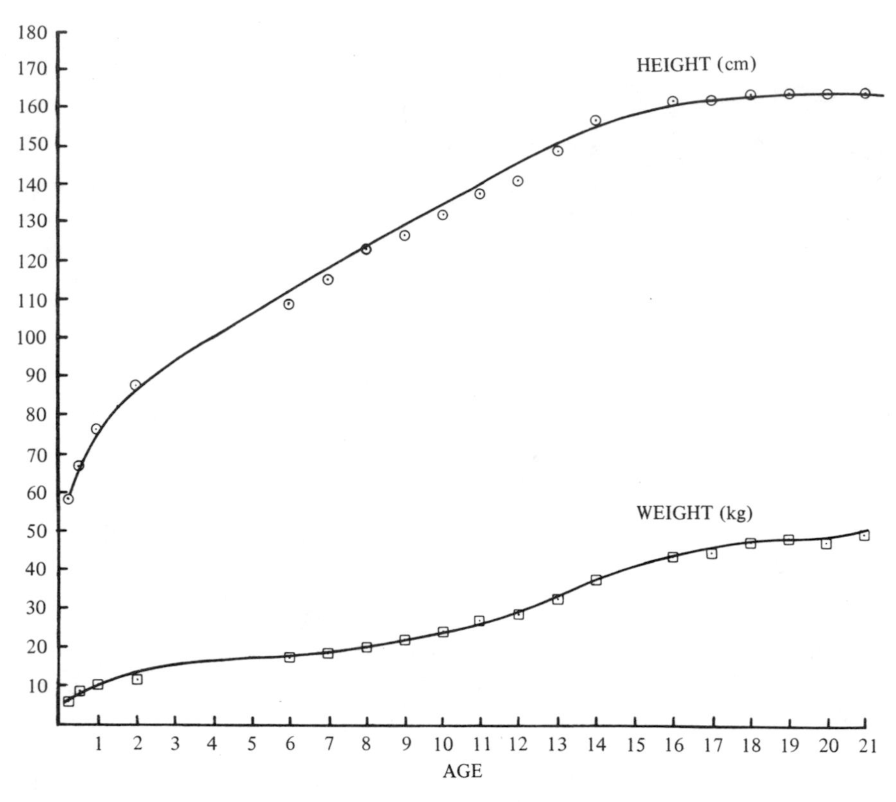

Figure 4.1. Growth in height and weight of an individual (data courtesy of Theresa Martino).

formance become more difficult, and growth again becomes slow. It is often important to forecast the timing of these changes in growth rate, both to determine for how long a given technical approach will still be competitive and to plan for the conversion to a new technical approach.

2. Substitution Curves

Frequently one is interested in forecasting the rate at which a new technology will be substituted for an older technology in a given application. Initially the older technology has the advantage in that it is familiar, its reliability is probably high, and both spare parts and technicians are readily available. The new technology is unknown and its reliability is uncertain; spare parts are hard to come by and skilled technicians are scarce. Hence the initial rate of substitution of the new technology is low. As the initial problems are solved the rate of substitution increases. As the substitution becomes complete, however, there will remain a few applications for which the old technology is well suited. The rate of substitution slows as the older technology becomes more and more difficult to displace.

The substitution of a new technology for an older one, then, often exhibits a growth curve. Behavior of this type is shown in Figure 4.3, which shows the substitution of mechanical power for sail power in the U.S. Merchant Marine.

Growth curves are frequently used to forecast the substitution of one technology for another, as well as for forecasting improvements in technical approaches. The usual use is to forecast the share of the market or the share of total installations. If the size of the market is changing, this is usually forecast separately; that is, growth curves for market substitution are often combined with forecasts of total use to obtain forecasts of actual market share or actual numbers of installations. Here we will focus only on forecasting the rate of substitution. Combining this with forecasts of total installations will be taken up in Chapter 10.

Figure 4.2. Growth in efficiency of incandescent and fluorescent lamps.

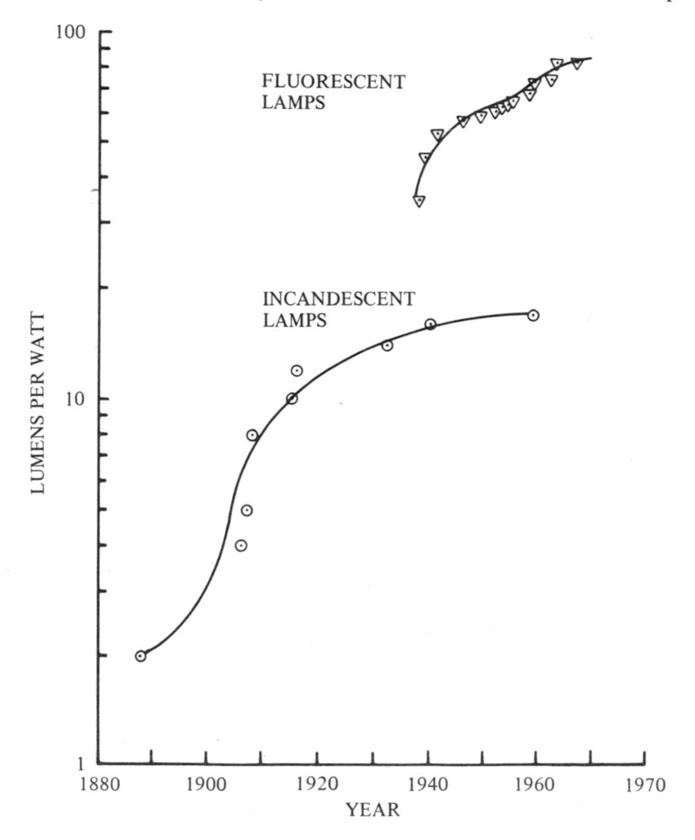

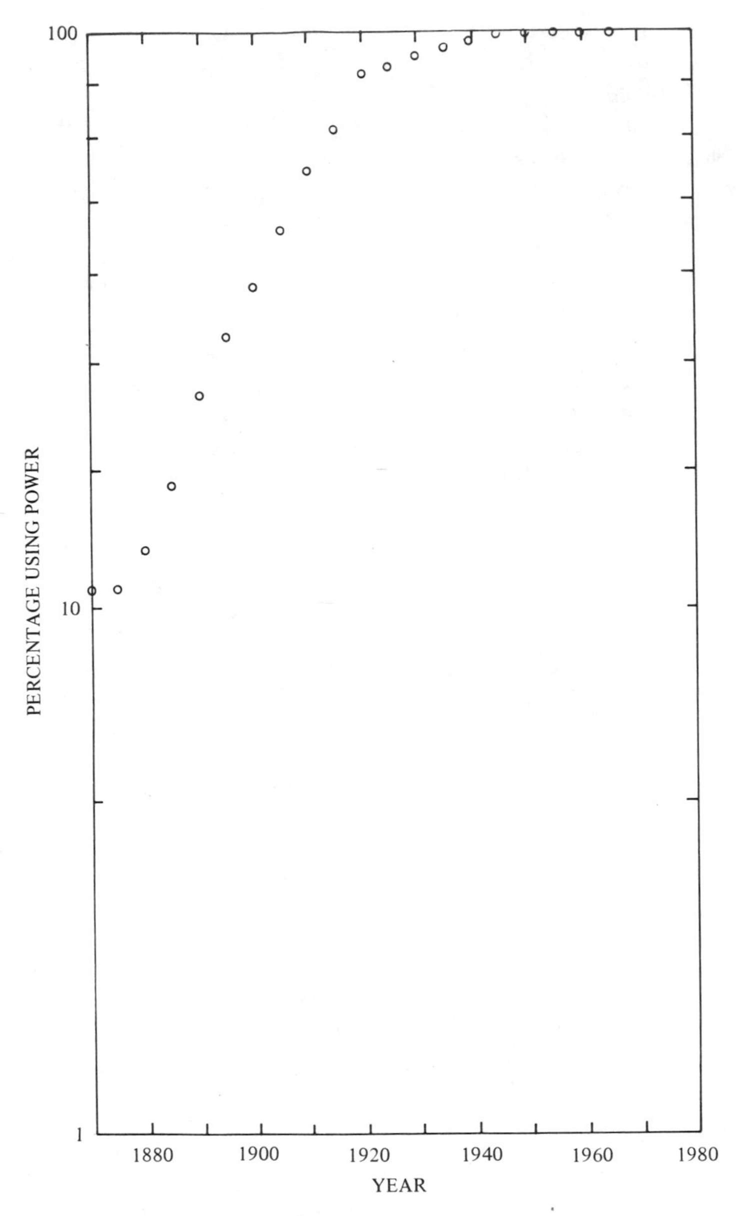

Figure 4.3. Percentage of the U.S. Merchant Marine using mechanical power.

3. The Pearl Curve

Forecasting by growth curves requires fitting a mathematical formula for a growth curve to a set of historical data. The fitting is usually done using linear regression (least squares). Once a mathematical formula has been fitted to the data, the forecast is made by extrapolating that curve to the years beyond the historical data.

One common formula for a growth curve is the Pearl curve, named after the American demographer Raymond Pearl, who popularized its use in demographic forecasting. The formula for the Pearl curve is

$$y = L/(1 + ae^{-bt}), \tag{4-1}$$

where L is an upper limit to the growth of the variable represented by y, a and b are the coefficients evaluated by "fitting" the curve to the data, and e is the base of the natural logarithms. The curve has an initial value of zero at time $t = -\infty$ and a value of L at time $t = +\infty$. (Note that if the initial value for a technology is nonzero, it can be added as a constant to the right-hand side). The inflection point of this curve occurs at $t = (\ln a)/b$, when $y = L/2$. The curve is symmetrical about this point, with the upper half being a reflection of the lower half.

The Pearl curve is unusual among growth curves in that the shape (steepness) and location can be controlled independently. Changes in a affect the location only; they do not alter the shape. Conversely, changes in b affect the shape only; they do not alter the location. This property makes the Pearl curve useful in a variety of mathematical applications besides forecasting.

Fitting a Pearl curve to a set of data by linear regression can be done only if the curve is first "straightened out"; that is, it must be transformed so that it is no longer S shaped but a straight line. This requires the following algebraic transformation of the Pearl curve formula:

$$Y = (L - y)/y = \ln a - bt; \tag{4-2}$$

That is, the difference between y and the upper limit L is divided by y. This can then be regressed on the time t. The constant term from the regression is $\ln a$, and the slope term from the regression is b. The coefficient b is taken to be intrinsically positive, so the line representing the transformed variable Y slopes down and to the right. Once a and b are obtained from the regression, they can be substituted back into the Pearl curve formula. The formula can then be extrapolated to future values of time by substituting the appropriate value for t.

Appendix 4 contains a growth curve program that can be used to fit a Pearl curve to a set of data. However, it is also possible to use a programmable pocket calculator. The transformation in equation (4-2) combined with the regression equations given in Appendix 1 can be programmed in about 200 steps. This is well within the range of many calculators now available. For modest-sized data sets, use of a calculator may be even more convenient than use of a computer. Moreover, the calculator can be used anywhere, while the forecaster may not always have access to a computer.

4. The Gompertz Curve

Another frequently used growth curve is the Gompertz curve, named after Benjamin Gompertz, English actuary and mathematician. The equation for the Gompertz curve is

$$y = Le^{-be^{-kt}}. \tag{4-3}$$

Like the Pearl curve, the Gompertz curve ranges from zero at $t = -\infty$ to the upper limit L at $t = +\infty$. However, the curve is not symmetrical. The inflection point occurs at $t = (\ln b)/k$, where $y = L/e$. Just as with the Pearl curve, it is necessary to straighten out the Gompertz curve before linear regression can be used to obtain the coefficients b and k. This is done by taking the logarithm twice, obtaining the linear form

$$Y = \ln[\ln(L/y)] = \ln b - kt. \tag{4-4}$$

When Y is regressed on t, the constant term is $\ln b$ and the slope term is k. As with the Pearl curve, the slope term is taken to be intrinsically positive. The straight line obtained by this transformation slopes down and to the right. Once b and k are obtained from the regression, they can be substituted in the formula. Future values of t may then be substituted into the formula to obtain forecasts of the variable y.

The growth curve program in Appendix 4 can also be used to fit a Gompertz curve. In addition, what was said above about using a pocket calculator for Pearl curve fitting applies equally well to the Gompertz curve.

5. Choosing the Proper Growth Curve

Given a set of historical data on some technical approach, which growth curve should the forecaster select to fit to the data, Pearl or Gompertz? Some forecasters fit both and choose the one with the largest correlation coefficient as the basis for subsequent forecasts. This is extremely bad practice. Forecasters are not concerned with how well a curve fits past data but, rather, with how well the curve will perform as a forecast.

The forecaster should select the growth curve that most closely matches the underlying dynamics which will produce the future behavior of the technology. The Pearl and Gompertz curves have completely different underlying dynamics. This can be seen by taking their derivatives: The slope of the Pearl curve is proportional to $y(L - y)$ and the slope of the Gompertz curve is (for $y > L/2$) proportional to $L - y$ only. That is, the slope of the Gompertz curve is a function only of distance to go to the upper limit, whereas the slope of the Pearl curve is a function of both distance to go and distance already come.

Consider the diffusion of a new technology. At the beginning there may be few suppliers, few repair facilities, and little information available to the would-be adopter. As adoption progresses, however, suppliers and repair facilities become more widespread and information more widely available. Thus increasing substitution makes further substitution easier. On the other hand, substitution will take place first in the easiest applications; the more difficult applications will remain for later. Thus the rate of substitution depends on both the success to date and the number of applications for which it has not yet been adopted (these being the toughest). In short, under these circumstances the rate of substitution is a function of both the distance to go and the distance already come, and the Pearl curve is the appropriate choice.

On the other hand, if success in substitution does not make further substitution easier, then the rate of substitution will be proportional only to the number of applications that are still holdouts. The farther the substitution progresses the more difficult further progress will be. Under these circumstances the Gompertz curve is the appropriate choice.

If the forecaster is concerned with the performance of a technology rather than with its substitution for an old one, the same considerations apply. Clearly progress is going to be harder the more the upper limit on performance is approached. But is there any offsetting factor such that some progress makes further progress easier? If there is, then the rate of improvement is a function of both the progress made so far and that remaining before the upper limit is reached, and the Pearl curve is appropriate. If there is no offsetting factor, then progress is a function only of the difference between current performance and the upper limit. In this case the Gompertz curve is the proper choice.

The correct way to choose a growth curve, then, is to determine the underlying dynamics of the situation. This requires some examination of the factors that hasten or retard growth in the performance of the technology. When the underlying dynamics have been determined, the appropriate growth curve can be readily selected. The choice of growth curve must be made on the basis of the factors driving the change in technology, not on the basis of mathematical goodness of fit to historical data.

6. The Base 10 Pearl Curve

The formula given earlier for the Pearl curve involved the base of the natural logarithms, e. However, the formula can also be given as follows:

$$y = L/(1 + 10^{A-Bt}). \tag{4-5}$$

With some algebraic manipulation similar to that applied to the standard version of the Pearl curve, this can be linearized as

$$Y = \log[y/(L - y)] = -A + Bt. \qquad (4\text{-}6)$$

The coefficients are denoted by capital letters to show that their numerical values are different from those in the standard Pearl curve; however, they play the same role. Coefficient A controls location and coefficient B controls the shape of the curve. The advantage of this formula for the Pearl curve is that the transformed variable $y/(L - y)$ forms a straight line when plotted against t on semilog paper. Of course, one does not have to express the Pearl curve in this form; all that is necessary is to transform the data as in equation (4-6). Not only is it then suitable for linear regression, but it can be plotted directly on semilog paper.

Note that the ratio $y/(L - y)$ increases by a factor of ten for every increase in t of $1/B$. This is the "tenfolding time" or "T time," and it is characteristic of a base 10 Pearl curve with shape coefficient B. It takes four T times for the ratio $y/(L - y)$ to go from 0.01 to 0.99, or for the ratio y/L to go from 0.0099 to 0.990099. A base 10 Pearl curve can be characterized in terms of its T time and the value of t at which the ratio $y/(L - y)$ achieves some specified value such as 1.0.

7. The Fisher–Pry Curve

Fisher and Pry (1971) developed an equation for a growth curve that seemed to fit a great many cases of technological substitution. Their equation was

$$f = \tfrac{1}{2}[1 + \tanh a(t - t_0)]. \qquad (4\text{-}7)$$

In this equation f is the fraction of applications in which the new technology has been substituted for the old, t_0 is the time for 50% substitution, and the coefficient a is the shape coefficient for the curve.

Fisher and Pry found that by plotting the ratio $f/(1 - f)$ on semilog paper, the substitution curves for some 17 different cases fell very close to straight lines. What was not realized at the time was that their formula was simply an alternate version of the Pearl curve; it can be converted into the standard form simply by algebraic manipulation. In effect, however, Fisher and Pry were the first to use the base 10 Pearl curve as a substitution curve. Thus the Fisher–Pry curve can be characterized in terms of its T time and the time for 50% substitution.

It is now quite common to use the Fisher–Pry transformation to plot substitution curves: the fraction of market share captured divided by the fraction not yet captured is plotted versus time on semilog paper. A straight line is fitted to the plot and projected to forecast the future levels of substitution. An example of this will be shown later in this chapter.

Fisher–Pry curves can be used to illustrate successive substitutions. Marchetti (1977) provides examples showing the successive use of different energy sources.

8. Estimating the Upper Limit to a Growth Curve

Some forecasters use curve-fitting methods that extract from the historical data not only the two coefficients of the Pearl or Gompertz curve, but the upper limit L as well. This is bad practice and should not be done. During the early history of a technical approach the upper limit has very little effect on its growth in performance. Thus data points from this period contain little information about the upper limit. Values for L obtained by such means are certain to contain a large error component.

Instead, the upper limit should be estimated on the basis of, for example, the physical and chemical limits imposed by nature on the technical approach to be forecast. These natural limits may exist in the form of a breakdown voltage, a maximum efficiency, limiting mechanical strength, a maximum optical resolution, a minimum detectable concentration of a chemical, and so on. Estimating these natural limits requires a detailed knowledge of the technical approach in question and is best done by the forecaster in conjunction with a specialist in the field. The question to be asked is, What physical, chemical, or other considerations set limits on the best that can be achieved by *this* specific technical approach? It is important not to confuse this question with that of what might be achieved by some different technical approach that might be adopted at some other time. Once these considerations are identified, their effects can be calculated and the ultimate limits determined. Calculations of this type are illustrated in Breedlove and Trammell (1970) and Keyes (1973).

Errors in the estimates of upper limits can have significant effects on the forecasts produced by fitting growth curves to the data. An example of this is shown in Tables 4.1, 4.2, and 4.3. A Pearl curve was fitted to the data on steam engine efficiency from Appendix 3. Five different upper limits were used: 45, 47.5, 50, 52.5, and 55%. Table 4.1 shows the variations in a and b that resulted from the different estimates of the upper

Table 4.1. Pearl Curve Coefficient Versus Upper Limit

Upper Limit (%)	a	b
45	9.464×10^{16}	0.02057
47.5	3.468×10^{16}	0.01997
50	1.630×10^{16}	0.01951
52.5	0.8989×10^{16}	0.01915
55	0.5079×10^{16}	0.01879

Table 4.2. Mid-point Year of Growth Curve for Various Upper Limits

Limit (%)	Midpoint Year
45	1899.89
47.5	1906.93
50	1913.16
52.5	1918.81
55	1924.98

limit. Table 4.2 shows the year in which the growth curve achieved half the upper limit for each of the estimated upper limits. Table 4.3 shows fitted versus actual values for the historical time period, and projections beyond the historical data. If the upper limit is underestimated, both a and b are too large; the curve rises too steeply and reaches the midpoint too soon. If the upper limit is overestimated, the curve rises too slowly and reaches its midpoint too late. It is therefore very important to estimate the upper limit correctly. The limit must be estimated on the basis of the natural limitations inherent in the technical approach, and not from the historical data points.

9. Selecting Variables for Substitution Curves

When one technology is replacing another, the upper limit to the substitution is usually 100%. In some cases there is a fraction of the applications where the new technology is obviously unsuitable, and substitution will not take place. In these cases the upper limit is clearly 100% of those

Table 4.3. Percentage Efficiency of Steam Engines, and Fitted Curves for Various Upper Limits

Year	Value (%)	Upper Limit				
		45%	47.5%	50%	52.5%	55%
1698	0.5	0.695844	0.720893	0.739921	0.755055	0.769692
1712	1.3	0.923358	0.948813	0.967849	0.982798	0.997088
1770	2.8	2.908092	2.895348	2.883963	2.873803	2.863003
1796	4.1	4.748043	4.672583	4.613907	4.566264	4.519272
1830	10	8.410136	8.410136	8.241532	8.107993	7.979042
1846	15	11.165011	10.853109	10.619986	10.436292	10.259681
1890	18	20.219151	19.772665	19.444459	19.189185	18.946511
1910	20	24.832104	24.477917	24.228496	24.041778	23.154308
1950	30	33.170092	33.378256	33.616564	33.862571	34.154308
1955	38	34.045017	34.348741	34.672615	34.996664	35.374092
1960		34.874986	35.276955	35.689572	36.095179	36.562821
1970		36.397894	37.000671	37.597085	38.173490	38.832195
1980		37.739429	38.543025	39.326328	40.079108	40.937928

applications for which the technology is suitable. For instance, Fisher and Pry examined the substitution of synthetic detergents for soap. It was clear that this was impractical in applications such as hand or bath soap; hence they took as the proper applications only laundry and dish soaps and examined the history of substitution in only those cases.

While determining the upper limit may be easy, measuring the degree of substitution is often difficult. For instance, the measure of substitution should not be dollar sales if cost is what is driving the substitution; this would underestimate the extent of the substitution. The measure of substitution should not be weight (pounds or tons) if one item is "denser" than another. For instance, in forecasting the substitution of nuclear for fossil-fueled electric power plants one would not use pounds of uranium substituted for pounds of coal as the measure of substitution. Instead, the proper measure would be in some units of output or capacity, such as the number of kilowatt-hours generated or the kilowatts of capacity installed. (Each of these would be appropriate for specific purposes).

The basic principle is to identify some function that is common to the two technologies and choose a variable which measures the amount of this function that each performs. For instance, Fisher and Pry examined the substitution of plastic tile flooring for wood flooring. Since plastic is much lighter than wood, weight would have been inappropriate as a measure of substitution. Instead, they used the number of square feet of each type of flooring installed annually. Choosing a variable in this manner provides a meaningful measure of the degree of substitution, making the resulting substitution forecast more useful.

10. An Example of a Forecast

Table A3.42 of Appendix 3 shows the percentage of American homes subscribing to cable television (CATV). Let us forecast the rate of adoption of this technology in American homes. Adoption of this technology is akin to substitution, and a growth curve would be an appropriate means of forecasting.

What is the upper limit to the adoption of CATV? Initially CATV was adopted only in small towns; regulations of the Federal Communications Commission (FCC) kept CATV out of large cities because of its potential threat to commercial television broadcasting. CATV was not adopted in rural areas because the low population density made the cost per customer too high. However, more recent FCC regulations have opened up large cities to CATV. Thus the way is open for CATV to be adopted in all "urban areas." If this were the only factor operating, we could establish the upper limit for CATV adoption as 100% of homes in urban areas.

However, two other considerations have widened the potential market. First, the FCC has authorized telephone companies to operate CATV systems in rural areas. Since these companies already have connections to virtually all homes in rural areas, they could extend CATV to these homes at low cost. Second, the new technology of fiber optics promises to reduce the cost of CATV distribution, making it economically feasible even in areas of low population density. The upper limit for the adoption of CATV is then 100% of all U.S. homes.

What is the appropriate form of the growth curve, Pearl or Gompertz? We must determine whether the rate of adoption of CATV depends only upon the portion of the market not yet reached, or whether it also depends upon the share already reached.

Ordinarily a new technology is adopted first in those applications where it offers the greatest advantage or lowest cost. However, FCC regulations kept CATV out of large cities, where distribution costs would be lowest. CATV was concentrated in small cities, where distribution costs were somewhat higher than they would have been in large cities. Nevertheless, changes in regulations now allow utility and cost to be the dominant factors. Thus we can expect CATV to be adopted first in those areas where the costs and benefits make it most valuable; the less advantageous applications will come later. Thus the rate of adoption is proportional to the number of "holdouts," going down as the number of "holdouts" declines.

However, the more widely adopted CATV is, the more attractive it will be. More and more potential customers will be within its reach; special programming will expand to make it more attractive. Thus the rate of CATV adoption will also be proportional to the number who have already adopted it, increasing with this number.

The Pearl curve is then more appropriate for this forecast than the Gompertz curve, since the rate of adoption is a function of both the number of adopters and the number of nonadopters. Thus we will fit a Pearl curve to the historical data, using an estimated upper limit of 100%.

The results of using the growth curve program of Appendix 4 to fit a Pearl curve to the historical data are plotted in Figure 4.4 as a Fisher–Pry plot. The straight line represents the Pearl curve fitted to the data and projected to the year 2010. The results are shown in detail in Table 4.4, which presents the upper and lower confidence limits as well as the numerical values of the fitted curve for various years. According to this projection, CATV will have been adopted by 63% of U.S. homes by 1990, 90% by 2000, and 98% by 2010.

Note that although the Fisher–Pry plot appears to be a good fit to the data, 13 of the 18 data points actually lie outside the 50% confidence limits. Nine of these are above the upper 50% limit and four are below the lower 50% limit. As can be seen from the plot, these departures from the fitted curve are not completely random; instead, they are systematic excursions above and below the fitted curve. This indicates that some systematic factors are at work in addition to random variation.

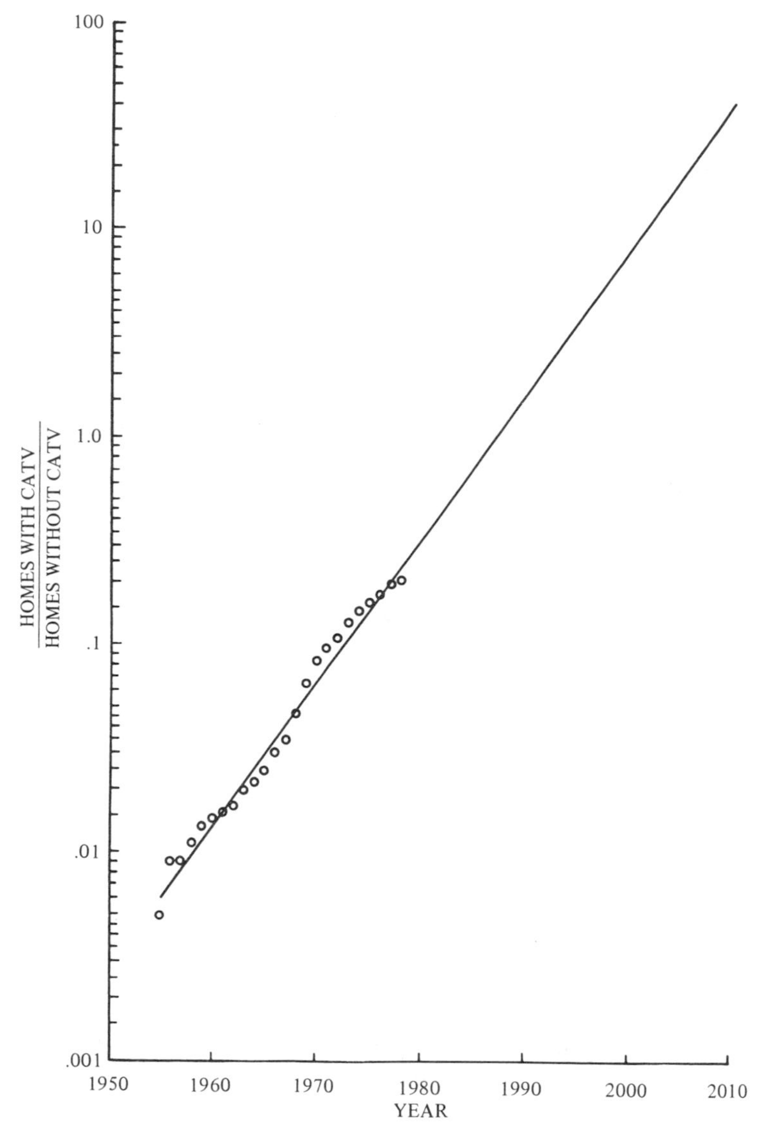

Figure 4.4 Fisher–Pry curve showing the adoption of CATV by U.S. homes.

Instead of the confidence limits showing the usual "trumpet" shape, the values in the table appear to be "pinched in" at both ends. This apparent distortion arises from the fact that the Pearl curve formula prevents these bounds from exceeding the upper limit of 100% or going below the limit of 0%. While this is the correct result, it is different from the usual plot of confidence limits.

The example has briefly illustrated the use of growth curves in fore-

Table 4.4. Growth of CATV

Year	Households with CATV (%)	Pearl Curve	50% Confidence Limits	
			Upper	Lower
1955	0.5	0.607	0.675	0.545
1956	0.9	0.712	0.792	0.641
1957	0.9	0.836	0.928	0.754
1958	1.1	0.980	1.087	0.883
1959	1.3	1.149	1.274	1.037
1960	1.4	1.347	1.493	1.216
1961	1.5	1.579	1.748	1.427
1962	1.7	1.850	2.046	1.672
1963	1.9	2.166	2.349	1.959
1964	2.1	2.535	2.800	2.295
1965	2.4	2.965	3.273	2.685
1966	2.9	3.465	3.823	3.140
1967	3.8	4.046	4.461	3.668
1968	4.4	4.719	5.200	4.281
1969	6.1	5.499	6.054	4.991
1970	7.7	6.398	7.039	5.811
1971	8.8	7.433	8.171	6.756
1972	9.7	8.619	9.467	7.840
1973	11.3	9.975	10.945	9.082
1974	12.7	11.517	12.623	10.496
1975	13.8	13.262	14.517	12.101
1976	14.8	15.227	16.641	13.912
1977	16.1	17.423	19.009	15.944
1978	17.1	19.863	21.627	18.209
1990		63.121	65.939	60.210
2000		89.545	90.800	88.141
2010		97.720	98.058	97.325

casting. Fitting the growth curve to the data is actually the smallest part of the task and is largely computerized. The main task of the forecaster is to assure that the right data have been collected and that they have been processed properly; this means determining what variable is the best measure of the technology to be forecast, determining the appropriate upper limit, and selecting the most suitable form of the growth curve. The forecaster's major contributions to the quality of the forecast have very little to do with the actual fitting of the growth curve.

References

Breedlove, J. R., and Trammell, G. T. (1970). "Molecular Microscopy: Fundamental Limitations," *Science* 170, 1310–1312.

Fisher, John C., and Robert Pry (1971). "A Simple Substitution Model of Technological Change," *Technological Forecasting and Social Change* 3, 75–88.

Keyes, R. W. (1973). "Physical Limitations in Digital Electronic Circuits," *IBM Report* RC4572, October 11.

Marchetti, C. (1977). "Primary Energy Substitution Models: On the Interaction Between Energy and Society," *Technological Forecasting and Social Change* 10, 345–356.

For Further Reference

Bossert, Richard W. (1977). "The Logistic Growth Curve Reviewed, Programmed, and Applied to Electric Utility Forecasting," *Technological Forecasting and Social Change* 10 (4), 357–368.

Bundegaard-Nielsen, M. (1976). "The International Diffusion of New Technology," *Technological Forecasting and Social Change* 8 (4), 365–370.

Hurter, Arthur P., Albert H. Rubenstein, Mark Bergman, Jane Garvey, and Joseph Martinich (1978). "Market Penetration by New Innovations: The Technological Literature," *Technological Forecasting and Social Change* 11 (3), 197–222.

Mahajan, Vijay, and Robert A. Peterson (1979). "Integrating Time and Space in Technological Substitution Models," *Technological Forecasting and Social Change* 14 (3), 231–241.

Sahal, Devendra (1977). "The Multidimensional Diffusion of Technology," *Technological Forecasting and Social Change* 10 (3), 277–298.

Problems

1. Plot a Gompertz curve and a Pearl curve on the same graph. Select their parameters so that they have the same upper limit and their inflection points occur at the same value of time. What are the major differences between the two curves?

2. According to estimates of the U.S. Geological Survey, the upper limit for installed hydroelectric capacity in the United States is 161,000 MW. Using this as an upper limit, fit a growth curve to the data in Table A3.46 in Appendix 3 on installed hydroelectric capacity. Is a Pearl curve or a Gompertz curve more appropriate? Why?

For the following problems use data from the appropriate tables in Appendix 3. Justify your choice of the upper limit and the form of the growth curve (Pearl or Gompertz).

3. Growth in power-plant efficiency (kWh/lb of coal).

4. Transition from sail to mechanical power in the U.S. Merchant Marine.

5. Growth in the percentage of dwellings with electric power.

6. Number of telephones per 1000 population.

Problems *(continued)*

7. Substitution of coal for wood as an energy source.

8. Substitution of oil and gas for coal as an energy source.

9. Number of automobiles per capita.

10. Growth in the reliability of space launches (percentage successful).

11. Substitution of surface mining for the deep mining of coal.

Chapter 5
Trend Extrapolation

1. Introduction

A specific technical approach to solving some problem will have a maximum level of performance it cannot exceed. This upper limit is set by physical laws that govern the phenomena utilized in the technical approach. However, the upper limit to the performance of a technical approach does not present an absolute barrier to progress. When a technical approach is reaching its limit, a new one will be found that utilizes a different set of physical or chemical phenomena. This new approach will therefore be subject to its own ultimate limit, but this is higher than the limit on the approach it will replace.

Such behavior is illustrated in Figure 5.1, which shows the growth curve for the speed of propeller-driven aircraft reaching its limit, while the growth curve for jet-propelled aircraft is just beginning its progress toward a higher limit. Note that the initial jet aircraft had lower speeds than did contemporary propeller-driven aircraft. This is typical of a "successor" technical approach; while it is still in the experimental stage, it often shows performance that is lesser than the current standard approach. The important point about the new technology is that it has an inherently higher upper limit so that it will ultimately surpass the current approach.

If we are to forecast long into the future, some method is needed that can project progress beyond the upper limit of the current technical approach. Growth curves are not suited for this purpose.

In projecting beyond the limits of the current technical approach, however, the forecaster must distinguish between predicting a successor technical approach and inventing it. The forecaster must be able to estimate the performance of devices available at some future time without knowing

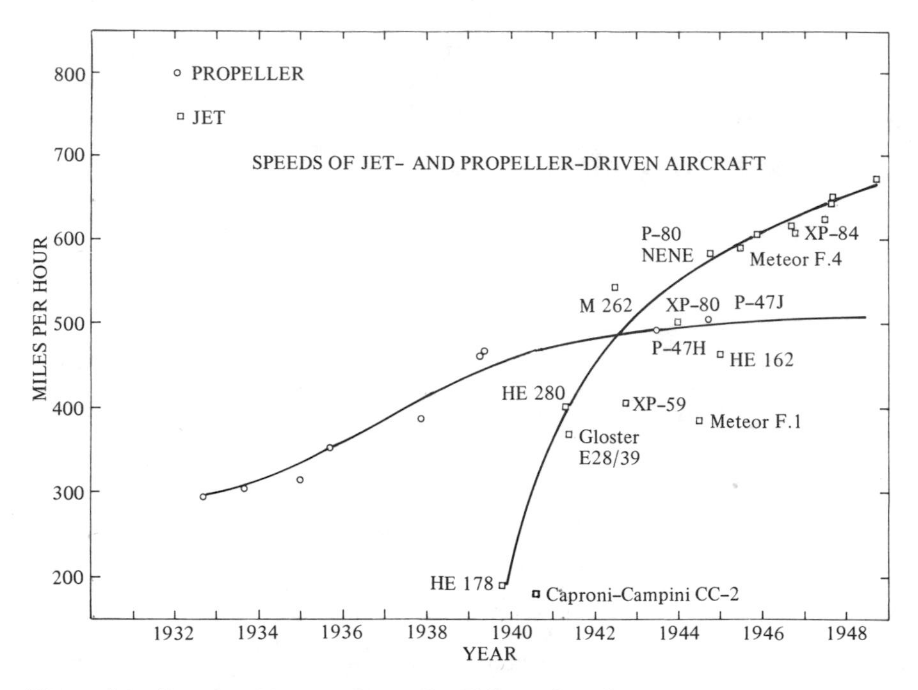

Figure 5.1. Speeds of jet- and propeller-driven aircraft.

the technical details of how they will operate. This chapter presents methods for making such long-term forecasts.

2. Exponential Trends

Figure 5.2 shows the top speeds of U.S. military aircraft from 1910 to 1968. During this period of time, many aircraft design innovations were introduced, including the enclosed cockpit, the all-metal fuselage, retractable landing gear, and other "streamlining" features, not to mention such innovations as the supercharged engine and the pressurized cabin. Each of these innovations allowed the aircraft designer to overcome some limitation of the prior technical approach. Nevertheless, the speed trend shown in Figure 5.2 appears to show a smooth, steady progression. The intersecting growth curves so apparent in Figure 5.1 are completely invisible in Figure 5.2. There is no hint from the long-term trend that any major changes in the technology have taken place.

This is actually a very common situation, and several of the problems at the end of this chapter display it clearly. The history of some technologies will be filled with a succession of technical approaches. Each successive approach will utilize a different set of phenomena to solve a specific, perennial problem or perform some continuing function. A plot of performance against time, however, will show a smooth trend, with none of the wiggles and jumps that might be expected as technical approaches with higher performance displace older approaches.

It is the existence of these long-term, smooth trends that makes it possible to forecast beyond the upper limit of the current technical approach. Moreover, the trends make it possible to forecast a successor approach without knowing what that approach will be. In fact, the forecaster's function has been fulfilled by providing warning that something new is coming; it is too much to ask for the invention as well. Moreover, the forecaster should not shrink from the implications of a long-term trend. If the past history of a technology has seen a succession of technical approaches that have cumulatively produced a trend, that trend must be projected into the future unless the forecaster can find some good reason

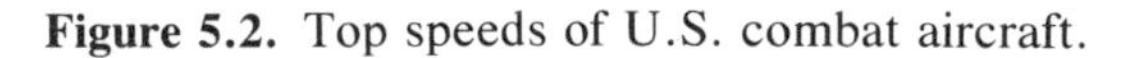

Figure 5.2. Top speeds of U.S. combat aircraft.

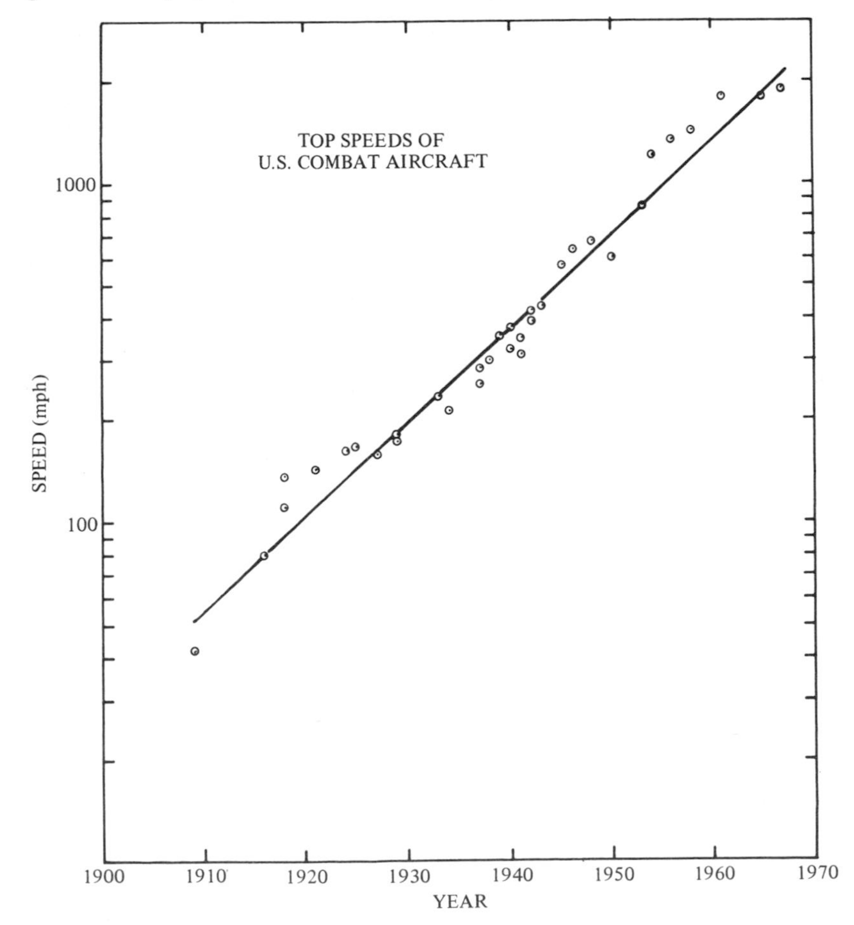

as to why the present is a point of discontinuity and the trend will not be continued. In particular, one should not shrink from projecting a trend simply because one does not see how it can be continued. It is not the forecaster's job to *see how* a trend can be continued but, rather, to *foresee* that it *will* be continued.

The most common long-term trend is exponential growth. This is growth by a constant percentage per unit time, that is, proportional to the value already reached. This is expressed mathematically as

$$dy/dt = ky, \tag{5-1}$$

where k is the constant of proportionality. Solving this differential equation, we have

$$y = y_0 e^{kt}. \tag{5-2}$$

This is equivalent to the growth of money under compound interest. The proportionality constant k can be viewed as an "interest rate." The only major difference is that interest is compounded periodically, while exponential growth is compounded continuously.

Exponential trends are conventionally plotted on semilogarithmic scales, as in Figure 5.2, where they then appear as straight lines. This can be seen by taking the logarithms of both sides in equation (5-2):

$$\ln y = Y = \ln y_0 + kt. \tag{5-3}$$

In fitting a trend to historical data, the forecaster regresses the logarithm of the actual trend values on time, using the linear form in equation (5-3). The constant term of the regression is the logarithm of the initial value y_0, and the slope term of the regression is the growth rate k.

Applying this to the data on combat aircraft plotted in Figure 5.2, we obtain the following equation:

$$Y = -118.30568 + 0.06404T. \tag{5-4}$$

This regression fit can be obtained using any of several standard fitting programs. All of them implement the least-squares fitting technique described in Appendix 1, although some will require the user to take the logarithm of the data prior to using the program. The trend program in Appendix 4 will fit either a linear trend or an exponential trend simply by choosing option 1 or 2, respectively. The user is not required to transform the data before using the program.

Linear and exponential trends can also be fitted quite easily with hand-held calculators, which sometimes have built-in regression programs. Some of these, however, require equally spaced data (e.g., monthly, weekly) and are therefore not suitable for time series of technological performance. However, a regression program can be written in about 200 steps. A programmable calculator is more convenient than a computer and for fitting only a small number of data points may actually take less time.

3. An Example of a Forecast

To illustrate the use of exponential trends for forecasting, let us forecast electric power production for the year 1970, using data from 1945 through 1965 (taken from Table A3.1 in Appendix 3).

A plot of the data is shown in Figure 5.3. The data alone suggest a straight line when plotted on semilog paper, hence an exponential trend is an appropriate method for forecasting. We fit a linear trend to the logarithms of the values for electric power production, using the regression program in Appendix 4, option 2. We obtain the equation

$$Y = 9.15434e^{0.075286T}, \tag{5-5}$$

where T is the year minus 1900. Substituting 70 for T, we obtain 1779.780 billion kW-h, which is about 8.5% greater than the actual value of 1639.771 billion kW-h. However, the actual value falls outside the 50% confidence limits, hence the comparatively small error is still much larger than most of the historically observed deviations from the fitted trend.

The fitted trend is shown along with the historical data in Figure 5.3, and the historical values, the fitted trend, and the confidence limits are given in Table 5.1.

As with any other forecasting technique, the use of exponential trends cannot be simply a mechanical process of fitting curves to data. Selecting the proper data and determining that an exponential trend is appropriate are part of the forecaster's task before the trend is fitted. Interpreting the trend is part of the forecaster's task after the trend is fitted. The use of confidence bounds makes the task of interpretation simpler and provides some guidance about the degree of deviation to be expected between trend and actual values.

4. A Model for Exponential Growth

Empirically, many technologies do grow exponentially. This idea is commonly accepted today. It is even widely deplored. However, it was not always recognized or accepted. Henry Adams in 1918 expressed the notion that the growth of technology is similar to the behavior of a mass introduced into a system of forces previously in equilibrium. The motion of the mass will accelerate until a new equilibrium is reached. Arthur Conan Doyle, in a story entitled "The Great Keinplatz Experiment," written shortly before the turn of the century, made the statement, "Knowledge begets knowledge as money begets interest." M. Petrov and A. Potemkin, writing in the Russian journal *Novy Mir* (*New World*) (June

1968, pp. 238–252), attribute the first formulation of the exponential law of the growth of science to Engels in his controversy with Malthus.

Thus the idea of exponential technological growth has a respectable intellectual history, as well as considerable empirical foundation. However, both the empirical evidence and intellectual acceptance present us

Figure 5.3. U.S. electric power production from 1945 to 1965, projected to 1970.

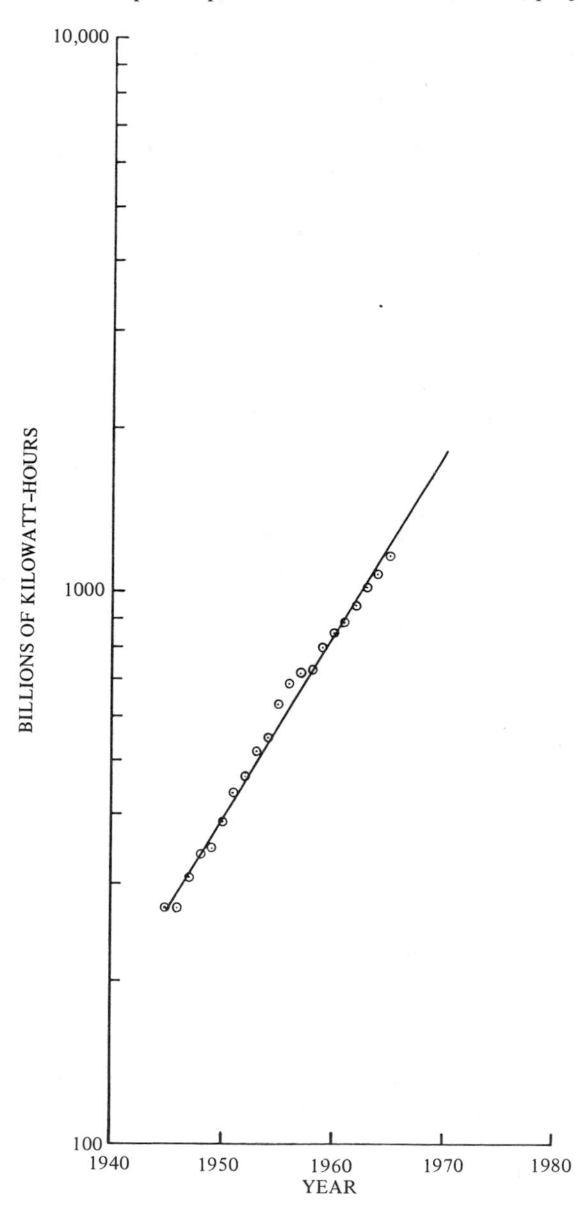

Table 5.1. U.S. Electrical Power Production (Billions of Kilowatt Hours)

			Confidence limits	
Year	Production	Fitted trend	Upper 50%	Lower 50%
1945	271	270.994	280.868	261.467
1946	270	292.184	302.715	282.018
1947	307	315.030	326.274	304.173
1948	337	339.663	351.680	328.057
1949	345	366.222	379.078	353.802
1950	389	394.858	408.627	381.553
1951	433	425.733	440.497	411.464
1952	463	459.022	474.871	443.702
1953	514	494.914	511.949	478.446
1954	545	533.613	551.945	515.889
1955	629	575.337	595.091	556.239
1956	685	620.324	641.635	599.721
1957	716	668.829	691.850	646.754
1958	724	721.126	746.025	697.058
1959	798	777.513	804.475	751.453
1960	844	838.308	867.540	810.061
1961	881	903.858	935.587	873.204
1962	947	974.532	1009.010	941.232
1963	1011	1050.730	1088.240	1014.520
1964	1084	1132.890	1173.730	1093.480
1965	1158	1221.480	1265.980	1178.530
1966		1316.990	1365.540	1270.160
1967		1419.970	1472.970	1368.870
1968		1531.000	1588.900	1475.200
1969		1650.710	1714.020	1589.740
1970		1779.780	1849.040	1713.120

with a brute fact: While exponential growth has been accepted, it has not been explained.

Attempts have been made to explain exponential growth through a competitive model. Holton (1962) was one of the first to advance such a behavioral explanation. Observing the exponential growth of the operating energy of nuclear-particle accelerators, he said,

> One research team will be busy elaborating and implementing an idea—usually that of one member of the group, as was the case with each of the early accelerators—and then work to exploit it fully. This is likely to take from two to five years. In the meantime, another group can look so to speak, over the heads of the first, who are bent to their task, and see beyond them an opportunity for its own activity. Building on what is already known from the yet incompletely exploited work of the first group, the second hurdles the first and establishes itself in new territory. Progress in physics is made not only by marching, but even better by leapfrogging. (Holton, 1962)

Holton need not have confined his remarks to the field of physics. Many other fields exhibit competition of this sort, with each participant attempting to get an edge on the others by leapfrogging them. Seamans (1969) has exploited this idea of competition in a model that appears to provide a satisfactory explanation of the phenomenon of exponential growth. This model will now be described.

Assume there are two competitors, A and B. They are in a situation, be it military, commercial, or scholarly, where each wants to be ahead of the other in the level of some functional capability. This may arise from the desire to meet a military threat, to obtain a commercial advantage, or to obtain priority for some piece of scholarly research. We assume that A desires to be $100m\%$ ahead of B, and B desires to be $100n\%$ ahead of A. We also assume that A has a response time S, from the time a decision is made to initiate a new project until that project has reached its goal; B has a response time T. A has a response fraction f, where f is a number between zero and unity; that is, after B has started a project to surpass A's current level, A is willing to wait a fraction f of B's response time before initiating his own leapfrogging project. We assume B has a response fraction g. A and B may be involved in a completely symmetrical situation, of course, where both have equal response times and response fractions and both desire the same percentage of lead over the other. The allowance for asymmetry does not unduly complicate the model and provides more generality.

We will assume that A starts at time $t = 0$ with a level of functional capability equal to unity.

At time $t = 0$, B initiates a project designed to overcome A's lead. This project will have the goal of achieving a level of $1 + n$, which will be reached at time T.

Once B initiates her project, A will wait until time fT to initiate a counterproject. This project will have the goal of achieving a level of $(1 + m)(1 + n)$, which will be reached at time $fT + S$.

B will then initiate a project at time $fT + gS$. This will have a goal of $(1 + m)(1 + n)^2$, which will be reached at time $(f + 1)T + gS$.

A will initiate a project at time $2fT + gS$. This will have a goal of $(1 + m)^2(1 + n)^2$, which will be reached at time $2fT + (g + 1)S$.

B will initiate a project at time $2fT + 2gS$. This will have a goal of $(1 + m)^2(1 + n)^3$, which will be reached at time $(2f + 1)T + 2gS$.

A will initiate a project at time $3fT + 2gS$. This will have a goal of $(1 + m)^3(1 + n)^3$, which will be reached at time $3fT + (2g + 1)S$.

Now that we see how this process of leapfrogging works, we can generalize the expressions above.

A will start one of his leapfrogging projects at a time given by $pfT + (p - 1)gS$, where p is any integer. The project will have a goal of $(1 + m)^p(1 + n)^p$, which will be reached at time $pfT + [(p - 1)g + 1]S$.

B will start one of her leapfrogging projects at a time given by $qfT + qgS$, where q is any integer. The project will have a goal of $(1 + m)^q(1 + n)^{q+1}$, which will be reached at time $(qf + 1)T + qgS$.

We can now see that the interval between the introduction of "new models" by A will be a time period $fT + gS$, and that the ratio between the level of functional capability of the new model and that of its predecessor will be $(1 + m)(1 + n)$. Similarly, B will introduce new models at intervals of $fT + gS$, each with a level of functional capability $(1 + m)(1 + n)$ times that of its predecessor. Part of the time A will have the superior capability, and part of the time B will have the superior capability. The fraction of the time that either is in the lead will depend on the respective response times and response fractions.

The buildup of capability described above is actually geometric in nature because of the discrete steplike nature of the process. If we imagine technology as growing smoothly, however, between the introduction of successive models, this process is equivalent to exponential growth. The exponent can be evaluated as

$$e^{a(fT+gS)} = (1 + m)(1 + n),$$

or

$$a = \ln[(1 + m)(1 + n)]/(fT + gS).$$

As would be expected, the shorter the response times and the smaller the response fractions of the two competitors, the larger the exponent, or, equivalently, the greater the rate of growth. Similarly, the greater the advantage each desires over the over, the greater the rate of growth.

The situation with many competitors is not significantly different. We can imagine each planning to outdo the current leader by a margin sufficient to provide a lead for a satisfactory length of time. The competitors will also try to keep track of the ongoing projects of all the others to avoid being leapfrogged shortly before or after their new model is ready. It could be argued that under these circumstances all the competitors would tend to adopt the same percentage increase for their goal, and the same response fraction. Those whose reponse time was much longer than that of their competitors would soon be forced out of the market; hence there would also be a tendency for all to achieve the same response time. Under these circumstances progress would be exponential, and all competitors remaining in the market would tend to take the lead alternately.

While this model provides a theoretical explanation for exponential growth and the only variables appearing in it are measurable (at least in principle), it still has some shortcomings. It implicitly assumes that the response times of the competitors are set by factors other than limitations in the possible growth of technology; that is, if a competitor sets a goal, it is assumed to be achievable, although possibly at some considerable expense and effort. Clearly, in practice neither the response times nor the percentage increments of improvement can be set arbitrarily, at the discretion of the competitors. Projects that are too ambitious, trying to achieve "too much too soon," will be failures or at best will slip, either in the performance actually achieved or in the actual delivery date. Thus this model provides an explanation of why in a competitive situation growth should be exponential so long as technology permits it; however, it does not explain why exponential growth should be possible.

As an illustration of the application of this model, we will consider the introduction of commercial passenger transport aircraft. Table 5.2, which is extracted from the transport aircraft table in Appendix 3, gives the year of introduction and the productivity, in passenger-miles per hour, of pace-setting transports introduced by three major manufacturers who have been in the market consistently for over 30 years. Table 5.2 also gives the interval in years between successive pace-setting models for each manufacturer, and the productivity ratio of the two successive models. For each manufacturer we can calculate the mean interval between the introduction of successive models, and the mean value of the natural logarithms of the performance ratios. We can then take the ratios of the average natural logarithm to the average interval between successive models. For the three manufacturers, these turn out to be the following: Boeing, 0.143; Douglas, 0.144; Lockheed, 0.131. The overall industry average of these three values is 0.139. This turns out to be not too much different from the exponent describing the overall growth of passenger-

Table 5.2. Pace-Setting Passenger Transports Introduced by Three Major Manufacturers

Year	Passenger-miles per hour	Interval (years)	Performance ratio
Boeing			
1933	2,000		
		5	3.8
1938	7,524		
		11	4.3
1949	32,250		
		10	3.5
1959	112,833		
		10	2.8
1969	313,600		
Douglas			
1934	2,982		
		1	1.6
1935	4,620		
		5	2.5
1940	11,550		
		7	1.9
1947	21,420		
		7	1.8
1954	38,855		
		4	2.8
1958	109,431		
Lockheed			
1934	1,920		
		6	2.0
1940	3,808		
		6	5.5
1946	21,056		
		4	1.6
1950	34,040		
		8	1.3
1958	44,100		

mile productivity for all transport aircraft, including those that were not pace-setters (see Problem 2 at the end of this chapter). Since we have three competitors, the model derived above is not strictly applicable, and the simple form of the exponent cannot be used. Furthermore, at least since World War II, the response time of the manufacturers has been governed more by the question of whether the airlines have paid off their last purchases, and are now ready to buy more aircraft, than by technological considerations. Nevertheless, the competitive aspects of the market are clear, with each of the competitors introducing successive aircraft models that exhibit about the same growth rate, and with different manufacturers holding the lead at different times. While the illustration of the model would be more satisfactory if we could derive the "right" exponent directly from a knowledge of the rate at which the airlines could absorb new aircraft (thus giving the interval between successive models), the illustration nevertheless shows the fundamental correctness of the model in a competitive situation.

5. Nonexponential Growth

The long-term trend of many technologies is exponential; however, some measures of technology grow in linear fashion. The example in Section 3 shows exponential growth. While many technologies do grow exponentially, this is not necessarily the case for all technologies or all technological parameters: Some may grow linearly or, more rarely, they may exhibit parabolic, cubic, or higher-powered growth laws. In most cases, however, if a technology fails to grow exponentially, it will behave in a linear fashion.

One example is aircraft span-to-length ratio. One of the measures of aircraft progress is the degree of streamlining, which is reflected in the ratio of the wing span to the aircraft length. Figure 5.4 shows the span-to-length ratio for bombers, fighters, and transport aircraft over four decades. The trends are not as clear-cut as some of those shown previously, but it is apparent that bombers tend to lag behind fighters in this respect and that the trend for transports actually intersects the trends for both bombers and fighters. This is a reflection of the different requirements for streamlining in the three classes of aircraft. It is interesting to note that the F-111 with wings folded falls near the fighter trend, but the same aircraft with wings extended falls in the transport trend. (The F-111 is a fighter-bomber.) The advent of swing-wing aircraft technology will likely render span-to-length ratios meaningless in the future. Regression analysis of the span-to-length data is as follows:

Transport aircraft. Regressing ratio of span to length on time:

$$Y = 31.72063 - 0.01566T, \qquad r = -0.86843. \tag{5-6}$$

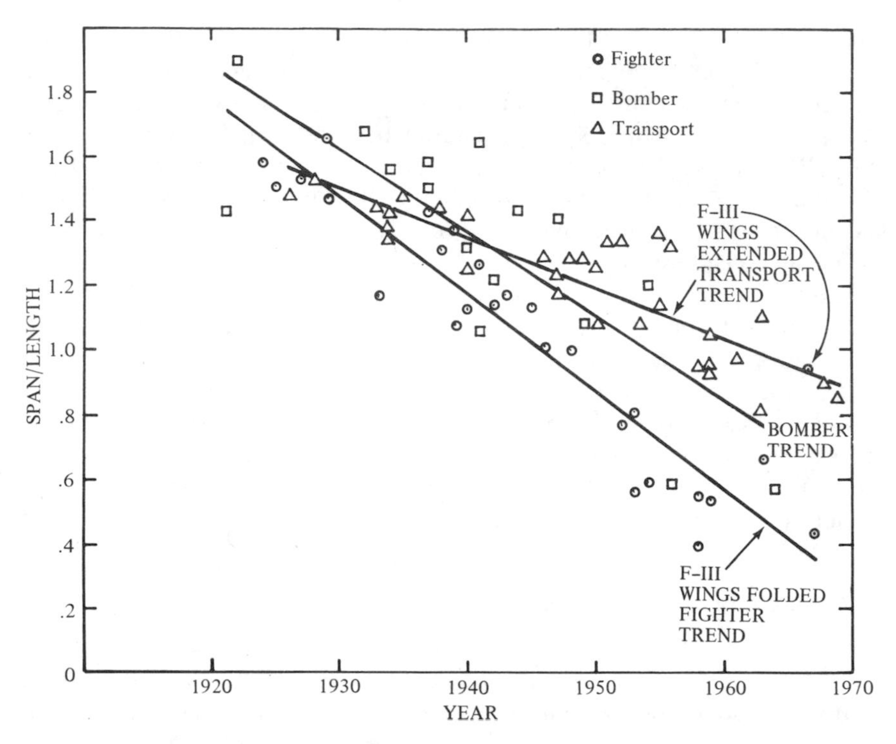

Figure 5.4. Ratio of wingspan to length for bombers, fighters, and transport aircraft.

Bombers. Regressing ratio of span to length on time:

$$Y = 51.8139 - 0.0260T, \qquad r = -0.81889. \tag{5-7}$$

Fighters. Regressing ratio of span to length on time:

$$Y = 60.1588 - 0.0304T, \qquad r = -0.94917. \tag{5-8}$$

As this example shows, the behavior of some technological parameters is linear. This example also indicates that in some cases it is more convenient to cast a technological parameter in a form where a decrease in the value of the parameter corresponds to technological progress. Thus, while in most cases the growth of a technological parameter will reflect the general exponential growth of technology, the forecaster should be alert to the possibility that a specific parameter may not grow exponentially. In such a case the forecaster is best advised to try to fit a linear law, keeping in mind that the proper parameter to forecast may be one that decreases with the increasing state of the art.

6. Qualitative Trends

In the preceding sections of this chapter we have discussed trends that can be described in quantitative terms. Not all trends, of course, can be thus described, but this does not mean they are any less real. It does mean, however, that it is harder to define them and that forecasts based on them will of necessity be less precise.

Table 5.3 (taken from Lamar, 1969) illustrates a qualitative trend. Since the time of the Wright brothers, aircraft have been designed so that more and more of the total aircraft can be adjusted, varied, or otherwise moved while in flight. The wing-warping capability of the original Wright aircraft was only the first step in a long sequence of such capabilities. Attempts to quantify this trend (e.g., in terms of the percentage of empty weight that can be varied in flight) may not be completely successful, since one movable portion may later be broken down into several movable portions, as illustrated in Table 5.3 by the step from simple flaps to complex flaps.

What can be done with a qualitative trend such as this? As with the trends we have been considering in previous sections, it is probably safe to extrapolate it. Aircraft will continue to be designed with more and more parts that can be moved or adjusted in flight. However, it does not appear to be possible to forecast just which parts will next be designed to be adjustable. In other areas where qualitative trends might be discernible, the same limitation will most likely apply. It will be possible to extrapolate the trend, in the sense that the type of phenomenon represented by the specific elements in a list (such as that shown in Table 5.3) will continue, but it probably will not be possible to forecast what specific elements will next be added to the list.

Nevertheless, in the absence of well-behaved quantitative trends such as those described earlier, there may be no alternative to the use of qualitative trends for obtaining a forecast. Forecasts based on qualitative trends may well be better than no forecasts at all, and they can certainly provide useful inputs to a decision by indicating that a particular qualitatively described phenomenon can be expected not only to continue at its present level but to increase or decrease in a continuation of past behavior.

7. A Behavioral Technology

In Chapter 1 it was claimed that behavioral technologies such as procedures and techniques are within the province of the technological forecaster. However, the examples shown so far have all been "hardware" technologies. It is worth showing an example of a behavioral technology that exhibits the same sort of behavior as do hardware technologies.

Table 5.3. Growth in Complexity of Variable Geometry on Aircraft[a]

				Thrust deflectors
				Thrust reversers
				Ejectable capsules
				Variable nozzles
				Variable inlets
				Flt fold wing tips
			Swing tails	Var. sweep wing
			Segmented elevons	Swing noses
			Eject seats	Droop noses
			Large cargo doors	Tilt wings
			Refueling booms	Tilt propellers
			Multistore combin.	Tilt engines
			External stores	→
		Slats	Speed brakes	→
		Flying tabs	Drag chutes	→
		Gear doors	Segmented spoilers	→
		Cargo doors	Lead edge flaps	→
		Multinspect doors	→	→
		Bomb bay doors	→	→
		Hatches	→	→
	Adj. stabilizer	Movable stabilizer	→	→
	Ground folding wings	→	→	→
	Arresting hooks	→	→	→
	Simple flaps	Complex flaps	→	→
	Retractable gear	→	→	→
	Trim tabs	→	→	→
Rudder	→	→	→	→
Elevator	→	→	→	→
Ailerons	→	→	→	→
			Time	

[a] *Source*: William E. Lamar, Air Force Flight Dynamics Laboratory. Originally published as Figure 5, p. 60, in the July 1969 issue of *Astronautics and Aeronautics*, a publication of the American Institute of Aeronautics and Astronautics.

For the last three centuries or so the scope of engineering projects has been increasing. One measure of the scope of an engineering project is simply dollar magnitude; however, there is more to such a project than simply spending money. A project must be managed so that the parts all come together at the right time and the tasks are accomplished in the proper order.

The techniques of engineering management have improved over the years, and it is now possible to manage projects that would have been beyond the capabilities of earlier managers and management techniques. Thus dollar magnitude is not only a measure of project scope but also a rough measure of the capability of engineering management techniques.

Figure 5.5 shows the dollar cost of a variety of engineering projects completed over the past three centuries. At any time there is a largest

Figure 5.5. Dollar costs of engineering projects.

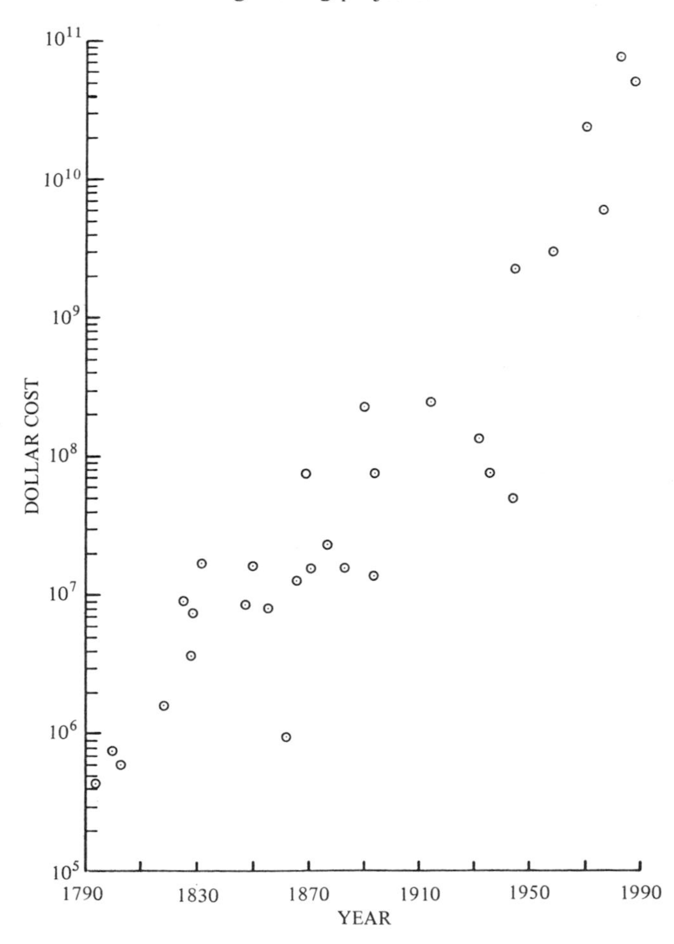

single project thus far completed. Many smaller projects are also carried out, however. Thus we are interested in the outer envelope of the plot of project size versus time, rather than, say, the average size of a project. The envelope of the plot of project magnitudes forms a long-term exponential trend, which indicates that engineering management capability exhibits the same kind of behavior as we have already seen for hardware technologies. Thus the methods of technological forecasting can be applied to behavioral technologies as well as to the more conventional technologies.

8. Summary

Any technical approach to achieving some functional capability will have an inherent upper limit beyond which that approach cannot go. Forecasting beyond this upper limit may still be possible, however. The forecaster cannot and need not state what technical approach will be used to achieve the increased level of capability; only the rate of innovation in the past need be observed, especially where there has already been a succession of technical approaches, and that rate can then be projected into the future. This can be done by fitting an appropriate trend (usually exponential but sometimes linear) to the historical data.

In some cases the overall technology will be approaching a fundamental limit, such as the speed of light, that affects all technical approaches alike. In such a case continuation of a historical trend is impossible and some alternative forecasting method is needed. However, in the absence of any such limit the historical trend should be projected and used as a forecast. Confidence limits can be placed on the forecast, where appropriate, to provide an estimate of the spread about the trend to be expected.

In any case, however, the forecaster cannot assume that an extrapolated trend will come about through passive waiting, especially when the past trend was established through aggressive action by the participants in the advance of the specific technology. The forecaster must assume, however, that the aggressive actions that shaped the trend in the past will continue. If it is known that the actions will not be continued, then some other method of forecasting is appropriate, since the trend will not continue either in the absence of the actions that produced it in the past.

References

Holton, G. (1962). "Scientific Research and Scholarship: Notes Toward the Design of Proper Scales," *Daedalus* 91, 362–399.

Lamar, W. E. (1969). "Military Aircraft: Technology for the Next Generation," *Aeronautics and Astronautics* (July).

Seamans, R. C., Jr. (1969). "Action and Reaction," *Technological Forecasting* 1, 17–32.

For Further Reference

Martino, Joseph P. (1971). "Examples of Technological Trend Forecasting for Research and Development Planning," *Technological Forecasting and Social Change* 2 (3, 4), 247–260.

Petroski, Henry J. (1977). "Trends in the Applied Mechanics Literature," *Technological Forecasting and Social Change* 10 (3), 309–318.

Problems

1. Take the data for the gross weight of U.S. single-place fighter aircraft from Table A3.9 in Appendix 3. Plot this data on rectangular-coordinate paper and fit a curve to it freehand. How easy would it be to extrapolate this curve? Plot the same data on semilog paper. Fit a straight line to it freehand. Also, fit an exponential trend to the data by regressing the logarithm of weight on time. How well does your freehand fit agree with your regression fit? What are the advantages of plotting data in a coordinate system where the trend appears as a straight line? What are the advantages of using an objective procedure such as linear regression for fitting a trend to data?

2. Fit an exponential trend to the data on commercial aircraft productivity in passenger-miles per hour for Table A3.4 in Appendix 3, using the data from 1926 to 1963. Project this trend to 1969. How well does the Boeing 747 fit the trend?

3. Fit an exponential trend to the data on the accuracy of measurements of mass from Table A3.39 in Appendix 3. Are you satisfied that the accuracy of measurement of mass improved exponentially over the time period of the data?

4. Consider the following innovations in automobiles:

Self-starter.	Power brakes.
Fluid drive.	Power steering.
Automatic shift.	Cruise control.

What did each of these innovations mean in terms of the strength and skill required of the driver? What might be some possible steps in the continuation of this qualitative trend?

5. From Table A3.47 of Appendix 3 on inanimate nonautomotive horsepower, plot the data for 1850–1880 and fit a trend to it. Project the trend to 1920. Now plot the data points for 1890–1920 on the same graph. How well did you forecast whese values? Now fit a trend to all the data points from 1850 to 1920 and project it to 1980. Plot the remainder of the data on the same graph. How well did you forecast these values this time? How do you explain the difference in the two forecasts?

6. Take the projection of electric power production in Table 5.1 and extend it to 1975. Compare this with the actual data from Table A3.1 in Appendix 3. How do you explain the results?

Chapter 6

Measures of Technology

1. Introduction

Previous chapters have described forecasting in terms of projecting a quantitative measure of technology. This chapter will describe means for selecting or developing appropriate measures of technology for forecasting purposes.

The starting point for a measure of technnology must be the function the technology performs. Any measure of the technology must be related to that function and quantitatively describe how well the technology performs it. Hence the first step in selecting a measure of technology is to identify clearly the function the technology performs. This is particularly true when the forecaster is dealing with a succession of technical approaches. Each of these will have its own technical measures of performance. The disparate technical approaches can be put on a common basis only in terms of the function they all perform. Once this function has been identified, the forecaster can then attempt to select one or more measures that meet the following criteria.

The selected technology measure must actually be measurable; that is, it must be possible to assign a numerical value to it on the basis of the performance of a device. The most desirable measure will be objectively measurable in terms of the characteristics of the device or procedure that embodies the technology being measured. In some cases an objective measure may not be possible; then the forecaster will have to make sure that the measure can be scaled judgmentally by someone familiar with the technology.

The selected measure must be a true representation of the state of the art; it must tell the full story of the way in which the technology performs its function. For a given state of the art a designer often has the freedom to make tradeoffs among various technical parameters. The selected measure of technology must capture these tradeoffs. For instance, in electronic amplifiers bandwidth can be increased if gain is reduced, and vice versa. Thus a widely used measure of performance is the gain–bandwidth product. Over a wide range of operating conditions the gain–bandwidth product of a transistor, vacuum tube, and so forth will be constant. This product represents the state of the art, and using it as a measure of technology captures the tradeoffs open to the designer.

The selected measure must be applicable to all the diverse technical approaches the forecaster will consider. This leads back to the starting point above: The measure of a technology must be based on the function performed; only then can diverse technical approaches be compared. See, for instance, the data on efficiency of illumination in Appendix 3 (Table A3-6). The output measure, lumens per watt, can be applied to a variety of illumination sources, even candles and gas lamps, after the rate of fuel consumption is converted to watts.

The measure of technology selected should be one for which data are actually available, and these should relate to a large number of devices, techniques, and so on, that are historical embodiments of the technology. If a large number of devices are included in the data sample, there is less chance for the results to be distorted by the peculiarities of a single device or class of devices. In addition, the data should cover as long a time span as possible. A long time span reduces the standard error of any curves fitted to the data by regression; in addition, it helps to eliminate distortions that might arise from circumstances peculiar to a particular time period (e.g., war, depression). However, it may not be possible to satisfy these criteria readily. The technology measure the forecaster prefers may be useless because data are simply not available, and some less desirable measure may have to be used simply because it is the only one for which adequate data can be obtained.

The measure chosen should be one for which the data are consistent with respect to the stage of innovation represented. If at all possible, all the data points should represent devices at the same stage of innovation. If the data come from devices at different stages of innovation, the forecast may be distorted. This is especially true if data from an early stage of innovation are grouped at one end of the time span and data from a late stage of innovation are grouped at the other end of the time span. Even if data from early and late stages are mixed randomly, the standard error of a regression fit will be larger than if consistent data had been used. The forecaster may not be able to obtain consistent data, in which case there is no choice but to use what is available; however, the possibility that inconsistent data may distort the forecast should be recognized.

In many cases the forecaster will use a measure of technology that is derived directly from the technical characteristics of the device. This may be a simple parameter such as speed or efficiency, or it may be a complex one such as a gain–bandwidth product, which captures an engineering tradeoff. The remainder of this chapter will present some examples of measures of technology that can be used when measures based on the technical characteristics of a technology are unavailable or unsuitable.

2. Scoring Models

The scoring model has long been used in operations research as a means of ranking or rating several alternatives. It is used when the alternatives have several characteristics and the value or importance of an alternative depends upon a combination of characteristics rather than on any single one. More recently the scoring model has been adapted to technological forecasting. It provides a means for obtaining a measure of technology when several technical parameters or characteristics are important and there is no analytical procedure for combining them into a composite measure.

The model can best be illustrated with an example. Delaney (1973) developed a measure of technology for aircraft hazard (fire and explosion) detectors. These devices are used in portions of aircraft such as fuel tanks and engine compartments, to detect the presence of fires or explosions. They may simply warn the pilot or, instead, automatically activate a fire extinguisher. Each hazard detector has several technical or operational parameters that describe specific aspects of its action. No single parameter gives a complete measure of hazard-detection technology; however, there is no theoretical basis on which these parameters could be combined to give a single measure of technology. Delaney solved the problem by using a scoring model, which is given by

$$T = SRMGKH/t, \qquad (6\text{-}1)$$

where the variables are as follows:

$S = AB$ is the specificity of the detector (its ability to react only to hazards), and A and B are given by the following relations.

$A = 50\lambda/(\lambda + \Delta\lambda)$, where λ is the wavelength (in angstroms) and $\Delta\lambda$ is the range of wavelengths to which the detector is sensitive.

B is determined by the electronics used:

$B = 1$ if discriminating electronics are not used;
$B = 2$ if discriminating electronics are used;
$B = 10$ if both optical redundancy and electronic logic are used.

R is the reset capability, that is, the ability to determine when the hazard is no longer present and be ready to detect another occurrence.

R = 1.0 for 4 or more cycles of reset capability;
R = 0.9 for 3 cycles capability;
R = 0.5 for 1 cycle capability;
R = 0.1 for $\frac{1}{2}$ cycle capability.

M is the reliability, in mean time between failures (MTBF, in hours) divided by a standard MTBF of 10^5 h.

G is the false alarm rate, in mean time between false warnings (MTBFW) divided by a standard MTBFW of 10^6 h.

K is the coverage factor, which is based on the detector's ability to give warning of a hazard throughout the volume protected.
K = 10 if volume sensors are used;
K = 8 if lineal sensors are used;
K = 5 if point sensors are used.
These factors were based on experimental data that showed that a point sensor would detect only 50% of the fires in an engine compartment; lineal sensors would detect 80% of the fires; and volume sensors would detect 100% of the fires.

H is the maximum temperature at which the sensor can still operate. There are two different upper limits to the maximum temperature, depending upon the use of the hazard detector: If it is used in an engine or structural compartment, the upper limit is 1255 °K, set by the structure itself; if it is used in a fuel tank, the upper limit is 483 °K, set by the fuel. To place detectors for both applications on the same basis, the actual upper limit for a fuel-tank detector was rescaled by 1255/493, or 2.6. Finally, to keep the number for this factor comparable to the numbers for the other factors, the maximum temperature is multiplied by 0.1.

t is the time in seconds required for the detector to respond to a hazard and provide an alarm. To keep this value comparable with the other values the response time is multiplied by 10.

A plot of the measure of technology for nine different hazard detectors introduced between 1949 and 1969 is shown in Figure 6.1. The measure of technology shows exponential growth over a 20-year period. Despite the highly subjective nature of some of the elements in the scoring model, the essential features of the technology have evidently been captured.

Next we take up a formal procedure for developing a scoring model for a measure of technology. The procedure has three steps:

1. Identify the factors to be included.
2. Weight the factors.
3. Construct the model.

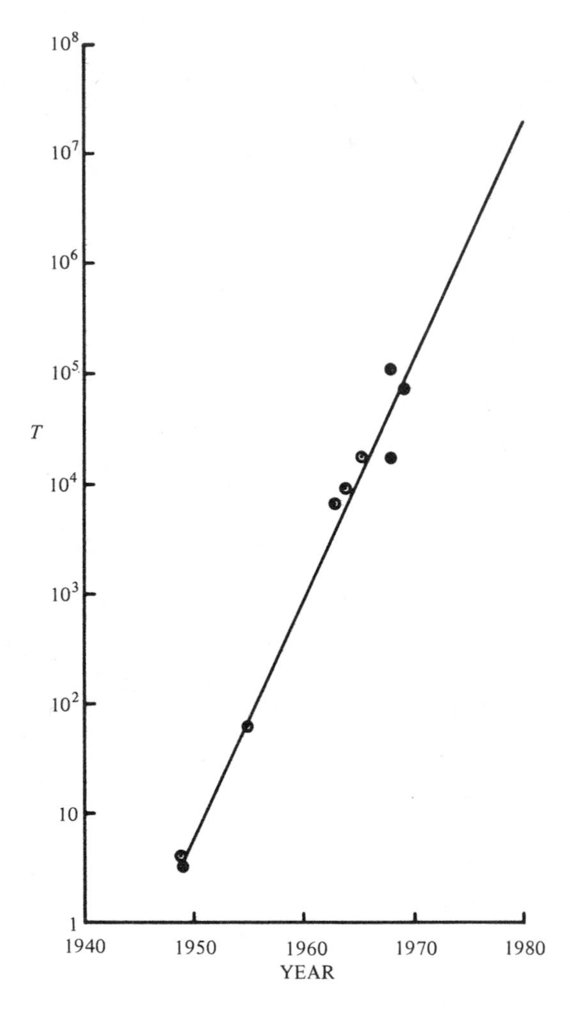

Figure 6.1. Measure of technology for aircraft hazard detectors.

Identifying the factors involves four steps in itself. The first of these is to list all the important factors that relate to how well the technology performs its function. These might include things like speed, weight, power consumption, efficiency, delay or waiting times, precision, and accuracy. The factors must be chosen to cover all the different technical approaches that may be represented among the devices to be compared (e.g., the hazard detectors described above included thermocouples, infrared detectors, and ultraviolet detectors). In general, it will be necessary to work with a specialist in the technology to identify the important factors.

The second step in identifying the factors is to refine the initial list to eliminate overlaps and double-counting. At the end of this step the forecaster will have a list of factors that are important to the technology and which measure independent aspects of it.

The third step is to verify that the factors can be measured or given judgmental ratings and that historical data or expert judges are available. If some factors cannot be measured or scaled, or data are not available, it may be necessary to repeat step 1 of the identification process. At the end of this step a list of factors will be available, and the forecaster will have either objective historical data or a means of obtaining judgmental data.

The fourth step in the identification process is to group the factors as is appropriate. One issue that always arises in the construction of scoring models is whether the factors should be multiplied or added together. This depends on the following considerations. If a factor is of such an overriding nature that it must be present, then it must multiply everything else in the scoring model. Thus if it receives a score of zero for some particular device, that device automatically has a technology measure of zero. On the other hand, some factors are not overriding, but subject to tradeoffs. A low value on one factor may be offset by a high value on another factor; hence a zero value on one of these factors would not require a zero for the overall measure of technology. Factors that can be traded against one another should be added together. Thus a group of such factors will enter the model as a sum. This sum may in turn be multiplied by one or more other group sums if it is the case that at least one factor in each group must have a nonzero value. Finally, the combination of sums and products may be multiplied by one or more overriding factors. Note that in some special cases a tradeoff may be multiplicative rather than additive. This is the case with the gain–bandwidth product mentioned earlier. However, in most cases these multiplicative tradeoffs involve factors that must have nonzero values if the measure of technology is to have a nonzero value; that is, an electronic device with zero gain or bandwidth would properly receive an overall technology measure of zero. Hence, even though there is a tradeoff involved, both factors should enter the model as multipliers.

Weighting the factors involves three steps. The first of these is to assign numerical weights to each of the factors, reflecting their relative importance. Consider a group of factors that are added together. One of these can be identified as being least important; it can be assigned a weight of one. Each other factor in the group can then be assigned a weight reflecting its importance relative to the least important factor. This weighting is carried out for each group of additive factors. If there are factors that multiply the entire model, these must be weighted next. Similarly, if groups of factors multiply other single factors or groups of factors, the groups must be weighted. Clearly, if we assign a coefficient to a multiplicative factor (or group), that coefficient can be factored out of the product and it simply becomes a multiplier for the whole model; it really has no effect in altering the weight of one factor as compared with another. Instead it is necessary to raise these multiplicative factors to powers that reflect their relative weights. Thus if one factor is twice as important as another factor it multiplies, it must be squared (or have twice the power of the other factor). This can be looked upon as weighting the logarithms of multiplicative factors in the same way as additive factors.

The second step in weighting the factors is to select a judgmental scale for those that cannot be measured but must be rated. A typical scale would range from 0 through 9; however, other ranges are possible, depending upon the degree of discrimination possible and desired. The important point is that the scale must be the same for all judgmental factors. If the scale for one factor ranges from 0 through 4 while the scale for another ranges from 0 through 9, the second is implicitly being given twice the weight of the first. Relative importance should be taken into account in the numerical weights assigned to the factors. The weighting and scaling of a factor should be kept separate.

The third step in weighting is to convert the measurable factors to a scale with the same range as that for the judgmental factors. A convenient way of doing this requires that the mean and standard deviation of the values be computed for each measurable factor. A scale is then devised for each factor in which the mean value lies in the middle of the scale range (e.g., if the scale runs from 0 through 9, the mean value for the factor may be scaled at the division between scale value 4 and scale value 5). The remaining scale ranges are then defined in terms of a suitable fraction of the standard deviation. For instance, if the distribution of the values for a factor is roughly normal, almost all the values will lie within three standard deviations of the mean. Thus if a scale from 0 through 9 is desired, intervals of half a standard deviation will divide the range of values conveniently. Scale value 4 should include all factor values from the mean to half a standard deviation below the mean; scale value 5 should include all factor values from the mean to half a standard deviation above the mean. Scale value 0 should include all factor values in the lower tail more than two standard deviations below the mean, and scale value 9 all those in the upper tail more than two standard deviations above the mean. If the distribution of factor measurements is not symmetrical about the mean, it may first be transformed (e.g., by taking the logarithm) to make it symmetrical before the mean and standard deviation are computed. By this procedure the measured and scaled factors can all be put on a common basis so that their relative importances are reflected in their numerical weights.

Finally, the scoring model must be constructed. This operation is almost automatic after the preceding steps have been completed. The model must have the groups of additive and multiplicative factors in the proper algebraic form. In addition, it is customary to construct the model as a fraction, with the desirable factors in the numerator and the undesirable factors in the denominator. Thus in the hazard-detector scoring model all the performance factors except for operating time were desirable and were placed in the numerator. The operating time appeared in the denominator.

Development of a scoring model using the procedure described above allows the forecaster to produce a measure of technology for a class of devices for which several parameters must be combined into a single number. With the scoring model the forecaster can integrate both objectively measurable variable factors and factors that must be scaled judgmentally, as well as combine factors in an intuitively correct manner even though there is no theoretical basis for an exact combination.

3. Constrained Scoring Models

The scoring model described in the preceding section can take any form and, in principle, might produce a technology measure that either grows exponentially (i.e., the aircraft hazard detector) or exhibits a growth curve or some other behavior. Here we take up a form of scoring model that assumes that every technology, in the long run, exhibits a growth curve. The model is based on a theoretical upper limit for the technology and thereby allows a comparison of the state of the art from one technology to another. This method was first described by Gordon and Munson (1981).

The scoring model is of the form

$$M_i = 100 \frac{C_i}{C^*} \left(K_1 \frac{X_{1i}}{X_1^*} + K_2 \frac{X_{2i}}{X_2^*} + \cdots + K_N \frac{X_{Ni}}{X_N^*} \right), \tag{6-2}$$

where the Xs are the different factors for the ith device and the X^*s are reference values to normalize the Xs. A reference value may be a theoretical upper limit on the corresponding X, or it may be some other reference value that will not be exceeded. The Xs are intended to increase with increasing performance. If some parameter decreases with increasing performance, such as the operating time of the hazard detectors described in the previous section, the corresponding X should be the reciprocal of that parameter, and the X^* the reciprocal of the lower limit or of the smallest value used as a reference. The Ks are the weights reflecting the importance of the various factors. The term C represents one or more overriding factors that must have a nonzero value. A factor of 100 normalizes the maximum possible value.

In principal, the Ks in this model could be obtained judgmentally, as were the weights in the scoring model of the previous section. However, the originators of this model instead obtained the Ks by regression on historical data for the set of devices for which the technology measure was desired. This was done as follows.

First, it was assumed that the technology measure would follow a Pearl curve with an upper limit equal to 100. The formula for the technology measure was thus incorporated in the following equation:

$$\tfrac{1}{2}[1 + \tanh[A(t - t_0)] = 100C'(K_1X_1' + K_2X_2' + \cdots + K_nX_N'), \quad (6\text{-}3)$$

where the Ks are as above but the primed variables represent the ratios in the previous form of the scoring model. It can be shown by algebraic manipulation that the expression on the left-hand side of the equation is the formula for a Pearl curve. In this version of the Pearl curve formula t_0 is the time at which the curve reaches 50% of its upper limit and A establishes the slope at t. In equation 6-3 there are $N + 2$ unknown coefficients: A, t_0, K_1, K_2, . . . , K_N. If there are at least $N + 2$ data points, these coefficients can be solved for. In the usual case there will be more than $N + 2$ data points, and a nonlinear regression method must be used to solve for the Ks.

The nonlinear regression method starts by choosing values for A and t_0; ordinary linear regression is then used to evaluate the Ks. A measure of the goodness of fit of the data to the growth curve is obtained, such as the standard error of estimate. Another pair of values for t_0 and A is chosen, and the regression repeated. This process is continued until a pair (A, t_0) is found that gives the smallest standard error of estimate for the fitted Ks. A nonlinear regression computer program can be developed that will carry out this search automatically, obtaining a numerical solution for the "best" values of A, t_0, and the Ks.

The advantages of this constrained scoring model are that it can be applied to a great variety of technologies and will allow direct comparisons of the degree of maturity of unrelated technologies. As with all scoring models, it allows the inclusion of factors that are judged to be relevant, even in the absence of any theory to guide the selection. By virtue of the assumption that the technology measure must follow a Pearl curve, however, the need for subjective weights for the factors is eliminated; the weights can be obtained by an objective means—nonlinear regression. This objectivity may well be an advantage when there is uncertainty about the relative importance of different factors in an overall measure of a technology.

4. Planar Tradeoff Surfaces

One of the reasons for using a composite measure of technology is the possibility of tradeoffs among several technical parameters of a complex device. The concept is that at any given state of the art the designer has the freedom to obtain more of one parameter at the expense of some other parameter. The specific application of a particular device will determine which parameters will be emphasized by the designer and which will be sacrificed. Thus several different devices representing the same state of the art may have different values for the various parameters, depending upon their particular applications. This leads to the concept of a tradeoff surface in the N-dimensional space defined by the N parameters that go to make up the state of the art and which can be traded off for one another. An advance in the state of the art means that more of one parameter can be obtained without sacrificing any of the remaining parameters. Thus an advance in the state of the art would mean moving to a "higher" tradeoff surface. Conversely, a design represented by a point "inside" the frontier for the current state of the art would be "inefficient" or "bad design," in the sense that the designer could have obtained more of at least one parameter without sacrificing any of the others.

One approach to defining tradeoff surfaces, described by Alexander and Nelson (1973), treats these surfaces as hyperplanes in the N-dimensional space of the parameters that define the state of the art. The tradeoff surfaces are given by the expression

$$M = K_1P_1 + K_2P_2 + \cdots + K_NP_N, \tag{6-4}$$

where the Ps are the values of specific parameters and the Ks are coefficients that define a hyperplane. Note that this reduces to an ordinary plane in three-dimensional space and to a line in two-dimensional space. Figure 6-2 illustrates this for a space of only three parameters. Each value of M corresponds to a plane representing a specific level of the state of the art. The planes are all parallel, and movement from one plane to another plane farther from the origin is equivalent to an increase in the state of the art. The ratio K_i/K_j is the rate at which parameter P_j can be traded off for parameter P_i; that is, if P_i increases by one unit, K_iP_i increases by K_i units. In order to keep M constant, P_j must be reduced by K_i/K_j units. Thus if plane M_3 in Figure 6.2 represents the current state of the art (the "frontier"), the designer has freedom to move about in that plane, trading an increase in one parameter for a decrease in either or both of the other two according to the tradeoff relationship given by the ratios of the Ks. If M_3 represents the state of the art, to move toward the origin, as back to plane M_2, would represent an inefficient design. A move away from the origin can happen only if the state of the art is increased. Note that the parameters should be selected so that an increase in the value corresponds to an increase in performance.

This approach, as developed by Alexander and Nelson, does not require that a numerical measure of the state of the art be developed; instead, the forecaster obtains historical data on devices that represent differing

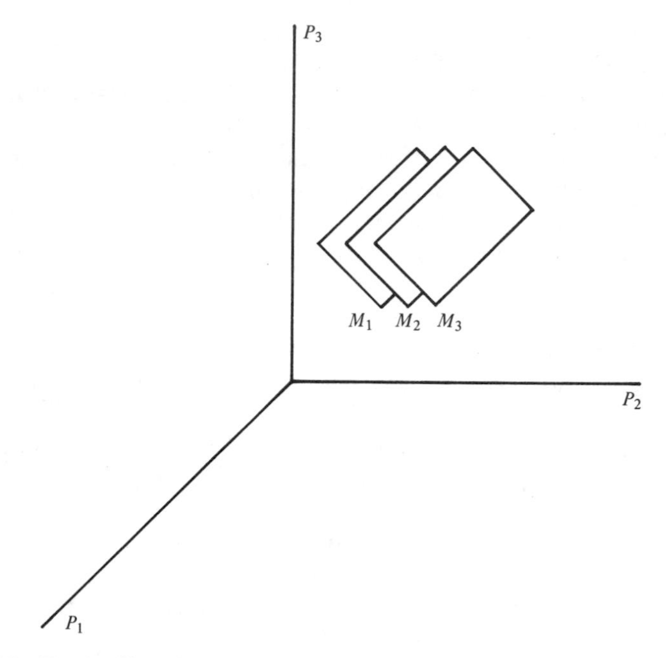

Figure 6.2. Tradeoff surfaces in parameter space.

levels of the state of the art. The tradeoff surface formula is obtained by regressing the year of introduction of each device on the parameter values for that device. The year of introduction is thus a "proxy" for the state-of-the-art measure M, and the Ks are obtained by multiple linear regression. Alexander and Nelson (1973) discuss the application of this approach to aircraft turbine engines, but it has also been applied to many other technologies.

As with any of the other composite technology measures, it is important to select the parameters that go into the equation carefully: They should be independent and should measure important aspects of the technology being measured. If the set of devices includes representatives of several technical approaches, the parameters chosen should be applicable to all the approaches.

This approach to tradeoff surfaces determines empirically, rather than theoretically, what tradeoff relationships were available to the designers. It can therefore be used in cases where tradeoffs are thought to exist but there is no theory available to guide the forecaster in combining the technical parameters of a device into an overall measure of technology.

The assumption of planar tradeoff surfaces may be unreasonable in some cases. Frequently it is more likely that the tradeoff surfaces are convex, so that trading off one variable for another becomes more and more difficult as further increments of improvement are demanded. Dodson (1970) describes a method similar to that described above, except that the tradeoff surfaces are ellipsoids (or hyperellipsoids in more than three dimensions). These ellipsoidal surfaces allow for an increasing difficulty of tradeoffs.

References

Alexander, Arthur J., and J. R. Nelson (1973). "Measuring Technological Change: Aircraft Turbine Engines," *Technological Forecasting and Social Change* 5, 189–203.

Delaney, Charles L. (1973). "Technology Forecasting: Aircraft Hazard Detection," *Technological Forecasting and Social Change* 5, 249–252.

Dodson, E. N. (1970). "A General Approach to Measurement of the State of the Art and Technological Advance," *Technological Forecasting and Social Change* 1, 391–408.

Gordon, Theodore J., and Thomas R. Munson (1981). "A Proposed Convention for Measuring the State of the Art of Products or Processes," *Technological Forecasting and Social Change* 20, 1–26.

Problem

1. Develop a scoring model for measuring automobile technology. Include the following factors: passenger capacity, luggage capacity, miles per gallon, range (miles per tankful of gasoline), time from 0 to 55 mph, and curb weight. Include any other factors you think are important. Put all the factors in a mathematical expression, with weights that you consider to be appropriate. Now take as many current-year automobile models as you have factors in your model, obtain the data on the factors, and score them. Derive the weights implicit in the manufacturers' tradeoffs by assuming that all the cars represent the same level of technology (i.e., set all scores to a common value and solve for the weights). Do your weightings agree with those implicitly used by the manufacturers?

Chapter 7
Correlation Methods

1. Introduction

The previous chapters have dealt with the direct forecasting of one or more measures of technological change. However, in some cases direct forecasting may not be the most appropriate means of gaining the information needed for a decision. The functional capability of interest may be highly correlated with something else that is easier to forecast or which can be measured directly. In such cases a forecast based on the correlation may be better than one obtained directly, since it makes use of information that is more readily obtained.

In this chapter we will discuss several forecasting methods that are based on correlations between the functional capability of interest and other factors which may be more readily forecast or measured. In all such cases the usefulness of the forecast depends upon the continued validity of the correlation on which it is based. This must be investigated for each case in which correlation methods are to be used.

The methods shown in this chapter are intended as examples only. In practice, many other correlations similar to the ones shown here can be found and used; the forecaster should be alert to discover such correlations in other cases and not assume that the ones shown here exhaust the possibilities.

2. Lead–Lag Correlation

There are cases where one technology appears to be a precursor to another. Thus the trend, growth curve, or other time history of one will be similar to that of the other, except one will lag behind the other. Forecasting the lagging technology is then done by determining the lag and projecting the values of functional capability *already achieved* by the leader.

An illustration of this situation is given in Figure 7–1, which displays a growth curve showing the increasing use of composite materials in aircraft (as a replacement for metals). The forecaster has estimated a 10-year lag between the first demonstration in an experimental aircraft and application in an operational aircraft, that is, a 10-year lag between the operating-prototype stage of innovation and operational use. Projecting a continuation of this 10-year lag, it is forecast that nearly 50% of the structural weight of an advanced fighter will be composite materials by the early 1990s (Hadcock, 1980).

The use of lead–lag correlations requires matching the levels of func-

Figure 7.1. Time between the development and application of advanced composites to aircraft (adapted from Hadcock, 1980).

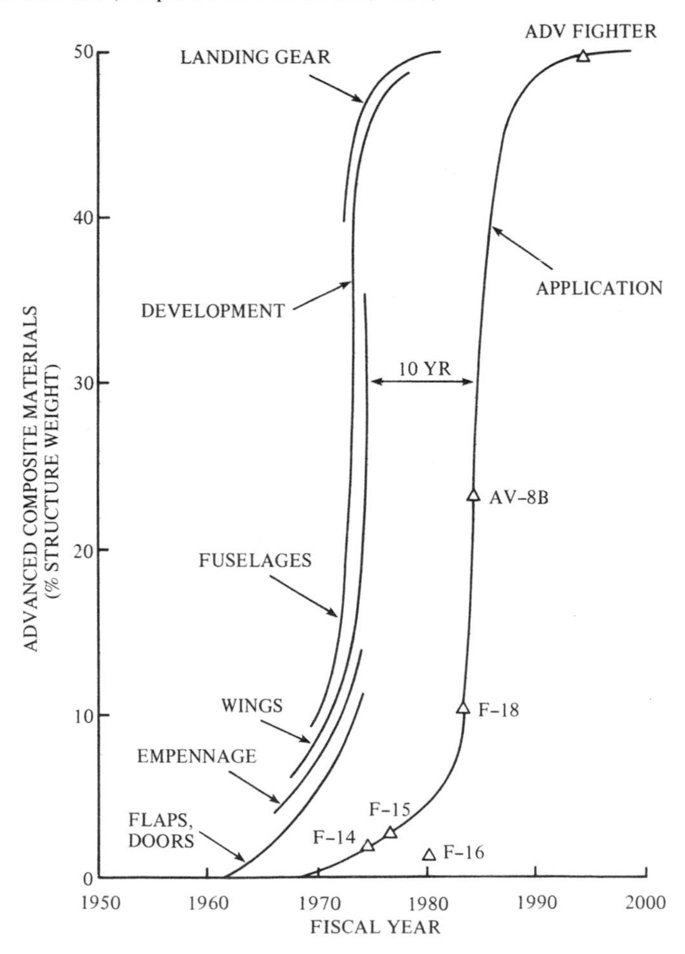

tional capability of specific devices in the leading and lagging technologies. For instance, if the leading technology had a level of functional capability X, and the lagging technology achieved level X ten years later, we could talk about a 10-year lag between the two. If there were several cases of matched pairs in the leading and lagging technologies, we might want to average them or determine whether there was any consistent trend in the lag time. Unfortunately, in practice there are likely to be few if any matched pairs. What actually happens is that both the leading and lagging technologies involve the introduction of devices with higher and higher levels of functional capability, but in most, if not all, cases the actual level of capability of a leading device is not exactly matched by that of a lagging device. Thus it is not possible to specify a lag time for devices of the same performance level.

Alternatively, we might look at the difference in functional capability between leading and lagging devices introduced in the same year. Rather than a lag time for given performance levels, this would give differences in performance levels at a given time. Unfortunately, the forecaster often finds that there are few, if any, cases where leading and lagging devices were introduced at essentially the same time.

Thus although the lead–lag relationship may be quite clear, the problem becomes one of matching pairs to obtain either a time lag for a same level of performance or a difference in the level of performance at a same time. What must be done in order to obtain the equivalent of matched pairs is interpolate between the performance of devices for either the leading or lagging technology. This is illustrated in the following example.

It is already known that much of the technology introduced into commercial transport aircraft during the period 1945–1970 was first demonstrated in military aircraft. Thus we would expect a lead–lag relationship between the performance of military aircraft and that of transport aircraft. One measure of functional capability is aircraft speed. So let us look for a relationship between the speeds of military aircraft and those of transport aircraft.

In equation (5-4) we have a relationship between the speed of military aircraft and time. This equation is repeated here, with X now designating leader performance:

$$X = -118.30568 + 0.06404T. \tag{7-1}$$

This equation permits us to interpolate between the speeds of actual military aircraft to obtain lag times for commercial transports, whose speeds do not correspond exactly with that of any actual military aircraft.

Let us make use of equation (7-1) to find a lag relationship for transport aircraft. Using Y to designate the performance of the lagging technology, the equation will be of the form

$$Y = A + B(T - D), \tag{7-2}$$

where A and B are the same as in the equation describing the leading technology [equation (7-1) is an example], and D is the lag. D need not be a constant but may itself be a function of T. Now consider one of the lagging devices, which was introduced in the year T_j with performance Y_j. Substituting this value of performance into equation (7-1), we obtain a time T_j', which is the interpolated time at which the leader technology achieved the level Y_j, Then $D_j = T_j - T_j'$. For instance, a transport aircraft was introduced in 1925 that had a speed of 95 mph. The natural logarithm of 95 is 4.55388. Substituting this into the left-hand side of equation (7-1) and solving, we obtain a T of 1918.452. This is the date (in year and decimal fraction) when the interpolated speed of combat aircraft was 95 mph. This gives a lag time of 6.548 years between combat aircraft achieving 95 mph and transport aircraft achieving the same speed. Without the interpolation equation we could not have determined this lag, since there never was a combat aircraft whose top speed was exactly 95 mph.

Once the lag for each device of the follower technology is obtained, we can regress the lag time on the year of introduction of that technology. This gives us an equation of the form

$$D = a + bT. \tag{7-3}$$

Substituting into equation (7-2) and rearranging terms, we have

$$Y = A - aB + B(1 - b)T, \tag{7-4}$$

which is the lagged trend equation for the follower technology, which we can then use to forecast Y on the basis of X values already achieved.

Table 7-1 gives the speed, the year of introduction, the interpolated year for combat aircraft, and the number of years of lag for pace-setting transports introduced between 1925 and 1959. (Transport aircraft introduced after 1959 have been omitted deliberately, since their top speeds were subject to an economic constraint.) Note that the number of significant figures given in Table 7.1 represents a spurious precision, since the times for introduction of both combat and transport aircraft were originally rounded off to whole years. The time lags can then be regressed on the year of introduction of the transport aircraft. Doing this, we obtain the equation

$$D = -443.79507 + 0.23225T. \tag{7-5}$$

Substituting this into equation (7-2), we obtain the lagged transport equation

$$Y = -89.88504 + 0.04917T. \tag{7-6}$$

Figure 7.2 shows the trend line for combat aircraft, the data points for pace-setting transport aircraft, and the lagged transport trend. Also shown are two subsonic transport aircraft, introduced in 1963 and 1969, that

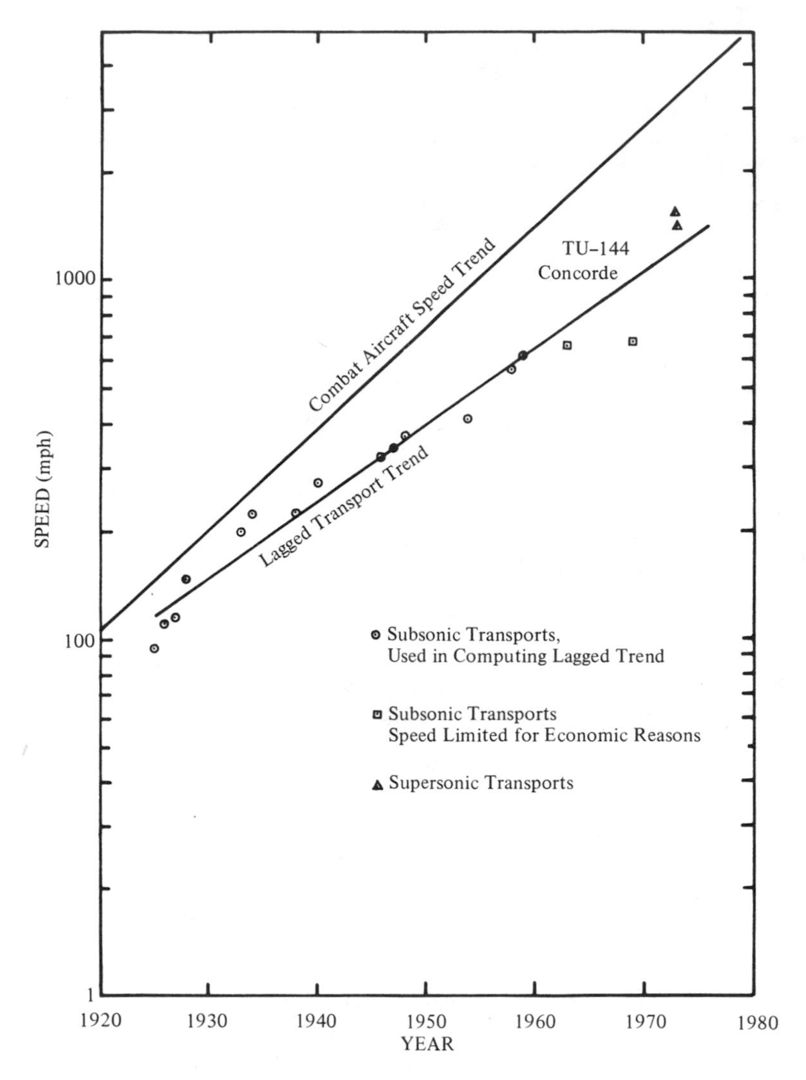

Figure 7.2. Correlation of combat aircraft and transport aircraft speeds.

were designed for speeds lower than technology would have permitted, because of economic considerations. In addition, data points for the Anglo-French supersonic transport Concorde and the Russian supersonic transport TU-144 are shown. These supersonic transports seem "early," in the sense that they are well above the lagged transport trend. However, an examination of Figure 5.2 will reveal that the military aircraft that had about the same speed as the Concorde and TU-144 were also above the combat aircraft speed trend. Since the representatives of the leading technology were "early," it does not seem unreasonable that the representatives of the follower technology would also be early.

Table 7.1. Lag of Transport Speed Behind Combat Aircraft Speed

Year	Speed (mph)	Natural logarithm of speed	Year for same combat speed	Lag in years
1925	95	4.554	1918.452	6.548
1926	111	4.710	1920.883	5.117
1927	116	4.754	1921.571	5.429
1928	148	4.997	1925.375	2.625
1933	200	5.298	1930.076	2.924
1934	225	5.416	1931.916	2.084
1938	228	5.429	1932.122	5.878
1940	275	5.617	1935.049	4.951
1946	329	5.796	1937.849	8.151
1947	347	5.849	1938.680	8.320
1948	375	5.927	1939.892	8.108
1954	409	6.014	1941.247	12.753
1958	579	6.361	1946.675	11.325
1959	622	6.433	1947.794	11.206

Whenever two technologies seem to show a leader–follower relationship, the method of this section can be applied. It is necessary to fit some curve (e.g., growth curve or trend) to the leader technology and use this to interpolate between the actual values achieved by its devices. The time interval between these interpolated values and the actual values for the follower represent the lags. These lags should then be examined for any consistent pattern of change with time. The time lag (either a constant or a function of time) is then introduced in the interpolation equation for the leader, and the resulting equation used to forecast the follower. The equation can be used to forecast the follower as far ahead as the lag time from the most recent value of the leader technology. If the leader technology can be forecast reliably, the lag equation can be used to forecast the follower even farther into the future than the lag time.

3. Technological Progress Function

The technological progress function (Fusfeld, 1970) is based on an observed correlation between performance and cumulative production. This is illustrated in Figure 7.3, which shows the improvement (decrease) in two performance measures of jet engines (specific fuel consumption and specific weight) with cumulative engine production. This phenomenon has been observed for a wide range of technologies.

While the correlation is interesting, little success has been achieved in placing it on a sound theoretical basis. Nevertheless, if this behavior is observed for a technology, it might be used for forecasting improvements. The problem, of course, is that it is necessary to forecast production in order to use the progress function. However, where good forecasts of

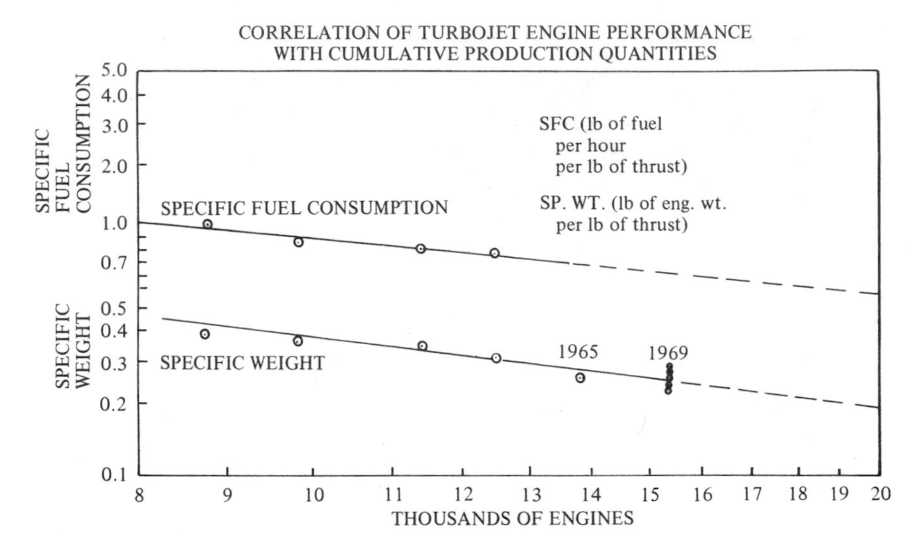

Figure 7.3. Correlation of turbojet engine performance with cumulative production quantities.

production can be made and a correlation is observed between cumulative production and performance, this method can be used.

The customary method for utilizing the progress function is to regress the logarithm of the level of functional capability on the logarithm of the cumulative production. The result should be a straight line when plotted on log–log paper. In effect, this means that a given percentage of increase in cumulative production brings about the same percentage of improvement in the technology. So as cumulative production grows, it becomes harder and harder to squeeze out the same increase in the performance of the technology. This relationship seems to make sense, which gives reason to believe that there is some basis for the progress function. Perhaps additional research will show why this method works as well as it does and for the technologies for which it does.

The regression program in Appendix 4, using option 3, can be used to fit a technological progress function to data on performance as a function of cumulative production.

4. Maximum Installation Size

In many industries there is a relatively constant relationship between total industry capacity and the size of the largest single installation. This relationship, which was apparently first observed by Simmonds (1972), can be used for forecasting maximum installation size if the total production or total capacity can be forecast.

Figure 7.4 illustrates this behavior: It shows the size of the largest steam turbine electric generator in the United States by year and the total installed steam turbine capacity (data from Appendix 3). The growth in size of the largest single unit roughly parallels the growth in total capacity. This parallelism is disturbed by rapid initial growth prior to 1918, as well as by the appearance of two "freak" units, one in 1915 and one in 1929. These were significantly larger than units introduced in the immediately previous years; they were not exceeded by other units for many years. In the case of the generator introduced in 1915, there was a lag of eight years before a slightly larger generator was introduced. In the case of the

Figure 7.4. Relation between total installed capacity and maximum single unit size for steam turbine electric plants.

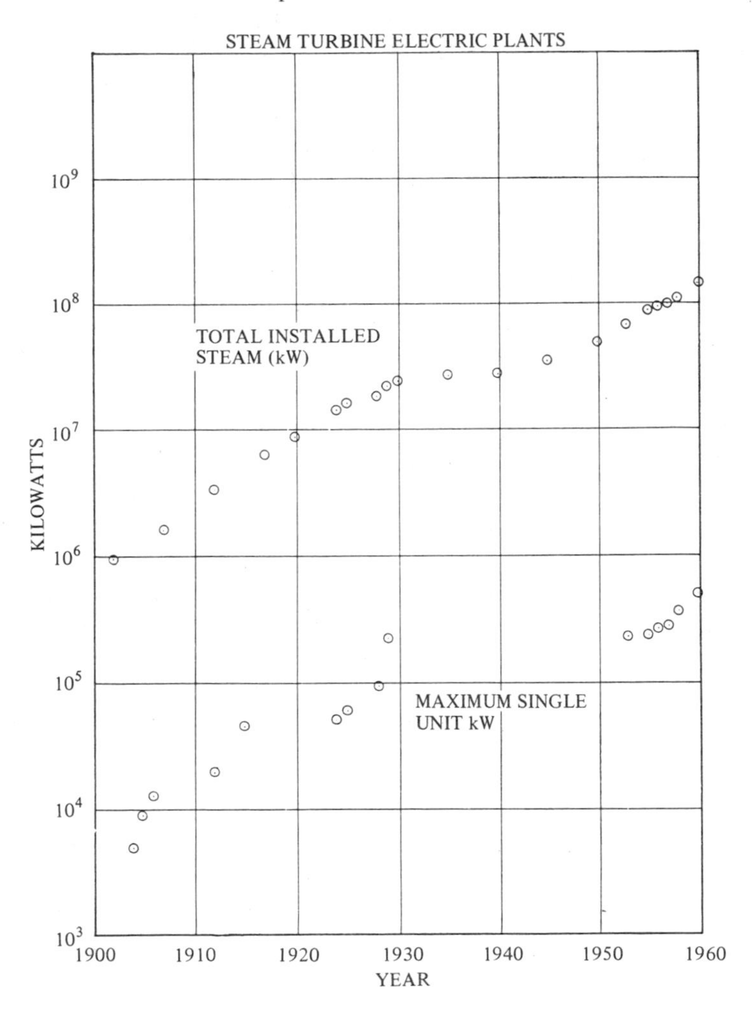

generator installed in 1929, there was a lag of 24 years before a slightly larger one was introduced. If the 15-year period of stagnation of the electric power industry (1930–1945) is subtracted, this lag becomes 9 years, not too much different from the lag following the 1915 generator. This is reasonable, since both units exceeded the prior state of the art by about the same percentage.

The ratio of largest single unit to the total capacity can be computed for the 11 units for which both items of data are available. The mean value of this ratio is 0.00415, and the standard deviation is 0.00192. If the 1929 unit is excluded from the calculation, the mean value becomes 0.00362, and the standard deviation 0.000975. It can be concluded that the maximum size of steam turbine does represent a fairly constant proportion of the total installed capacity of electric plants in the U.S. This proportion remained nearly constant for over 40 years.

There seems to be no theoretical basis for the correlation between total capacity and maximum single unit. Nevertheless, when it is observed and total capacity can be forecast, the correlation can be used to forecast the size of the largest single unit. The advantage of this correlation, for the technological forecaster, is that variables such as total capacity and total production are often comparatively easy to forecast on the basis of such things as population or the Gross National Product, whereas the maximum single unit may be much more difficult to forecast. Therefore this relation can be used advantageously by the technological forecaster when it is found to exist.

5. Correlation with Economic Factors

In some cases the level of functional capability of some technology may be related to economic factors. For many technologies related to communications or transportation the extent of development within a country is often closely correlated with the wealth of that country.

Consider a telephone system. One measure of its extent of development is the number of telephones per capita. This is an indicator of the degree to which the system allows the user to communicate with other people. The first telephone subscriber in an area has a problem: There is no one to call. The more telephone subscribers there are, the more useful a telephone is to each subscriber.

The Gross National Product (GNP) per capita is often used as a measure of national wealth. It is not a perfect indicator, since it may conceal sizable disparities in income; it can also be misleading in totalitarian countries, where the perquisites of the leaders do not show up as cash income. Despite its shortcomings, however, the GNP per capita is a useful measure of national wealth.

Appendix 3 includes data on the GNP per capita and the number of telephones per capita for 40 nations of the world. This data is plotted in Figure 7.5, which also shows a trend fitted to the data using the regression program in Appendix 4, option 3. The best-fit equation is

$$Y = (0.676786 \times 10^{-5})T^{1.22092},$$

where T is the GNP per capita in dollars.

Once the relationship between the GNP per capita and the number of telephones per capita is known, this equation can be used to forecast the development of the telephone system in a nation. First one forecasts the GNP and the population, obtains their ratio, and then substitutes this into the best-fit equation above. The advantage of this method is that there is often no direct way to forecast something like telephones per capita, except by extrapolation. Extrapolation, however, is unlikely to be an acceptable method of forecasting under conditions of rapid economic change, precisely when a forecast is most likely to be needed. It is then that this method can be most useful.

Two caveats must be entered: First, this method is cross-sectional, whereas the forecaster really wants longitudinal data. Second, the forecaster must watch out for correlating the economy with itself.

Figure 7.5. The number of telephones per capita versus the Gross National Product per capita for various nations.

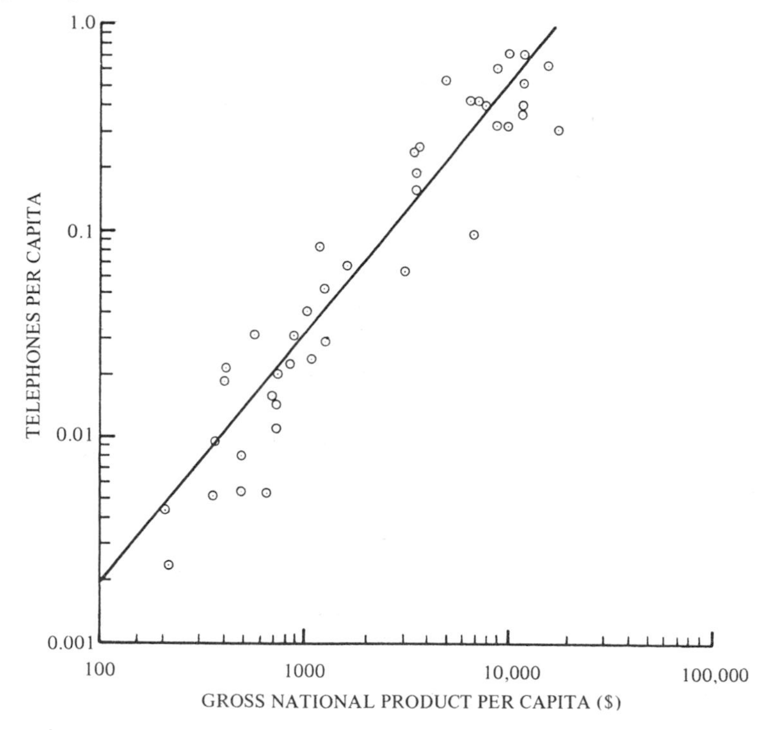

What the forecaster really wants is a longitudinal relationship, that is, one that can say something about the future on the basis of data about the past. However, this method does not offer any relationship between the past and the future; instead, it observes cross-sectional data and says that the future of some countries will be like the present of others. In using this method the forecaster risks subscribing to the "railroad track theory of history." This theory assumes that all countries follow the same track, passing through the same stations in the same order, but some having gotten an earlier start. Therefore the future of the late starters will be like the past and present of the early starters. There is probably little if any validity to this theory. The forecaster should be careful not to inadvertently fall into adopting it.

Nevertheless, in the absence of any other method of forecasting, the use of cross-sectional data in the place of longitudinal data can be defended as being better than throwing up one's hands in despair. Hence, while the forecaster should be aware of the traps involved, this method will nevertheless be found useful.

There are many known relationships like the one between telephones per capita and the GNP per capita. For instance, it is well known that both electric power and steel production per capita are closely correlated with the GNP per capita. However, items such as electric power and steel are really significant elements of an economy, and correlating these with the economy is to some degree correlating the economy with itself. The results are not particularly meaningful. Use of this method should be restricted to those cases that are not equivalent to correlating the economy with itself; that is, this method should be used only for technologies that do not represent a significant fraction of the nation's capital investment, or involve a significant fraction of the nation's productive activity. Obviously the value of the method depends upon there being a correlation between the technology of interest and the economy. But this correlation should result from some third factor relating the technology and the economy. The correlation should not arise simply because the technology *is* a significant fraction of the economy.

Despite these two caveats, this method is extremely useful in particular circumstances, and the forecaster should be aware of correlations with economic factors that can be used when direct forecasting is not possible.

References

Fusfeld, Alan R. (1970). "The Technological Progress Function: A New Technique for Forecasting," *Technological Forecasting* 1, 301–312.

Hadcock, Richard N. (1980). "The Cautious Course to Introducing New SDM Technology into Production Systems," *Astronautics and Aeronautics* (March), pp. 31–33.

Simmonds, W. H. C. (1972). "The Analysis of Industrial Behavior and Its Use in Forecasting," *Technological Forecasting and Social Change* 3, 205–224.

Problems

1. Take the data on speed records of rockets and experimental aircraft and the data on aviation speed records for the period since 1945 from the tables in Appendix 3. Assume that the experimental aircraft technology is a precursor to the military aircraft technology represented by the aviation speed records. Using the lead–lag correlation method, prepare a forecast of the speeds of future military aircraft. Does your forecast appear reasonable?

2. Using the data from Appendix 3 on the speeds of pace-setting commercial transport aircraft and the cumulative production of aircraft, fit a technological progress function and use it to forecast the speeds of commercial transport aircraft. How might you estimate the future production of commercial transport aircraft?

3. Using the data from Appendix 3 on the total installed hydroelectric capacity and maximum hydroelectric turbine capacity, estimate the relationship between the total installed capacity and the maximum single unit. Using the U.S. Geological Survey's estimate of the ultimate installed hydroelectric capacity in the United States (161,000 MW), forecast the largest single unit size that would be used in the United States. Do you consider your forecast to be reasonable? Why or why not?

4. Using the data from Appendix 3, correlate the number of radios per capita in various countries with the Gross National Product per capita. Assuming a 4% growth rate for the Spanish economy, when will Spain have 0.75 radios per capita?

Chapter 8
Causal Models

1. Introduction

The five preceding chapters have described forecasting methods that were rational and explicit, as opposed to the intuitive forecasting involved in Delphi. Forecasts produced by these methods can be analyzed in detail and critiqued by persons other than the forecaster. However, these methods still have shortcomings.

Analogies, growth curves, and trends all assume that whatever has been causing the growth of technology in the past will continue to cause that growth; they require no knowledge of these causes of growth.

The ability of these methods to make forecasts on the basis of past information is very useful; however, the fact that they ignore the causal factors producing technological change means that they have the following serious shortcomings:

1. They are unable to warn the forecaster that there has already been a significant change in the conditions that produced past behavior and that therefore this behavior will not continue.
2. They are unable to predict the future when the forecaster knows an important factor affecting technological change will be altered.
3. They are unable to give policy guidance about which factors should be changed, and by how much, to produce a desired technological change.

In this chapter we will discuss forecasting models that take causal factors into account. These relate technological change to the specific factors that produce it, and they can warn one of the effects of change, as well as provide policy guidance about the changes needed to produce desired effects on the rate of technological change.

The development of causal models requires an understanding of what causes technological change. Here we face a major shortcoming: Existing theories of technological change are not adequate to allow us to develop elaborate and precise models. Nevertheless, enough theoretical work has been done to allow some useful models to be developed.

In the remainder of this chapter we will discuss three types of models: The first type attempts to forecast technological change on the basis of factors internal to the system that produces technological change; the second type assumes that technology is driven by economic factors; and the third type includes the relevant portions of the social and economic system in which the technology is being developed.

2. Technology-Only Models

This class of models assumes that technological change can be explained fully by factors internal to the technology-producing system. This does not mean that technology-only models assume technology to be "autonomous" or "out of control"; it means only that the external factors affecting technology are reflected in factors internal to the technology-producing system, and hence a knowledge of the latter is sufficient for forecasting purposes. We will cover two growth curve models.

The Growth of Scientific Knowledge. This model explains the growth of scientific knowledge or information on the basis of factors within science and technology. The model asserts that the rate of increase of scientific knowledge in some field is a function of the information already known, the maximum that can be known in that field, the number of people working in the field, and the amount of information transferred among these workers. This relationship can be expressed as

$$dI/dt = K(1 - I/L)[N + \tfrac{1}{2}mfN(N - 1)], \qquad (8\text{-}1)$$

where I is the amount of information available, L is the maximum amount of information knowable, and N is the number of people working in the field. There is a maximum of $\frac{1}{2}N/(N - 1)$ information-transferring transactions possible per unit time among the workers in the field; the factor f represents the fraction of these that actually take place. The factor m represents the relative productivity resulting from one of these transfers, as compared with a researcher working alone; this may be smaller or larger than unity. K is a proportionality constant.

If N is much greater than one, which is the case in most fields of technology, equation (8-1) can be simplified to

$$dI/dt = K(1 - I/L)(\tfrac{1}{2}mfN^2). \qquad (8\text{-}2)$$

If we take N to be constant, this can be solved to give

$$I = L - (L - I_0)e^{-KmfN^2t/2L}. \tag{8-3}$$

This is a growth curve that has value I_0 at time $t = 0$. It reaches an upper limit L as $t \to \infty$; however, unlike the Pearl curve, it has a value of $I = I_0$ at a finite value of t.

If this model is applied to a new field of science, I is such a small fraction of L that their ratio can be ignored. However, once the field has more than ten people working in it, N becomes negligible with respect to N^2. Therefore we can write

$$dI/dt = \tfrac{1}{2}KmfN^2. \tag{8-4}$$

If we now assume that the growth of the number of people in the field is exponential, that is, $N = N_0e^{ct}$, we have

$$dI/dt = \tfrac{1}{2}kmfN_0^2e^{2ct}, \tag{8-5}$$

which can be rewritten as

$$\ln(dI/dt) = \ln(\tfrac{1}{2}KmfN_0^2) + 2ct; \tag{8-6}$$

that is, the logarithm grows linearly at a rate that is twice that of the number of people working in the field. Figure 8.1 illustrates this behavior. It shows the production of papers in the field of masers and lasers from 1957 to 1968. The growth rate from 1958 through 1964 is 0.5349, that from 1964 through 1968, 0.1901. In terms of the model we can interpret this as saying that the production of papers grew faster than did the number of people, until the field became too large for effective communication. Afterward, that growth tapered off to equal the growth rate of the number of workers. Note that in this example we have taken the number of papers published as the measure of information available. This is not a precise measure, but the two quantities are strongly correlated and it is the best measure available.

There are many possible variations of these information growth models; however, they are probably most useful at the two ends of the size spectrum, very new and small fields and very large fields, where the actual number of interactions that take place is only a small fraction of the total possible. At these two extremes these models and their variations can be useful in forecasting the growth of knowledge.

A Universal Growth Curve. This model was developed by Floyd (1968), and it represents an attempt to explain growth toward an upper limit on the basis of effort expended. Floyd's model is given by the following equation:

$$P(f, t) = 1 - \exp\left(-(F - f)\int_{-\infty}^{t} KNW\,dt'\right), \tag{8-7}$$

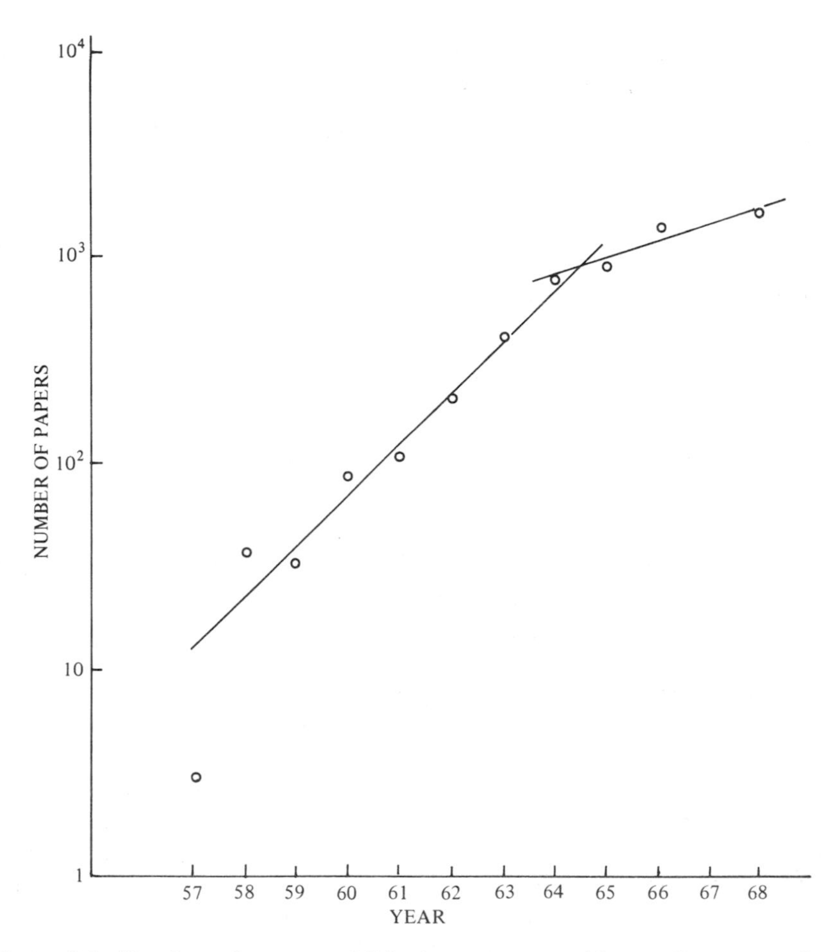

Figure 8.1. Number of papers published on masers and lasers, by year, as listed in *Physics Abstracts*.

where f is the functional capability (Floyd calls it "figure of merit"), F is the upper limit on f, W is the number of researchers working in the field to advance f, N is the number of attempts per unit time per worker to advance f, K is a constant, and $P(f, t)$ is the probability of achieving level f by time t.

By making some assumptions about the rate of growth of the number of workers in the field, Floyd converted equation (8-7) to the following form:

$$P(f, t) = 1 - \exp\left(-(F - f)\int_{-\infty}^{t}(f - f_c)T(t')\,dt'\right), \qquad (8\text{-}8)$$

where f_c is the functional capability of the competitive technology and $T(t)$ is a slowly varying function of time representing the total activity to advance f.

After some manipulation the integration in the exponent is carried out, with the following result:

$$P(f, t) = 1 - \exp(-\{-\ln(2)(C_1 t + C_2)/[\ln(Y - 1) + Y + C_2]\}, \quad (8\text{-}9)$$

where

$$Y = (F - f_c)/(F - f). \quad (8\text{-}10)$$

If we set $P(f, t) = 0.5$, equation (8-9) can be manipulated to yield

$$\ln(2) = \frac{\ln(2)(C_1 t + C_2)}{\ln(Y - 1) + Y + C_2} \quad (8\text{-}11)$$

or

$$\ln(Y - 1) + Y = Ct, \quad (8\text{-}12)$$

where the subscript on C has been dropped.

If the upper limit F and the functional capability f_c of the competitive technology are known, two historical data points are sufficient to solve for C. If more historical data are available, a best-fit value for C can be obtained.

Floyd originally developed a nomogram to solve equation (8-12), since at that time a solution by numerical means would have been quite tedious. However, with the advent of hand-held calculators it is actually easier to solve the equation numerically than with Floyd's nomogram.

The expression $F - f_c$ is constant and needs to be calculated only once. The expression $F - f$ needs to be calculated for each value of f. The value F itself is a constant. Thus a calculator with three storage registers is adequate for solving the problem without repetitively entering the constant terms. One register stores F and one stores $F - f_c$ once this has been computed. It is then necessary only to key in each value of f to compute $F - f$. Y is then computed and stored in the third register. Next $\ln(Y - 1)$ is computed and added to the stored value of Y.

We will work an example using Floyd's model. The first two columns of Table 8.1 are an abbreviated list of speed records set by propeller-driven aircraft. All but the last two entries are drawn from the speed record table in Appendix 3. The last two entries are unofficial records set by the XP-47H and the XP-47J.

Assume a value of 600 mph for F as the maximum speed for propeller-driven aircraft. Take f_c to be 100 mph, about the maximum speed for a train. Using equation (8-10), we obtain the values of Y in the third column; using equation (8-12), we obtain the values for Ct in the last column.

We can now plot Ct versus t, the year in which the record was achieved, to obtain a value for C. This is shown in Figure 8.2, where the first seven points from Table 8.1 are plotted. The line is fitted to these seven data

Table 8.1. Aircraft Speed Records

Year	Speed	Y	Ct
1912	108.2	1.017	−3.077
1913	126.7	1.056	−1.819
1920	194.5	1.233	−0.233
1921	205.2	1.266	−0.056
1922	222.9	1.325	0.205
1923	266.6	1.499	0.806
1924	278.5	1.555	0.967
1932	294.3	1.636	1.182
1933	304.9	1.694	1.330
1934	314.3	1.750	1.463
1935	352.4	2.019	2.039
1937	379.6	2.269	2.507
1939	469.2	3.823	4.860
1943	490.0	4.545	5.811
1944	504.0	5.208	6.645

points by linear regression, and extended to 1945 as a forecast. From the regression, $Ct = 0$ in 1920, and $C = 0.299$.

Using the regression equation, we obtain the values for Ct shown in Table 8.2; these must be converted to aircraft speeds. To do this we must first obtain the Y values corresponding to each value of Ct. Unfortunately, equation (8-12) cannot be solved for Y; however, we can take advantage of the fact that both terms on the right-hand side increase with increasing Y. We can estimate values for Y and, using a hand-held calculator, determine the corresponding Ct. If the Ct value is too large, we need to reduce the value of Y, and conversely if Ct is too small. Let us determine the speed corresponding to the fitted Ct value of 1.253, corresponding to 1925. Take 1.7 as the first estimate of Y. Solving equation (8-12), Ct is 1.343, which is too large. Thus the next estimate for Y should be smaller. The following table shows one sequence of estimates that converges on a good estimate for Y.

Y	Ct
1.7	1.343
1.6	1.089
1.65	1.219
1.66	1.244
1.67	1.269
1.665	1.257

In the same manner we can fill in the remainder of the second column of Table 8.2. Solving equation (8-10) for f, we can use the estimated Y values to obtain the final column of Table 8.2.

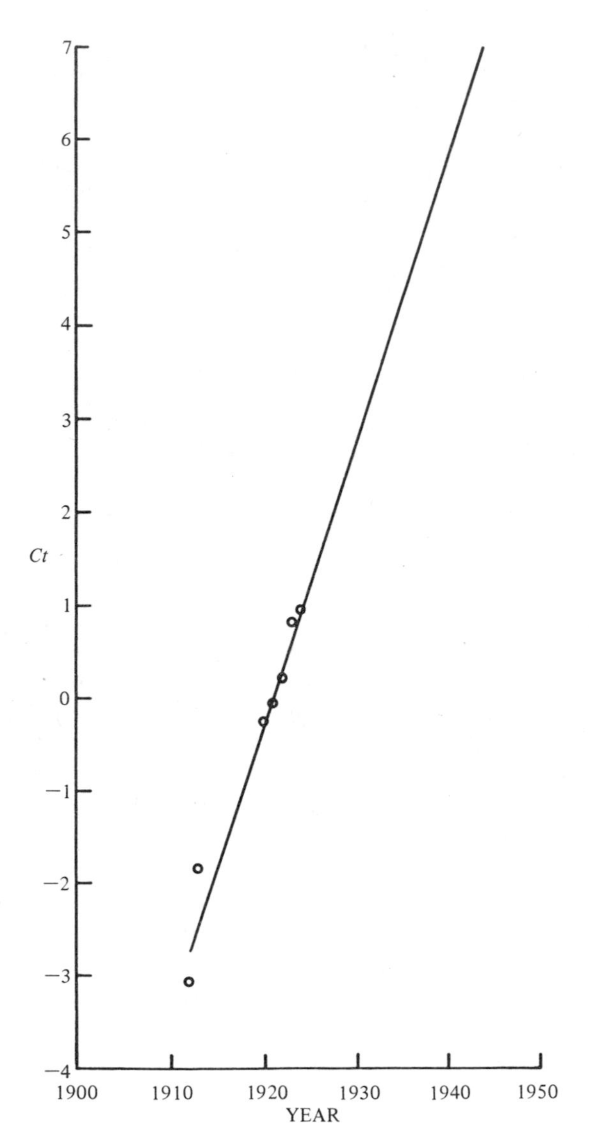

Figure 8.2. Plot of *Ct* versus time for aircraft speed.

These results are plotted in Figure 8.3. The plotted curve is taken from the last column of Table 8.2. The data points shown as circles are those used to fit the curve, and those shown as triangles are the remainder of the data points from Table 8.1. It is striking that a model based on aviation progress from 1912 to 1924 gives a remarkably good forecast for performance in the late 1930s and the 1940s. However, the model does not do well at all for the four speed records set in the early 1930s. It is possible

Table 8.2. Floyd Curve Fitted to Aircraft Speed Records

Year	Ct	Y	Speed
1915	−1.738	1.061	128.75
1920	−0.243	1.229	193.17
1925	1.253	1.665	299.70
1930	2.748	2.405	392.09
1935	4.244	3.375	464.86
1940	5.739	4.490	488.64
1945	7.234	5.688	512.09
1950	8.730	6.950	528.06

that the Great Depression slowed down aviation progress, contributing at least in part to the overoptimistic forecast for 1932–1935.

The Floyd model seems suitable for forecasting the progress of a single technical approach. The upper limit in performance and the performance of the next-best competing technology must be known. At least two historical data points must be available to permit the evaluation of a constant. With the use of a calculator, fitting the curve is straightforward and fairly rapid.

3. A Technoeconomic Model

In Chapter 4 we used a Pearl curve to forecast the substitution of one technology for another; however, we needed some data from the early years of the substitution to forecast the remainder. Mansfield (1968) introduced a technoeconomic model for forecasting substitution. He obtained data on the adoption history of three innovations in each of four industries.

The innovations are shown in Table 8.3. n_{ij} is the number of firms that ultimately adopted the innovation. p_{ij} is a measure of the profitability of the innovation; it is the ratio of the payback time for the innovation to the payback time required by the industry before an innovation can be justified. S_{ij} is a measure of the cost of adopting the innovation; it is the initial investment required by the jth innovation divided by the average assets of the n_{ij} firms that adopted the innovation.

The parameters l_{ij} and ϕ_{ij} are the location and shape parameters, respectively, obtained by fitting a Pearl curve to the substitution history (these were termed a and b, respectively, in Chapter 4). The parameter r_{ij} is simply the correlation coefficient for the regression fit, and the last column is the RMS error of the fit.

The rapidity of adoption is given by the parameter ϕ_{ij}. Mansfield hypothesized that the more profitable the innovation, the more rapidly it would be adopted (larger ϕ_{ij}), and the more costly the innovation to adopt, the more slowly it would be adopted (smaller ϕ_{ij}). He therefore regressed

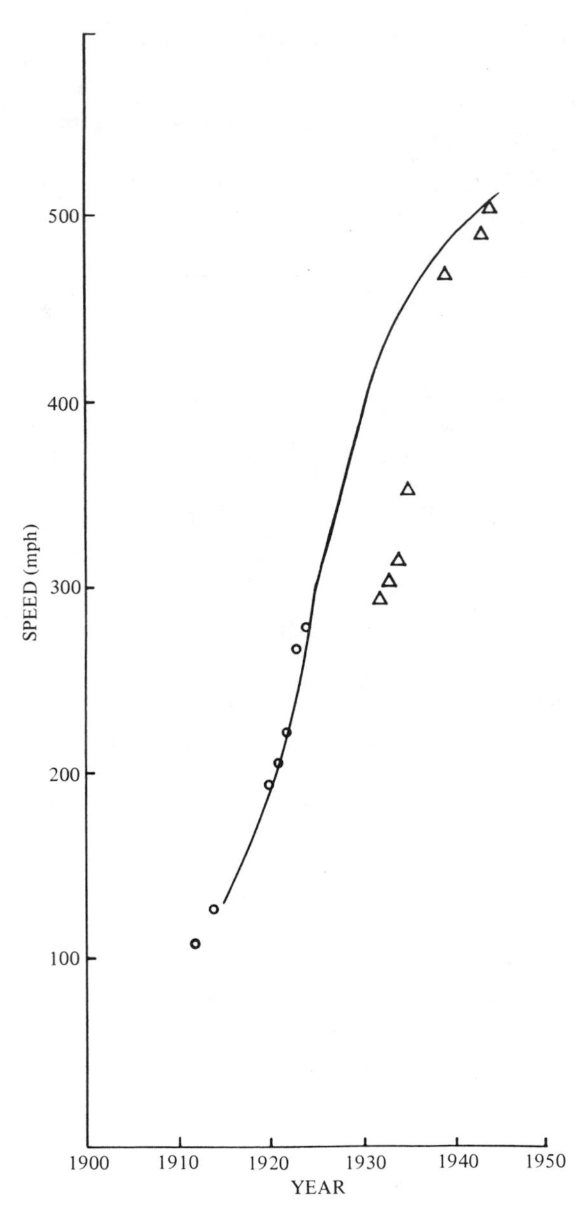

Figure 8.3. Floyd curve of aircraft speed versus time.

Table 8.3. Innovations in Three Major Industries[a]

Innovation	Sample data			Parameter estimates			RMS error
	n_{ij}	p_{ij}	S_{ij}	l_{ij}	ϕ_{ij}	r_{ij}	
Diesel locomotive	25	1.59	0.015	−6.64	0.20	0.89	2.13
Centralized traffic control	24	1.48	0.024	−7.13	0.19	0.94	1.52
Car retarders	25	1.25	0.785	−3.95	0.11	0.90	5.02
Continuous wide-strip mill	12	1.87	4.908	−10.47	0.34	0.95	0.90
Byproduct coke oven	12	1.47	2.083	−1.47	0.17	0.98	0.84
Continuous annealing	9	1.25	0.554	−8.51	0.17	0.93	1.42
Shuttle car	15	1.74	0.013	−13.48	0.32	0.95	2.03
Trackless mobile loader	15	1.65	0.019	−13.03	0.32	0.97	1.66
Continuous mining machine	17	2.00	0.301	−24.96	0.49	0.98	2.22
Tin container	22	5.07	0.267	−84.35	2.40	0.96	3.00
High-speed bottle filler	16	1.20	0.575	−20.58	0.36	0.97	0.95
Pallet-loading machine	19	1.67	0.115	−29.07	0.55	0.97	1.58

[a] *Source*: E. Mansfield, *Industrial Research and Technological Innovation* (New York: W. W. Norton), 1968, p. 142; reprinted by courtesy of W. W. Norton & Co.

the computed values of the ϕ_{ij} on the S_{ij} and p_{ij} for the innovations. The actual regression equation he used was

$$\phi_{ij} = b_1 d_1 + b_2 d_2 + b_3 d_3 + b_4 d_4 + a_5 p_{ij} + a_6 S_{ij}. \tag{8-13}$$

This equation makes use of a trick frequently employed in situations such as this: The d_i are dummy variables that are assigned the value of 1 when the data in the equation apply to the ith industry; otherwise they are zero. This trick allows us to use all the data from all the industries to estimate a_5 and a_6, while still obtaining separate values of b_i for each industry. Mansfield obtained the following values from this regression:

$$\phi_{ij} = \begin{Bmatrix} -0.29 \\ -0.57 \\ -0.52 \\ -0.59 \end{Bmatrix} + \underset{(0.015)}{0.530} p_{ij} - \underset{(0.014)}{0.027} S_{ij}. \tag{8-14}$$

The constants in the large braces apply, from top to bottom, to the brewing industry, the bituminous coal industry, the steel industry, and the railroad industry. The coefficients of p_{ij} and S_{ij} have the expected signs; that is, an increase in profitability increases ϕ_{ij}, whereas an increase in the required initial investment decreases ϕ_{ij}. The numbers in parentheses below the coefficients are the standard errors of the coefficients. Both coefficients are significantly different from zero.

We can interpret the regression results as follows. The effects of the profitability and size of the initial investment required are the same for all industries, but for innovations with comparable profitabilities and sizes of investment, some industries will adopt more rapidly than others. In particular, the brewing industry is far more prone to adopt innovations rapidly than are the other three. We can now compare the values of ϕ_{ij} given by equation (8-14) with those obtained by fitting a Pearl curve to the data for adoptions. The result is shown in Figure 8.4; where the ϕ_{ij} values computed from the Pearl curve are plotted against the ϕ_{ij} values computed from equation (8-14). The 45° line indicates equality; that is, if the regression equation agreed perfectly with the Pearl curve fits to the data, all the points would lie on the line. The goodness of fit indicates that the profitability and size of an investment do explain a great deal of the variation in the rates of substitution for innovations in different industries. In addition, some industries seem inherently more innovative than others.

These results can be used to develop technoeconomic models for forecasting the rate of adoption in other industries. Using Mansfield's approach, it is first necessary to obtain the industry innovativeness coefficient [the term in large braces in equation (8-14)] from data about the adoption of prior innovations in the industry. Then the profitability and investment cost can be used in equation (8-14) to the predict rate of adoption. However, Blackman et al. (1973) have developed an extension that allows the forecaster to estimate the industry innovativeness coefficient from other data on R&D expenditures and similar factors, making

Figure 8.4. Actual versus computed ϕ_{ij} for Mansfield's study of the diffusion of 12 innovations (courtesy of W. W. Norton & Co.).

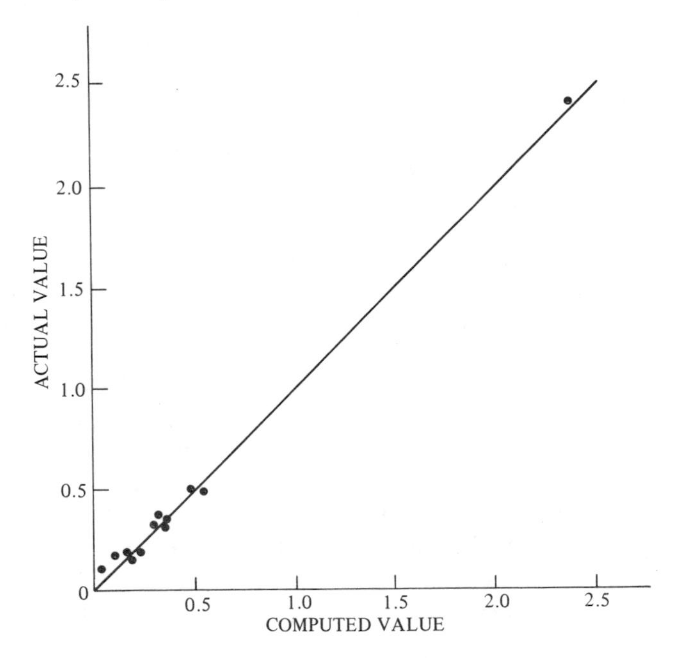

it unnecessary to use data from previous innovations. Thus, by either repeating Mansfield's research on a particular industry or using Blackman's extension, it should be possible to develop a technoeconomic model for forecasting the rate of adoption of a specific innovation in a particular industry. A more elaborate extension of Mansfield's model is given by Martino et al. (1978), who investigated 41 innovations in 14 different industries. Their model explained 50% of the variance in the diffusion rate among the 41 innovations on the basis of the economic and technical data available at the time the innovation was introduced. This more elaborate model should be usable for forecasting purposes with reasonable accuracy.

4. A Simulation Model

Sometimes a causal model cannot be expressed in the closed analytical form of the models presented earlier in this chapter. In such a case the relationships expressed in the model must be simulated on a digital computer, which would then produce a forecast by calculating the changing values of the variables, and the relationships among them, with the passage of time in the simulation.

Most simulation languages are complex, and the more powerful they are, the more complex. However, Kane (1972) has presented a simulation language, KSIM, that is simple to use and yet powerful enough to provide a meaningful treatment of real problems. KSIM is actually a specific simulation model rather than a simulation language; therefore not all real-world situations can be modeled by it. Nevertheless, it does provide a useful tool for analyzing many moderately complex real-world problems.

KSIM has the following properties:

1. Variables are restricted to the range 0–1. Actually, this is not a severe restriction: In the real world, all variables must be bounded above and below, and by a change of scale, any real-world variable can be converted to one within the range 0–1.
2. A variable increases or decreases, depending upon whether the net impact of the other variables in the system is positive or negative. A variable may impact on itself, and this is accounted for in computing the net impact.
3. The response of a variable to impacts from other variables decreases as the variable reaches its extremes of 0 or 1 and is a maximum near values of 0.5. Put another way, a variable is most sensitive to impacts from other variables in the midrange of its values, and least sensitive near the extremes.
4. Other things being equal, the impact produced by a variable will be larger, the larger the value of the impacting variable.

5. Complex interactions are decomposed into pairwise interactions, and only the pairwise interactions are specifically incorporated into the model.

It is this last property that produces the most serious limitation on KSIM's ability to model real-world problems. It says that the strength of impact of one variable on a second depends only on the magnitudes of those two variables and is not affected by the magnitude of any other variable. This is not true for many systems. For instance, many ecological systems include interactions that involve the product of two variables. KSIM cannot be used to model systems for which such interactions exist. Kane has developed more advanced models that are capable of handling such interactions; however, they are beyond the scope of this discussion.

KSIM operates by starting each variable with an initial value, stepping forward a time increment Δt, and computing a new value for each variable on the basis of the impacts from all the variables during that time increment. The change in a variable x_i during a time increment is given by

$$x_i(t + \Delta t) = x_i(t)^{p_i}, \tag{8-15}$$

where the exponent p_i is obtained from

$$p_i(t) = \frac{1 + (\Delta t/2) \sum_{j=1}^{N} (|\alpha_{ij}| - \alpha_{ij})x_j}{1 + (\Delta t/2) \sum_{j=1}^{N} (|\alpha_{ij}| + \alpha_{ij})x_j}. \tag{8-16}$$

Here the α_{ij} are the impacts of x_j on x_i, and Δt is the time increment. The form of equation (8-16) assures that p_i will always be positive. Raising a variable whose values lie between 0 and 1 to any positive power will keep the values within that range. Equation (8-16) can be expressed as follows, to make its purpose more clear:

$$p_i(t) = \frac{1 + \Delta t\,|\text{sum of negative impacts on } x_i|}{1 + \Delta t\,|\text{sum of positive impacts on } x_i|}. \tag{8-17}$$

When the negative impacts are stronger than the positive impacts, the numerator of p_i will be larger than the denominator, p_i will be larger than 1, and $x_i(t + \Delta t)$ will be *smaller* than $x_i(t)$. Note that the effect of adding the cross-impact factors to their absolute values doubles the negative impacts (in the numerator) and cancels the positive impacts. In the denominator, the effects are just the opposite, which explains the need for the factor of $\frac{1}{2}$ in both the numerator and the denominator: to restore the impacts to their original size.

From equation (8-15) it can be seen that a given magnitude p_i will have less effect when x_i is near either 0 or 1 than when x_i is near 0.5. From equation (8-16) it can be seen that the larger x_j, the stronger its impact on x_i will be. Finally, condition (4) above means simply that all the effects of the variables on each other are expressed by the α_{ij}.

To use KSIM it is necessary to specify the initial values for each of the variables, their impacts on the others (the α_{ij}), and the impact of the "outside world" on each variable. The outside world is a variable that affects but is not affected by the variables in the system. Formally, this means that the matrix of impact coefficients has one more column than it has rows. The last column represents impacts from the outside world but does not correspond to a variable that can receive impacts from the variables in the system.

Before using KSIM it is necessary to identify the variables that will go into the model, and once these are identified, the interactions among them have to be specified. This procedure is really the heart of the modeling process. KSIM then simply traces out the implications of whatever has been built into the model by specifying the impacts among the variables.

To illustrate this process let us use KSIM to prepare a forecast of the use of solar energy in homes for space heating and hot water. The system will be described by the following variables:

1. The proportion of homes using solar energy for space heating and hot water. The current value of this variable is 0.1 of its possible maximum.
2. The price of fossil fuel. We take this as currently being 0.3 of its possible maximum.
3. The performance of solar technology for a given cost of installation and operation. This is taken as 0.3 of its possible maximum value (i.e., when fully developed, the performance of a unit of a given cost will be over three times today's performance).
4. R&D on solar home-heating technology. This is taken as being 0.2 of its possible maximum value.

Next we consider the impacts on each of these variables.

The proportion of homes using solar energy is affected by itself, in that greater use implies a more widespread availability of technicians, sales outlets, and advertising. It is also affected by the price of fossil fuel, since the higher the price, the more incentive there is to install a solar energy unit. The proportion of homes using solar energy is also affected by higher performance, as this increases its cost effectiveness and makes it more attractive.

The price of fossil fuel is affected negatively by the use of solar energy, since greater use of solar energy decreases the demand for fossil fuel. The price of fossil fuel is affected by itself, as increased price decreases demand and reduces the pressure for further price increases. The price of fossil fuel is affected by demand in the "outside world," since home heating is not the only use for fossil fuel. Its price would rise with increasing scarcity, even without the demand for home heating.

The performance–price ratio of solar energy installations is affected by increased use, as cumulative production brings design improvements. It is also affected by R&D, as cumulative R&D brings design improvements.

R&D on solar energy equipment is affected by the degree of use, since firms often tend to spend a fixed percentage of sales on R&D. Hence the greater the cumulative sales, the greater the cumulative expenditures for R&D.

The impacts among the variables are shown in Table 8.4, where they have been assigned numerical weights on a scale from −10 (maximum unfavorable impact) to 10 (maximum favorable impact). For use in plotting the results of the simulation, the variables are assigned the plotting symbols *S* for solar energy use, *F* for fossil fuel price, *P* for the performance of solar installations, and *R* for R&D expenditures. The initial values given above, the impact factors in Table 8.4, and these plotting symbols have been included in the DATA statements in the KSIM program in Appendix 4. Figure 8.5 shows the results of the simulation (note that here the KSIM program was modified to print out every 25th time increment instead of every 10th). As can be seen in Figure 8.5, all four variables increase towards their maximum values, but within the time horizon of the simulation the fossil fuel price did not quite reach the maximum possible value, being only about 94% of the maximum.

Several points need to be mentioned about this simulation run.

First, the model chosen is quite simple. Other variables might be included; however, they can readily be added to the model if desired.

Second, there may be some disagreement about the initial values of the variables. Is solar energy today really used in 10% of the maximum possible number of houses? It is relatively simple to test the sensitivity of the outcome to this value by rerunning the model with some other initial value.

Third, there may be disagreement about the structure of the interactions as reflected in the matrix of impacts. Again, the numbers can be changed and the model rerun to determine whether the changes produce significantly different results.

Table 8.4. Matrix of Impacts for KSIM Model

Variable	1	2	3	4	5[a]
1	3	5	4	0	0
2	−2	−2	0	0	6
3	3	0	0	7	0
4	8	0	0	0	0

[a] Outside world.

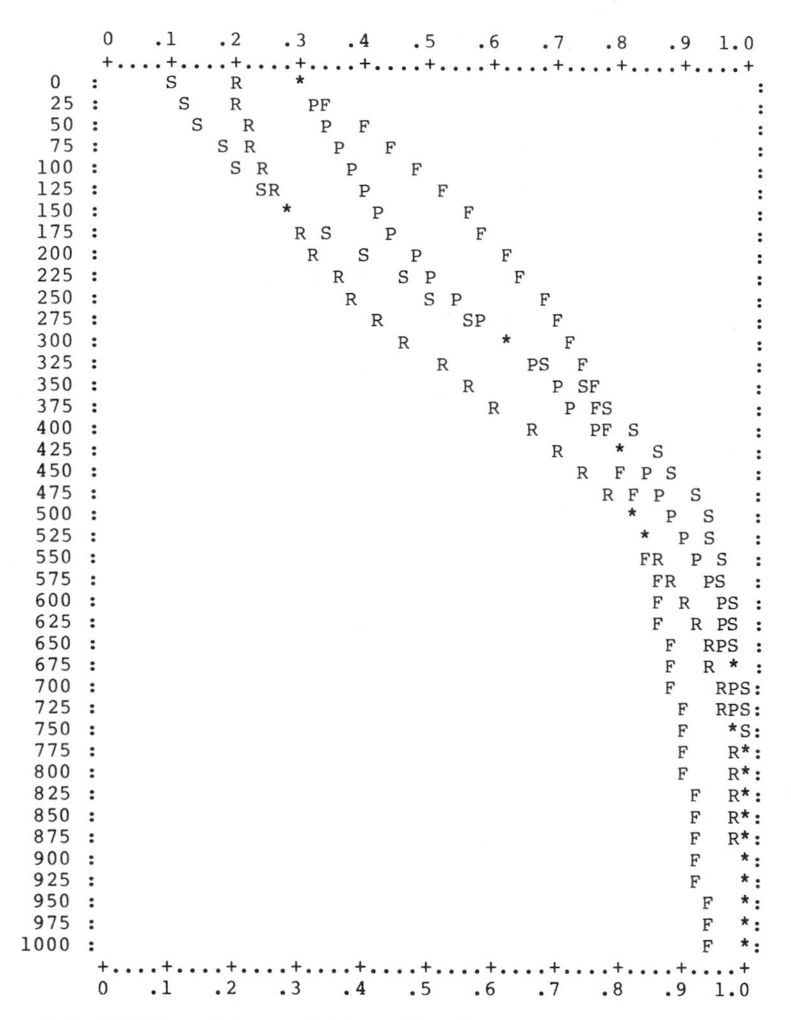

Figure 8.5. KSIM model run of the problem in text.

Fourth, KSIM gives results that are only qualitative. The time scale on the printout plot is not meaningful in terms of a specific number of years. The purpose of KSIM is to explicate the consequences of the structure of the system; it is not capable of providing a precise numerical forecast.

Fifth, while the initial values can reasonably be scaled to objective data, it appears that the numerical values of the impacts are completely subjective. This is true, but it is not a significant shortcoming of KSIM. As Jay Forrester of MIT has noted, most of the knowledge in the world is not written down; it exists in people's heads, in their intuitive understanding of how the pieces of the world fit together. The only way this knowledge can be made "public" is in assigning numerical values according to some scale. This quantified knowledge is superior to the original mental model from which it is derived, since it can be compared with the models produced by other people, and the reasons for the differences explored. Hence the subjectivity of the impact matrix in KSIM is not justification enough for rejecting the entire concept.

Despite the shortcomings of KSIM in comparison with other simulation languages, it allows the user to "get inside" a problem quickly and gain some understanding of the consequences of the interactions perceived. Thus KSIM is a useful tool for the technological forecaster. It does not replace more complex simulation models, but it can often be used as a means of gaining the understanding necessary to develop a more complex model.

5. Summary

Causal models are patterned after the models of the "exact" sciences. The intent is to go beyond simple correlation or extrapolation and to identify causal linkages among events. However, a major problem with causal models of socioeconomic activities is that there is no essential reason for people to behave in a fixed way. In particular, there is no reason why technologists must behave in the ways presumed by causal models of technological change. Thus, although an explanatory model fits past behavior well, there is no guarantee that it will fit future behavior.

In addition to the lack of "necessity," models of technological change suffer from another shortcoming: They are based on a hypothetical structure for the workings of the technology-producing system, which is described in a mathematical model. Coefficients of the model are then determined from past data, and the completed model used for forecasting. At the very least, this past data contains errors of rounding and aggregation. It may also contain errors of mistranscription or miscopying. Finally, some data may be missing through distortion or deliberate refusals to disclose them. Hence the coefficients of the model are in error to the degree that the data were in error. Even if the theory behind the model is correct, the model will produce erroneous forecasts to the extent that it was fitted to erroneous data.

Schoeffler (1955) describes these and other errors of models in detail. Although he writes of economic models, what he says applies equally well to causal models of technological change.

Despite these shortcomings, causal models remain an important tool of the technological forecaster. They permit one to take into account known changes in the factors that have produced technological progress in the past; they also allow an exploration of the consequences of alternative policy choices and an examination of the sensitivity of forecasts to particular factors whose values are uncertain.

References

Blackman, A. Wade, Jr., Edward J. Seligman, and Gene C. Sogliero (1973). "An Innovation Index Based on Factor Analysis," *Technological Forecasting and Social Change* 4, 301–316.

Floyd, A. L. (1968). "A Methodology for Trend-Forecasting of Figures of Merit," in *Technological Forecasting for Industry and Government,* James R. Bright (ed.) (Englewood Cliffs, NJ: Prentice-Hall).

Kane, Julius (1972). "A Primer for a New Cross-Impact Language—KSIM," *Technological Forecasting and Social Change* 4, 129–142.

Mansfield, E. (1968). *Industrial Research and Technological Innovation* (New York: W. W. Norton).

Martino, Joseph P., Kuei-Lin Chen, and Ralph C. Lenz, Jr. (1978). "Predicting the Diffusion Rate of Industrial Innovations," University of Dayton Research Institute Technical Report UDRI-TR-78-42, available from National Technical Information Service as PB-286693.

Schoeffler, S. (1955). *The Failures of Economics: A Diagnostic Study* (Cambridge, MA: Harvard University Press).

For Further Reference

Technological Forecasting and Social Change 9 (1, 2), 1976. The entire issue is devoted to the use of models in technological forecasting, and many advanced applications are discussed.

Technological Forecasting and Social Change 14 (4), 1979. This issue is devoted to structural modeling in technological forecasting, including several papers that treat applications of KSIM.

Problems

1. Using Floyd's model and the data on the efficiency of incandescent lamps from Appendix 3, fit the model to the data from 1888 through 1940 and forecast the efficiency for 1959. Use a value of 19 for the maximum possible efficiency and the efficiency of a candle (from the table on the efficiency of illumination, Appendix 3) as the competitive efficiency.

2. Using Floyd's model and the data on the efficiency of power plants in Btu per kilowatt-hour from Appendix 3, fit the model to the data from 1925 to 1940 and forecast the efficiency for 1960. Use a value equivalent to 100% for the maximum possible efficiency and the efficiency of a thermocouple (about 2%) for the competitive efficiency. Note that in this case increasing efficiency means fewer Btu per kilowatt-hour; thus F will be smaller than f, and f_c larger than f. However, note from equation (8-10) that this change amounts

Problems *(continued)*

to multiplying the numerator and denominator of the right-hand side by -1, which leaves the value of the fraction unchanged. Hence the same formulas can be used as though F were larger than f, and f_c smaller than f.

3. Your company is attempting to introduce an innovation into the railroad industry. The estimated payback period for the innovation is two-thirds the average maximum payback period required by firms in the industry. The initial investment in the innovation, per firm, will be 50% of the average assets of firms in the railroad industry. How many years will elapse between the time when 10% of the firms will have adopted the innovation and when 90% of the firms will have done so? [Hint: to avoid the need to obtain the constant term l_{ij}, first show that for percentage points P_1 and P_2, with $P_1 < P_2$, the time span from P_1 to P_2 is given by

$$T = (1/\phi_{ij}) \ln[(1 - P_1)P_2/P_1(1 - P_2)].$$

4. Use KSIM to forecast the rate of adoption of the electric car (fraction of total number of cars on the road). Include the following variables in your model:
Price of gasoline.
Price of electricity.
Cumulative R&D on batteries.
Price of electric cars.
Price of gasoline-powered cars.
Real per-capita income (i.e., with inflation removed).
Fraction of population living in urban areas.
Add other variables if you believe they are important. Estimate the current (initial) values for each of these variables. Estimate the entries in the matrix of impacts and run this base-line model to see the consequences of your estimates. Next test the sensitivity of the results of your estimates by making the following changes, one at a time, and rerunning the model: Change the initial value of one variable by a factor of 2 (double it if less than 0.5, halve it if greater than 0.5). Change the impact of one variable on another by a factor of 2. Incorporate a policy intervention (impact from the outside world) on battery R&D. Change your initial estimate by a factor of 2 if you initially incorporated this impact; incorporate it with a value of 5 if you did not initially incorporate it. (This problem may be done as a group or class project.)

Chapter 9

Forecasting Breakthroughs

1. Introduction

Forecasts based on trends or growth curves seem to require continuity between the past and the future. These methods therefore seem inherently incapable of predicting "breakthroughs," whose very name seems to imply surprise and lack of advanced warning. Thus is it true that breakthroughs are inherently unpredictable?

Webster defines breakthrough as "a sensational advance in scientific knowledge in which some baffling major problem is solved." This definition does not inherently require surprise or unpredictability in the breakthrough; however, it is not precise, since it is not clear how great an advance must be to qualify as "sensational," nor how bad a problem must be to qualify as "baffling."

For the sake of precision we will use the following definition of a technological breakthrough: An advance in the level of performance of some class of devices or techniques, perhaps based on previously unutilized principles, that significantly transcends the limits of prior devices or techniques. This definition has several implications: One is that simply moving up the growth curve of an established technical approach does not qualify as a breakthrough; another is that the predictable limits of a specific technique must be surpassed by the breakthrough. This may come from finding that the previously accepted limit was in error, or it may come from the use of some new technique or discovery to perform a particular function. In the latter case it is not necessary that the discovery be new in an absolute sense; it may have been widely used for some other purpose before and have been adapted to the new function. Finally, the definition implies that adopting a successor technique that has a level of inherent capability higher than that of the prior technique represents a breakthrough. That is, a breakthrough means a move to a new growth curve with an upper limit higher than that of the old growth curve.

This last point is particularly important. A breakthrough may be identified as such simply because it is on a new growth curve, even though the current level of performance is inferior to existing devices on the old growth curve. This clearly implies that at least some breakthroughs are predictable, since the appearance of the first low-performance device on a new growth curve is a precursor event for the device that will later surpass the limits of the old growth curve. Such a situation is shown in Figure 5.1, which depicts the growth curves for propeller-driven and jet-propelled aircraft. Even though the early jets were inferior to their propeller-driven contemporaries, the jet engine was clearly a breakthrough, because it had the inherent capability to surpass the performance of the older technical approach.

2. Examples of Breakthroughs

Almost any list of major "breakthroughs" of the current century will include atomic energy, the transistor, and penicillin. According to technological folklore, each of these burst unheralded on a startled public. However, even a cursory glance at the history of each of these breakthroughs shows many precursor events. To illustrate this point, a review of the history of each is worthwhile.

A. Atomic Energy

Table 9.1 shows some major events in the history of atomic energy. The earliest event was a scientific breakthrough and was totally unpredictable; the equivalence of mass and energy was a violation of accepted scientific theory. Once this equivalence was recognized, however, there was at least the theoretical possibility of converting matter into energy.

The next major event was the discovery that not all chemically identical elements were physically identical. Before this it was believed that mass was the most important characteristic of an atom and two atoms with the same mass were identical in all properties. However, Boltwood in 1906 and McCoy and Ross in 1907 showed that the radioactive elements ionium and radiothorium were chemically identical with the known element thorium, although they had different masses (and therefore had been assumed to be three distinct elements). Soddy coined the name *isotopes* for chemically identical elements with different masses. In 1913 J. J. Thompson found that neon also had two isotopes, demonstrating that nonradioactive elements could have isotopes as well as radioactive elements.

Prior to 1911 it had been assumed that the matter in an atom was uniformly distributed throughout the space it occupied. However, Ruth-

Table 9.1. Major Scientific Events in the History of Atomic Energy

1905	Mass–energy equivalence. Publication of a paper by Einstein establishing the equivalence of mass and energy.
1906	Isotopes of radioactive elements. Discovery of chemically identical elements with different radioactive properties.
1911	Atomic structure. Experiments by Rutherford show that the mass of an atom is concentrated in a positively charged nucleus.
1913	Isotopes of nonradioactive elements. Discovery of isotopes through differences in physical properties.
1919	Ejection of protons from nitrogen. First artificially induced nuclear reaction.
1919	Mass spectroscopy. Accurate determination of the masses of isotopes.
1920s	Mass defect (packing fraction). Discovery that the mass of a nucleus is less than the sum of the masses of the constituent particles.
1932	Discovery of the neutron.
1938	Fission of uranium nucleus.
1939	Chain reaction hypothesized.
1942	Chain reaction produced.

erford showed that virtually all the mass of an atom was concentrated in a small nucleus. As a result of his experiments, it became clear that atoms had equal numbers of positively and negatively charged particles, and the positively charged particles made up most of the mass of the atom.

Rutherford continued his study of the structure of the atom. In 1919 he bombarded nitrogen atoms with alpha particles and found that a nuclear reaction took place. The nitrogen emitted a proton (nucleus of a hydrogen atom) and turned into an oxygen atom; moreover, the emitted proton had more energy than did the alpha particles, indicating that somehow energy was being obtained from the nucleus of the atom.

The source of this energy was determined with the development of mass spectrometers, which allowed researchers to determine the masses of atoms. It had been customary to quote atomic masses in terms of multiples of the mass of a hydrogen atom; however, most elements had computed masses that were nonintegral multiples of the mass of hydrogen. With the discovery of isotopes it was assumed that these nonintegral masses were simply the result of mixtures of isotopes; however, better mass spectrometers showed that individual isotopes still had nonintegral masses. In all cases the mass was less than it "should have been" if the atom were made up of protons and electrons. This "mass defect" was explained as mass that had been converted into "binding energy" to hold the atomic nucleus together. It was discovered that the average binding energy per particle in the nucleus increased with increasing atomic mass, up to a maximum for atomic masses around 90; above this the average binding energy per particle declined again. This clearly meant that energy could be obtained from a nucleus by either combining two light atoms to make a moderately heavy one or splitting a heavy atom into two moderately heavy ones. However, there appeared to be no way to do this on a practical scale. Even atomic particle accelerators ("atom smashers") "missed" their target atoms too often. The efficiency of any process using a particle accelerator would be incredibly low.

In 1932 Chadwick discovered the neutron. It was quickly discovered that neutrons were much more efficient at causing atomic transmutations than were other types of particles. All during the 1930s many scientists throughout the world conducted experiments in which they bombarded various elements, including uranium, with neutrons.

In 1938 Hahn and Strassman conducted such an experiment with uranium and discovered that they had split the uranium atom into smaller fragments which gave off a great deal of energy. When they published their results, many other researchers realized they had just missed discovering uranium fission themselves because they failed to interpret the results of their experiments properly.

At this point it became clear to many people that it was possible to obtain energy from the fission of uranium, even if many details still needed to be resolved. At the very least, one could imagine a power plant containing some uranium and a neutron source. The neutrons would fission the uranium, which would release energy. This energy could be extracted and used to perform work. Such a power plant was described by science fiction author Robert A. Heinlein in *Blowups Happen*. The technical feasibility of such a plan was certain; only economic feasibility remained in doubt.

In 1939 Fermi hypothesized that uranium atoms might give off neutrons when they were split by neutrons. If so, a "chain reaction" was possible, which would eliminate the need for a neutron source. This chain reaction was demonstrated in 1942. The remaining problems took another 14 years to solve, with the first large-scale commercial production of electrical power from atomic energy being achieved in 1956 at the Calder Hall Station in England. This was 18 years after the discovery of fission and 51 years after mass–energy equivalence was discovered.

The point of all this is that there were a great many precursor events between the unpredictable scientific breakthrough of 1905 and the eventual commercial use of atomic power in 1956. Not all these precursor events provided positive signals: Some, such as the impracticality of atomic energy plants using particle accelerators, were false negative signals; some, such as the possibility of fusing light atoms into moderately heavy ones, pointed in the wrong direction. Nevertheless, atomic energy was not an unheralded event. Many writers, both of fiction and of popularized science, discussed atomic energy from 1914 to 1944 on the basis of their knowledge of contemporary scientific findings.

B. The Transistor

In 1945 the Bell Telephone Laboratories (BTL) established a group to conduct research on a solid-state amplifier. The intent was to draw on the research done on semiconductors during World War II and on Bell's pre-war research in quantum mechanics. The group made a false start on a solid-state amplifier patterned after a vacuum tube; it then switched to what would now be called a field effect transistor (FET). Their attempts to build a FET were guided by a theory that A. H. Wilson had published in 1931 which described the theory of semiconductors. They built several devices on the basis of this theory. None of them worked. While attempting to determine the reasons for failure, the group discovered the point contact transistor. This was in 1947. With the additional insight into semiconductor behavior this provided, they developed the theory of the junction transistor and fabricated one in 1951. Finally, in 1952 they achieved the FET they had initially been looking for.

It is important to note that the BTL group working on the solid-state amplifier were confident that they could build one, on the basis of existing theory. The major contribution they expected to make to the development was the improved semiconductor materials that had been developed during the war. The existence of the theory itself and the improvements in materials were both precursor events that gave advance warning of the possibility of a solid-state amplifier.

However, there were other precursor events. In 1925 J. E. Lilienfeld applied for a Canadian patent on a solid-state amplifier. In 1926 he applied for a U.S. patent on the same device and was granted U.S. patent number 1,745,175 in 1930. In 1935 Oskar Heil of Berlin was granted British patent number 439,457 on a solid-state amplifier. In retrospect, the devices of both Lilienfeld and Heil are recognizable as FETs. Neither inventor could have built a working device, however, because the materials available to them were not adequate.

The transistor was clearly a breakthrough. But it did not come unheralded; on the contrary, it was the result of a deliberate search by people who took the advance warning given by the precursor events seriously.

C. Pencillin

Penicillin was isolated by Chain and Florey in 1939. They traced their work to the observation by Fleming in 1928 that penicillium mold had an antibacterial effect. This observation was in itself an important precursor event.

However, there was an even more important precursor event, which was completely lost sight of. E. A. C. Duchesne entered the French Army

Medical Academy in 1894, at age 20. He became interested in the possible antagonism between molds and bacteria and did his thesis research on the topic. He experimented with a penicillium mold, observing that under certain circumstances the growing mold would kill bacteria. Next he grew the mold on a nutrient solution. He infected guinea pigs with virulent bacteria and injected half the experimental animals with the nutrient broth on which he had grown the mold. Those receiving the "penicillin" injections survived; the remainder did not. He reported his results in his thesis prior to his graduation in 1897. After graduation he began his intended career as a doctor in the French Army, serving at various posts where it was impossible to continue his research. In 1902 he contracted what seems to have been tuberculosis, and he died from the disease in 1912. His thesis was not published and was lost until a librarian accidentally discovered it around 1944. Once it was rediscovered, Fleming and others involved in the World War II work on penicillin recognized its importance, particularly Duchesne's methodical research that led him directly to the application of his observations about the antagonism between molds and bacteria.

3. Monitoring for Breakthroughs

This examination of breakthroughs shows that there were many precursor events that would have made it possible to forecast the eventual development of these technologies. The same is true of all other technological breakthroughs (though not necessarily of all scientific breakthroughs). Forecasting a technological breakthrough, then, requires that precursor events be identified and used to provide advance warning.

Monitoring for breakthroughs is a systematic means for identifying precursors and using them to provide advance warning. The monitoring process is designed to help the forecaster answer two questions: Which events are precursors? What do the precursors signify?

Monitoring involves four steps: collecting, screening, evaluating, and threshold setting. Each step has a specific purpose. Taken together, the steps provide a system that will help the forecaster extract the meaningful signals from a welter of noise and irrelevant events.

Collection brings information into the system. It must be designed to cover all likely sources of information. Thus the first step in the collection process is to decide which sources to monitor and to what depth. The approach should be flexible enough to permit the addition or deletion of sources as experience dictates. For instance, initial coverage of legislative activity might have included only the U.S. Congress. Later experience might indicate that one or more state legislatures should be included, either because the initial coverage was inadequate or because conditions have changed. The design of the collection activity should not be so rigid that changes cannot be made.

Once the forecaster has selected the sources to be monitored, the next step is to screen the items found. It is a serious mistake to let everything into the system in the hope that it can later be located through some sophisticated retrieval procedure. Too much effort will be spent cataloging and coding items that will never be used and which must eventually be purged from the system. The only items that should be allowed to enter the system are those which are relevant to the concerns of the forecaster's organization; all others should be screened out. The forecaster will normally need the help of other members of the organization to carry out the screening. In an industrial organization help might be solicited from people in R&D, finance, marketing, or production to determine whether an item should be allowed in or screened out.

Once an item has been accepted as relevant, it must be given further evaluation to determine its significance. The evaluation seeks answers to questions such as the following: What does this mean to my organization? If it represents the start of a trend or pattern, would it affect our mission? Would it make a product obsolete? Would it alter a production process? Will it have an impact on a customer? A supplier? The evaluation must examine the item not only by itself but in the context of the other items already allowed into the system and evaluated. Does it confirm a pattern suggested by earlier signals? Does it indicate a shift away from a previous pattern? Does it indicate a new pattern? In evaluation, just as in screening, the forecaster should seek advice from others in the organization who are better informed about the possible significance of each item.

Evaluation does not stop with the review of each item as it enters the system; as additional confirming signals appear, it includes sketching out the patterns that appear to be building up. Thus the information in the system includes not only a collection of events, but a set of hypotheses about what they mean. These hypotheses, in turn, lead to a set of statements about what additional confirming (or disconfirming) evidence might be expected.

Finally, evaluation requires purging the system of events and hypotheses that have been discredited by subsequent events. This purge is required to assure that the system contains only active and currently valuable material. Material that is of only historical interest can be retired to the archives. This material should not be thrown away; it should be used, from time to time, to review how well the monitoring activity has performed in the past and to identify means for improvement.

As evaluation continues, the evidence for one or more hypotheses will become stronger and stronger. A hypothesis cannot be confirmed, of course, until the breakthrough actually occurs, but by then it is too late to issue a forecast that gives advance warning. Hence the forecast must be made before the hypothesis is confirmed but when it is "strong enough." This requires that the forecaster set a threshold for each hypothesis. When the confirming signals show that the hypothesis has exceeded its threshold, it is time to make the forecast. However, the hypothesis may still be disconfirmed after the forecast is given; therefore the threshold should be set to balance the risks of acting too soon with the risks of acting too late. Here again, it is not the forecaster's role to select the proper balance, others in the organization should be consulted to determine the balance of risks, notably the managers responsible for the success of the organization. It may well be that several thresholds can be set. When a hypothesis passes one of the lower thresholds, it might trigger some low-cost hedging action that does not commit the organization too deeply; passing a higher threshold might trigger more significant actions.

These four steps of the monitoring process cannot be carried out in a simple linear fashion. In practice, a great deal of looping back is required. For instance, evaluation may show that screening has been done improperly, allowing too much or too little material to enter the system, and thus a revision of the screening criteria may be in order. From time to time it may be necessary to reset thresholds as additional evidence is evaluated. The monitoring system cannot be allowed to "run itself." It is supposed to be an aid to the forecaster, not a substitute for thought. The forecaster should be actively involved in operating the system and must revise it whenever circumstances indicate that changes are needed.

4. Where to Look for Signals

The examples given earlier show that breakthrough signals often occur in areas remote from the final application. The first signal indicating atomic energy, for instance, came from an esoteric branch of physics. Thus a systematic monitoring program must cover a wide range of sources. How should the sources be selected? How can the forecaster minimize the chance of overlooking something?

The list of sectors of the environment, introduced in Chapter 3, provides one means of assuring that important sources are not overlooked. The forecaster should cover each sector of the environment to the depth that seems appropriate for the particular pattern being tracked. In certain cases particular sectors can be omitted. However, the decision to not cover a sector should be made deliberately, not simply by default. Some questions pertinent to each sector of the environment are given below.

The Technological Sector. What performance level has the technology already achieved? Will performance improve? At what stage of innovation is the technology? What else is needed to make it feasible? What would have to be changed to make it a threat (or an opportunity)? What changes are needed to make it compatible with complementary technologies? What changes in supporting technologies would make the technology easier to produce, operate, or maintain?

The Economic Sector. What is the market size? What are production and distribution costs? Will these change? Are licensing agreements possible? Are innovative financing methods possible?

The Managerial Sector. Have new managers been assigned to a project dealing with the technology? Have new management techniques been devised that extend the scope of the "manageable?" Can new management procedures reduce technical risk? Shorten development time? If new management techniques were in use, what external signs would be present? Have these signs been observed?

The Political Sector. Have new laws been passed, or regulations issued? Is there agitation for changes in laws or regulations? Has an incumbent been defeated by someone with a significantly different attitude on matters affecting the technology? Has there been a court decision that will have a bearing on the technology? Has the funding of an agency been increased or decreased? Has an agency been given expanded jurisdiction? Has a new agency been established with jurisdiction that covers the technology? Are there signals from the economic, social, cultural, or intellectual sectors that might presage political changes?

The Social Sector. What changes are taking place in the structure of society, such as shifts in age distribution, geographic distribution, level of education, marriage age, family size, and income level? Are there any shifts in custom or tradition? Do younger groups have significantly different customs or traditions than did their elders?

The Cultural Sector. Have any groups acquired new values or abandoned previous ones? Has the number of people who subscribe to a value increased or decreased? Is a particular value being emphasized or deemphasized? Have the relative rankings of some values changed? Is the acceptable or tolerable level of realization of some value changing? Is the range of cases or situations held to be within the purview of some value expanding or contracting? Is the desired level of implementation of some value being retargeted upward or downward? Is the cost or benefit of maintaining some value shifting under the impact of external influences?

The Intellectual Sector. Are the values of the intellectual leadership of society shifting in any of the ways indicated in the preceding paragraph? Are these value shifts being reflected in television programs, movies, novels, essays, newspaper columns, or magazines of opinion?

The Religious–Ethical Sector. Have there been expansions or contractions of the application of some doctrine by one or more major religious bodies? Are there precursors of this expansion or contraction in discussions by theologians or prominent lay people? Have smaller denominations made expansions or contractions that might be precursors of similar moves by major denominations? Have there been changes in the professional standards of scientists and engineers? Have there been precursors of such changes in discussions by leaders in the profession?

The Ecological Sector. Have the unintended effects of some technology been observed? Has new knowledge shown that the previously accepted effects of some technology may no longer be acceptable?

In each sector the forecaster will have to identify the questions to be answered. Once this has been done, the forecaster must select sources that are likely to provide signals should these signals occur in the sectors being monitored.

The forecaster should take advantage of the help of others in the organization. People who regularly read some journal or monitor some source for their own reasons are often willing to provide the forecaster with information they happen to come across. By drawing on efforts people are already undertaking for their own purposes, the additional workload imposed by systematic monitoring may be kept quite low.

5. An Example

An example will serve to illustrate the monitoring procedure and make it more concrete. We will consider a historical breakthrough, the jet engine, which replaced the piston engine–propeller combination for propelling aircraft. Although the jet engine did not bring about an immediate jump in aircraft speed, it transcended the limitations of prior propulsion methods and thus qualifies as a breakthrough.

To provide a context for the example we will assume that the monitoring is being done by the Powerair Engine Company, which manufactures airplane engines. Naturally, Powerair is interested in the possibility that something else might replace the piston engine as a propulsion method for airplanes. Hence it has established a monitoring program that screens potential signals which might warn of changes in its business.

The year is 1930. A British patent is granted to Flying Officer Frank Whittle, of the Royal Air Force (RAF), for an aircraft engine based on the jet principle. In the engine described in the patent, air is drawn in through a turbine compressor, fuel is burned in the compressed air, and the combustion gases are used to drive a turbine, which in turn drives the compressor. There is sufficient energy left in the combustion gases, after driving the turbine, to provide some net thrust for aircraft propulsion.

Powerair's monitoring program receives notice of the granting of the patent to Whittle. The patent is treated as a potential signal and survives the initial screening. It is then subjected to a more thorough evaluation. The significance is clear: If the Whittle jet engine were to become practical, it could pose a threat to Powerair's product line. It represents a potentially competing product. Powerair is therefore interested in determining the extent of the threat. If the threat is of sufficient magnitude, Powerair may wish to get into the jet engine business itself or initiate an improvement program in its own engines to keep them ahead of the jet engine, if possible.

Once the signal has been allowed into the monitoring system, the first questions the Powerair forecaster asks are, Is this a new idea? Has anything similar been proposed before? Historical research, which identifies previous instances of essentially the same signal, can help in evaluating the significance of the signal, in identifying a pattern that may already be forming, and in identifying useful sources of additional signals. The forecaster's historical research is indeed fruitful, and several prior instances of similar ideas are found.

In 1910 Henri Coanda proposed an aircraft propulsion system somewhat similar to the Whittle system. In the Coanda system a piston engine would drive a compressor, which would provide compressed air to a combustion chamber. The combustion gases would be ejected through a nozzle, providing thrust. In 1913 Lorin, a French engineer, proposed a jet engine in which the compressor is eliminated completely, with the necessary compression being derived from the aircraft's forward speed. In 1921 Maxime Guillaume was granted a French patent on a jet engine with a turbine-driven compressor very much similar to Whittle's design. In 1929 Dr. A. A. Griffith of the Royal Aircraft Establishment (RAE) proposed that a turbine engine be used to drive a propeller to provide aircraft propulsion. Thus the forecaster finds that the idea is not a new one and that it has considerable history behind it. Furthermore, this research has turned up not one potential threat, but three: the ramjet, the turboprop, and the turbojet. The ramjet does not appear to be an immediate threat; it requires an aircraft speed sufficient to provide compression by a ram effect. In 1930 such speeds are not yet within sight. However, both the turboprop and the turbojet may be more immediate threats. There are enough similarities between them, however, that the practicality of the one essentially implies that of the other. Hence it is as easy to monitor for both as for either of them.

Next, Powerair's forecaster asks, If the idea is that old and has occurred to that many people, why has nothing ever come of it? Perhaps it is no threat at all. Further historical research uncovers what can be considered a denying signal. In 1922 the U.S. Army Air Service had become interested in jet engines and had requested Dr. Edgar A. Buckingham of the

National Bureau of Standards to investigate their feasibility. Buckingham's report, published in 1923 by the National Advisory Committee for Aeronautics (NACA), had showed that at a speed of 250 mph the fuel consumption of a jet engine would be four times that of a piston engine. Hence at that speed the jet engine would not be competitive with the piston engine. Since as of 1930 the world's record aircraft speed was 278.48 mph, set in 1924, it is easy to see why the jet engine was not then a competitor with the piston engine.

However, the forecaster now has one environmental factor to monitor: If aircraft speeds significantly higher than 250 mph become a possibility, the jet engine might become a threat. In fact, repeating Buckingham's work, it can be calculated that jet engines become competitive with piston engines at speeds in the neighborhood of 500 mph. Hence Powerair's forecaster decides to monitor not just the current speed records for aircraft, but the speeds of proposed aircraft and speed trends in general.

What else should be monitored? Whittle's calculations can be repeated to determine that if the jet engine is to be practical, the compressor must achieve a compression ratio of at least 4 to 1, and the compressor efficiency must be at least 75%. The forecaster finds that Dr. A. A. Griffith has been at work in this field as well and in 1926 developed a theory that can be used to design efficient compressors. However, the best compressors currently available, those used in turbosuperchargers for aircraft engines, have compression ratios and efficiencies much below those required for a practical jet engine. (A turbosupercharger is a compressor that enables an aircraft engine to operate at high altitudes by compressing the air entering the engine to sea-level pressure; it derives its power from a turbine driven by engine exhaust gases.) Hence the forecaster decides to monitor the compression ratio and efficiency of turbosuperchargers.

In addition, one can calculate, as Whittle did, that the turbine blades in the jet engine would have to withstand a temperature of about 1500 °F while turning at several thousand rpm. Not only is this temperature above the melting point of most steel alloys, but most alloys are subject to "creep" at high temperatures and stresses ("creep" describes a stretching or distortion to which some metals are subject when placed under continuous stress). The turbine blades in turbosuperchargers are subject to stresses similar to those in a jet engine, hence the forecaster decides to monitor the temperature resistance and creep properties of metals used in turbosuperchargers. Exhaust valves in aircraft engines are also subject to high temperatures and are required to have low creep to prevent distortion and leakage in operation. Hence the properties of metals used for this application will also be monitored.

Our hypothetical forecaster has repeated calculations that actually were or could have been made in 1930 and has identified several factors to monitor. The next task is to set thresholds. At what value of compressor ratio or efficiency should the threat of the jet engine be recognized as having arrived? At what level of temperature and operating life for turbosuperchargers or exhaust valves should the alarm be sounded?

At this point the hypothetical example begins to break down. There is always a certain amount of judgment involved in setting a threshold, and it is difficult to duplicate now a judgment that would have been rendered in 1930. In fact, there is considerable likelihood that any threshold calculated for this example would reflect a great deal more hindsight than it would 1930-style foresight. Suffice it to say that we have identified the factors to monitor and have determined values for these factors at which the threat is no longer impending but has arrived. The warning threshold should be set somewhat below these values. Rather than try to duplicate a judgment that would have been made in 1930, we will stop short of actually setting thresholds.

To complete the example for illustrative purposes, however, we will look at some developments subsequent to 1930. In 1931 a turbosupercharger was available with a compression ratio of 2 to 1 and an efficiency of 62%. By 1935 these had been raised to 2.5 to 1 and 65%. The values achieved in 1935 represented a significant fraction of the values needed for a practical jet engine; however, they were not yet adequate. There was still some hard work to be done to bring them up to the levels needed.

After 1935 events came more rapidly. In that year Hans von Ohain obtained a German patent on a turbojet almost identical with Whittle's design. In 1936 Whittle began construction of a jet engine at his own company, Power Jets Ltd. In the same year von Ohain received support from the Heinkel Company in Germany for the fabrication of a jet engine. In 1938 the U.S. Air Corps laboratories at Wright Field (Dayton, Ohio) initiated a five-year program of development of gas turbines for jet propulsion; the NACA began a program of compressor development; and the RAE initiated work on a turbocompressor based on Griffith's 1929 design for a turboprop engine.

The important point here is that by 1938 or so it was obvious to just about everyone in the aircraft engine business that jet propulsion was on the verge of being practical, and that the time had arrived to get started developing it. If a monitoring program were to be of any help, it should have sounded the alarm before this. Clearly, 1930 was too early to sound the alarm; likewise, 1938 was too late. Probably 1935 was the earliest year in which it would have been reasonable to sound the alarm. This implies that the warning thresholds for compression ratio and efficiency should have been about 50% and 80%, respectively, of the required values. However, there is no precise formula by which these thresholds can be determined. The proper threshold can be set only by asking, How hard is it to go from a compression ratio of 2.5 to 1 to a ratio of 4 to 1, and from 65% efficiency to 75% efficiency? If the answer is, Not very, then the warning should be sounded. By 1935 both Whittle and von Ohain had reached the conclusion that the required levels could be achieved.

History proved that they were right. But it is not possible to divest ourselves of knowledge of the historical outcome and decide what thresholds should have been set in the 1930s.

In this example we have shown that sufficient signals were available in the 1930s to have given advance warning of the development of the jet engine. We have shown how the jet engine could have been forecast through a well-designed environmental monitoring procedure. However, this example should not be taken as a criticism of those who failed to foresee the jet engine in 1930. One major reason that people in the past failed to forecast breakthroughs is that they did not realize it was possible. The lesson to be drawn from this example is that once the forecaster is aware of the possibility of monitoring for signals, these can in fact be identified in a fairly straightforward fashion. The signals we identified for the period of the 1930s did not demand any special gifts of foresight, they demanded only knowledge that was available then and calculations which were actually made by people of the time. If any inference about culpability is to be drawn from this example, it is that forecasters of today are culpable if they fail to learn from the mistakes of the past and fall into the trap of repeating them.

6. Summary

The forecasting methods given in previous chapters seem to require continuity between the past and the future in order to forecast the future. The question then arises, Can a breakthrough be forecast? The answer is yes, but it depends upon another kind of continuity between the past and the future. Breakthroughs in technology do not come as "bolts from the blue"; on the contrary, they are the end result of a chain, or even a network, of precursor events, and these events give warning that a breakthrough is coming.

Forecasting breakthroughs, then, involves a systematic search for these precursors, coupled with an evaluation of the significance of the precursors found. The forecaster seeking advance warning of a breakthrough must search all the relevant sectors of the environment in order not to miss important signals of coming breakthroughs. Moreover, the signals found must be synthesized into possible patterns of change, and the forecaster should continue to search for additional signals that will confirm the patterns.

References

Kahn, Herman, and Anthony Weiner (1967). *The Year 2000* (New York: Macmillan).

Taylor, Gordon R. (1968). *The Biological Time Bomb* (New York: New American Library).

For Further Reference

Heiman, G. (1963). *Jet Pioneers* (New York: Duell, Sloan and Pierce).

Neville, L. E. (1948). *Jet Propulsion Progress* (New York: McGraw-Hill).

Whittle, F. (1953). *Jet: The Story of a Pioneer* (London, England: Frederick Muller).

Problems

1. Which of the following devices or techniques represent breakthroughs as defined in the text? What functional capability was advanced? What limitations of the prior techniques were surpassed by the breakthrough?
   a. Bessemer steel-making process.
   b. Logarithms.
   c. Typewriter.
   d. Vaccination.
   e. Canning.
   f. Magnetic recording.

2. Review the history of the invention of the airplane from the standpoint of its inevitability. What were the achievements and contributions of Sir George Cayley, Sir Hiram Maxim, Otto Lilienthal, Octave Chanute, Samuel P. Langley, and the Wright brothers?

3. Review the history of the self-propelled road vehicle, which culminated in the development of the automobile. What were the three major contending power sources during the period 1890–1915? What were the advantages and disadvantages of each for automobile propulsion? What might have been the future course of automobile development if a boiler capable of rapid warmup had been invented prior to the invention of the automobile self-starter?

4. The following items are adapted from a list in Herman Kahn and Anthony Wiener's book *The Year 2000* (1967):
   a. Use of lasers for destructive (defensive) purposes.
   b. Extensive commercial application of shaped-charge explosives.
   c. Extensive use of "cyborg" (cybernetic–organic) techniques (mechanical aids or substitutes for human organs, senses, limbs, or other components).
   d. Three-dimensional photography, illustrations, movies, and television.
   e. Practical use of direct electronic communication with and stimulation of the brain.
   f. Capability to choose the sex of unborn children.
   g. Permanent inhabited undersea installations and perhaps even colonies.
   h. Automated universal (real-time) credit, audit, and banking systems.
   i. Recoverable boosters for economic space launching.
   j. Practical large-scale desalinization.

   Each of these items can be considered as a forecast based on at least one precursor event. For each item, answer the following questions:
   a. What additional signals would tend to confirm or deny the forecast?
   b. What sources should be monitored for these signals?

Problems *(continued)*

c. Have any confirming or denying signals appeared since the original forecast was made in 1967?

5. Gordon R. Taylor, in *The Biological Time Bomb* (1968) has forecast the availability of the following capabilities (on the basis of research already completed):
   a. Creation of life by entirely artificial means from nonliving substances.
   b. Nonsexual human reproduction, including "test-tube babies," parthenogenesis, and "cloning" (production of duplicate individuals starting with a single cell of the duplicated person).
   c. Practical development of organ transplantation on a large scale, with the establishment of "organ banks" for the temporary storage of usable organs.
   d. Control of aging and eventual elimination of death from old age.
   e. Chemical control or modification of intelligence, mood, memory, judgment, and other mental processes.
   f. "Genetic engineering," the growing of plants, animals, and humans to meet prescribed specifications on their heredity.

   For each of these items answer the same questions as in the preceding problem.

Chapter 10

Combining Forecasts

1. Introduction

Each individual forecasting method has its own strengths and weaknesses. In many cases a forecast can be improved by combining separate forecasts obtained by different methods; by doing this the forecaster avoids problems of trying to select the one best method for a particular application. In addition, the strengths of one method may help compensate for the weaknesses of another.

The chapter will cover some of the more important methods for combining forecasts. It will include combined forecasts of a single technology and means for combining forecasts of several different items into a composite whole. Particular emphasis will be placed on the problem of consistency in combining forecasts of different items into a single overall forecast.

2. Trend and Growth Curves

The history of a technology will often show a sequence of different technical approaches. Each new approach usually represented a breakthrough, since it surpassed the upper limit of the previous approach. Thus a plot of the performance of individual devices in the history of the technology might show a succession of growth curves, one for each technical approach, while the overall performance of the technology is given by an exponential trend.

A forecast of such a technology should involve projecting both the overall trend and the relevant growth curves: The trend will give an estimate of future levels of performance to be expected; projecting the growth curve for the currently used technical approach will provide additional information. If that growth curve can "keep up" with the trend for some time yet, this indicates that a near-term threat to the technical approach is unlikely. However, if the growth curve is nearing its upper limit and must soon fall below the trend (or has already begun to), this is an indication that a successor technical approach will eventually replace the current technical approach. If such a successor approach can be identified, even if it is only in its earliest stages of development, projecting its growth curve can help predict the time when it will compete successfully with the current approach.

Using both trend and growth curves, then, can provide much more information about the future course of some technology than can the use of either by itself.

3. Trend and Analogy

External events can cause a deviation, temporary or permanent, from a long-term trend. Many of the technologies described in the data of Appendix 3 show the effects of the Great Depression and of World War II. Other technologies show the effects of external events much more localized than those two major social upheavals. When important external events are identified, either in prospect or in retrospect, analogies can be used to forecast the effects they will have on long-term trends.*

The purpose of combining analogies and trends is to forecast the magnitude and timing of the deviation from a trend that some external event will cause. (In all this discussion it is assumed that the requirements of a formal analogy have been met.) The simplest case is that of an event that is analogous to some earlier event which has already impacted on a trend. The results of the impact are then forecast as analogous to the results of the previous event. A more complex situation involves analogies not only between impacting events but between technologies as well. The impact of the expected event on the technology of interest is then taken to be analogous to the impact of the model event on the model technology.

The combination of trends and analogies can be applied in reverse as well. When a deviation from a trend has been observed in the history of some technology, the forecaster may try to identify the cause of the deviation. In some cases this will be fairly obvious (e.g., wars, depressions); in other cases the cause will be specific to the technology and it will be necessary to carry out a historical investigation. Once the cause of the deviation is determined, however, it can be used as the basis for an analogy to forecast future deviations from analogous causes.

---

* Note that external events can impact growth curves in the same way they do trends. However, growth curves tend to have shorter time scales than do trends, hence the impacts are not so easily observed. Nevertheless, keeping in mind the differences in time scale, what is said here about trends applies equally well to growth curves.

4. Components and Aggregates

The forecaster often deals with an aggregate that can be decomposed into components. This raises the question of whether the components should be forecast and added to obtain the aggregate, or the aggregate forecast and the components derived from it. Sometimes the aggregate appears to have "lawful" behavior, while the components are to some extent interchangeable or in competition with one another. In such cases a "top–down" forecast is appropriate, with the aggregate being forecast first and the total allocated among the components. For instance, consider a forecast of the amount of food preserved by each of several technologies such as canning, freezing, and dehydration. The total amount of preserved food consumed in the United States is determined by population and income. Hence this total should be forecast first. It should then be allocated among the different preservation technologies on the basis of cost and other competitive factors. It would make no sense to forecast each technology separately and add them to get the total. On the other hand, some aggregates exist in a statistical sense only. They are made up of independent components that should be forecast separately, in a "bottom–up" manner. For instance, the number of aircraft engines manufactured in the United States is made up of engines for general aviation, for airlines, and for military use. Each of these components can change more or less independently of the others, and they must be forecast separately to get the total production.

In some cases, however, forecasts of components and aggregates can be combined fruitfully. Consider the problem of forecasting the number of households in the United States with cable television (CATV). The forecaster could attempt to forecast this total directly; however, finding a suitable "growth law" would be difficult. So long as the population continues to grow, there is no obvious upper limit. Hence growth curves are not suitable. Ultimately, the number can grow no faster than the population; at that time whatever growth law fits population would also fit CATV. However, the real value of a forecast is during the time when the number of households with CATV is coming into equilibrium with the population.

The situation is typical of cases in which forecasts of components and aggregates can be combined. In Chapter 4 a forecast was made of the fraction of households with CATV. Since a fraction has an obvious upper limit, a growth curve was a suitable approach. If the total number of households is then forecast as an aggregate, the number of households with CATV can be calculated as the product of the two forecasts.

There are many other cases in which it is better to forecast some total separately from the forecast of the fraction that will utilize or involve a particular technology. In these cases combining forecasts of the total and of the fraction will be better than attempts to directly forecast the number using the technology.

It should be noted that when a technological forecaster requires forecasts of populations, households, and so on, the best source is the U.S. Bureau of the Census. First, the use of Census Bureau forecasts avoids the need to defend the forecasts; no one else has more credible forecasts. Second, population forecasting is a specialized topic with its own problems and techniques. Technological forecasters are well advised to stick to their own specialties; the expertise of other forecasters in their fields of specialization should be utilized.

5. Scenarios

Sometimes the forecaster has a set of forecasts that are related in some way, or they all bear on a particular situation. Taken together, they provide an overall picture of an environment, as opposed to a small segment of the environment as captured by each of the forecasts individually. Scenarios are used to combine these individual forecasts into a composite whole.

A scenario is a written description of a situation. Kahn and Wiener defined scenario as follows:

> Scenarios are hypothetical sequences of events constructed for the purpose of focusing attention on causal processes and decision-points. They answer two types of questions: (a) Precisely how might some hypothetical situation come about, step by step? and (b) What alternatives exist, for each actor, at each step, for preventing, diverting, or facilitating the process? (Kahn and Wiener, 1967)

Scenarios have three purposes:

1. To display the interactions among several trends and events in order to provide a holistic picture of the future.
2. To help check the internal consistency of the set of forecasts on which they are based.
3. To depict a future situation in a way readily understandable by the nonspecialist in the subject area.

Technological forecasters frequently make use of scenarios for one or more of these purposes. Scenarios can be very effective in providing a composite picture of some future for use as the background for a decision. The scenario can define the alternatives and the choices that must be made among them, as well as portray an environment into which a technological development must fit, including the effects of the technology on the environment and vice versa. A range of scenarios can be used to test the "robustness" of some technological strategy by depicting a variety of environments, from favorable to unfavorable, with which the strategy must cope.

Because of their varied uses, no single scenario can serve all purposes; therefore there is no single "correct" way to write one. Writing a scenario is as much of an art as is writing a novel. Nevertheless, there are procedures that can help the forecaster go about his task of writing scenarios in a systematic manner. Following the procedures will not guarantee a good scenario; however, it will usually help improve it.

The forecaster should follow these steps in writing a scenario:

1. Develop a framework. What happens in each sector of the environment? What trends should be considered? Will they continue or change? If there is change, when and in what way? Are there decision points for critical decisions? Who will make the decisions?
2. Forecast the technology (or technologies) to be considered. When will it be deployed? What is the scale of adoption? What are the impacts from and on the framework trends and events? What are the specific decision points? Who are the decision makers?
3. Plot the scenario. Identify events that could trigger decisions. Select a small number of sequences of events and decisions. Check for consistency.
4. Write the scenario. Fill in the outline developed in the previous steps by a verbal narrative describing the events.

The following are the main problems the forecaster must solve in writing a scenario, and special pains must be taken to overcome them:

1. Transition from the present. The trajectory by which the future situation is reached must be developed carefully, especially if the scenario is to serve as a guide to action. The sequence of events and the necessary choices must be portrayed clearly.
2. Plausibility. A scenario will be plausible only if the chain of events that leads to it also appears plausible. A chain of very unlikely events ("acts of God") does not make for a plausible scenario. Even though individual events may have low likelihoods, they should be members of plausible classes; that is, they whould be the sort of thing that could plausibly happen, even if the specific events themselves may not be particularly likely.
3. Reversal of trends. Some scenarios will require that historical trends be reversed. Major trends have in fact been reversed, but the scenario writer must provide the reader with some reason to believe that the reversal of a trend is reasonable.
4. Convincing linkages among events. Frequently the scenario writer has to provide a reason for why a particular event will take place rather than some other, or why one decision is made instead of

another. These choices are supposed to be triggered by other events in the scenario. The linkages between causal events and subsequent events must be convincing. The reader should be satisfied that the causal event really would have the effect claimed.

5. Motivation of actors. Major actors will make certain decisions during the development of the scenario. The writer must provide proper motivation for these decisions. This is often particularly difficult in "best-case" or "worst-case" scenarios. It is often hard to imagine someone knowingly and deliberately taking the actions needed to produce these extreme scenarios.

Despite these problems, the scenario is a popular means for combining forecasts into a composite whole. This popularity is due in no small measure to the power a well-written scenario has to make an otherwise dull forecast come to life and appear in vivid colors. The technological forecaster should be prepared to take advantage of this power of the scenario when it is appropriate.

6. Cross Impact Models

Two elements of every forecast are the time of the event and its probability. However, the forecast of the time and probability for any event is based on assumptions about other events that have not yet occurred, but which might occur between now and the time forecast for the event of interest. A particular forecast will assume certain outcomes for these intermediate events. If the intermediate events have different outcomes, the original forecast becomes irrelevant. It may be a correct deduction from its assumptions, but it no longer applies because its assumptions are no longer valid.

Cross Impact models are a means of taking the dependency of some forecasts on other forecasts into account. A Cross Impact model consists of a set of events, each assigned a timing and probability; in addition, it consists of a set of cross impacts among the events. If the forecast for event E_2 assumed a specific outcome for an earlier event E_1, a different outcome for E_1 is then said to have a cross impact on E_2. This cross impact shows up as a change in timing, in probability, or in both for E_2. The impact may come from nonoccurrence, if the earlier event was forecast to occur; conversely, the impact may come from occurrence if the earlier event was forecast not to occur. If the earlier event was not considered at all, both occurrence and nonoccurrence may have impacts.

In principle, there may be cross impacts between every pair of events in a Cross Impact model. For N events in the model there may be $N(N - 1)$ cross impacts. In practice, the number is much smaller than this. If the outcome of E_1 will be determined before the earliest E_2 could possibly occur, then a cross impact from E_2 to E_1 is impossible and can be ignored. A typical Cross Impact model may have 200 events; there would be a maximum of 39,800 cross impacts. In practice, however, the number of cross impacts is unlikely to exceed 1000.

A Cross Impact model, then, looks upon the future as a collection of interconnected events. As time passes, the outcome of certain events is determined. The cross impacts from these events, as they are determined, cause the forecast timing of the remaining events to be earlier or later, and the forecast probability to be higher or lower. Ultimately, the outcome of every event is decided, and it either does or does not occur.

In use, the Cross Impact model simulates this future. The event with the earliest forecast date is selected and its outcome determined (for instance, by drawing a random number). Depending on the outcome, adjustments are made in the timing and probability of all the other events that receive cross impacts from this first event. The next event is selected, its outcome determined, and adjustments made in the remaining events. This process is completed when the outcome for all events has been determined.

This process will be illustrated by an example. Table 10.1 lists a set of events, each with a timing and a probability. Table 10.2 lists cross impacts among these events. For instance, if 4 million households have videodiscs by 1981, the penetration of CATV into 41 million households will be delayed from 1985 to 1987. [The forecaster(s) evidently believed that success of the videodisc would delay CATV, either by diverting consumer expenditures or by occupying viewer time or for some similar reason.] No cross impacts from nonoccurrence are shown in Table 10.2; however, some are implied. The failure of 4 million households to buy videodiscs clearly precludes having 8 million households with that technology. Therefore the cross impact from VDISC8MHSE on CATV41MHSE clearly includes the assumption that the latter has already been impacted by CATV4MHSE. Note that impacts may affect timing, probability, or both.

Table 10.1. Events for Cross Impact Model

Event	Short name	Year	Probability
CATV in 21 million households	CATV21MHSE	1980	0.95
CATV in 41 million households	CATV41MHSE	1985	0.90
Videodiscs in 4 million households	VDISC4MHSE	1981	0.70
Videodiscs in 8 million households	VDISC8MHSE	1985	0.50
380 Satellite broadband channels	380SAT CH	1980	0.95
1130 Satellite broadband channels	1130SAT CH	1985	0.90
1800 Satellite broadband channels	1800SAT CH	1990	0.95
61 Million long-distance calls	61MLDCALLS	1980	1.00
83 Million long-distance calls	83MLDCALLS	1985	1.00

Table 10.2. Cross Impacts Among Events

Impacting event	Impacted event	New year of occurrence	New probability
VDISC4MHSE	CATV41MHSE	1987	0.90
VDISC8MHSE	CATV41MHSE	1991	0.90
61MLDCALLS	1130SAT CH	1983	1.00
83MLDCALLS	1800SAT CH	1987	1.00
CATV41MHSE	1800SAT CH	1987	0.85

We will follow one possible sequence of outcomes from this sample Cross Impact model, which is shown in Table 10.3. The first three columns repeat information from Table 10.1, except that the events are arranged in chronological order. We start with the first event.

We draw a two-digit random number, either from the table in Appendix 2 or from a calculator. If the number is in the range 01–95, the first event is taken to occur. If the number is in the range 96–00, the first event does not occur. Assume that the number is favorable and the event occurs. The outcome is shown in the final column. Assume that the second event also occurs, as well as the third. However, this third event has a cross impact on 1130SAT CH. The adjustments to the latter are shown under the column First impact. Having determined the outcome of all events forecast for 1980, we move to 1981. Assume that VDISC4MHSE occurs; this delays CATV41MHSE as shown. The next event in order is now 1130SAT CH, which has been advanced to 1983. Assume that this also occurs. There are now two events forecast for 1985. Consider VDISC8MHSE first. This also occurs, further delaying CATV41MHSE to 1991. Assume that 83MLDCALLS also occurs in 1985; this advances

Table 10.3. One Possible Outcome of the Cross Impact Model

	Initial forecast		First impact		Second impact		
Event	Year	Probability	Year	Probability	Year	Probability	Outcome
CATV21MHSE	1980	0.95					1980
380SAT CH	1980	0.85					1980
61MLDCALLS	1980	1.00					1980
VDISC4MHSE	1981	0.70					1981
CATV41MHSE	1985	0.90	1987	0.90	1991	0.90	1991
VDISC8MHSE	1985	0.50					1985
1130SAT CH	1985	0.90	1983	1.00			1983
83MLDCALLS	1985	1.00					1985
1800SAT CH	1990	0.85	1987	1.00			1987

1800SAT CH to 1987 and increases its probability to 1. Assume that the remaining two events also occur.*

The results are shown in the last column of Table 10.3. All the events did occur, but not all took place on the dates originally forecast. Some dates were changed by cross impacts from earlier events.

This is only one of the possible outcomes implicit in the original set of data. If the simulation were repeated with a different set of random numbers, a different outcome might be obtained. There are actually $2^9 = 512$ distinct possible outcomes for this set of 9 events; each differs from all the rest for the outcome of at least one event.

With a realistic model of 200 events the number of distinct possible outcomes is the astronomical figure of $2^{200} = 1.607 \times 10^{60}$. It would be impossible to list even a significant fraction of these different outcomes. In practice, the Cross Impact model is used to generate a large number of synthetic "future histories" by repeating the simulation enough times. Usually 100 times is sufficient to obtain statistically repeatable results. Each of the resulting synthetic future histories could be considered the outline of a scenario, and built up with a verbal description.

The set of future histories is not converted into scenarios; instead, it is analyzed statistically. Usually the frequency of occurrence of each event is computed. Each event was initially forecast to have a certain probability of occurrence that was based on certain assumptions about prior events. The frequency of occurrence in the output of a model is an estimate of an event's probability of occurrence *in the context of all the other events,* rather than considered by itself. This frequency of occurrence can be compared with the original forecast, and tests of statistical significance applied to determine whether the other events make a significant difference in the probability of occurrence.

Other statistics that are customarily computed from the Cross Impact model's output include the frequency of occurrence by year and frequencies of the joint occurrence of pairs of events.

Nothing comes out of a Cross Impact model that was not implicit in the input. However, in a model of a practical situation many of the implications of the input will not be readily apparent; they may result from unobvious chains of cross impacts. Use of the model allows the forecaster to identify "surprise" futures that are implicit but not immediately obvious in the data. Searching the input data for these surprise futures is an unproductive use of the forecaster's time; it is much better to run the model on a computer and extract the results from the computer output.

---

* Note that when two or more events can occur in the same year, the event to be tested first for occurrence should be selected by drawing a random number. If the events are always chosen in the same order, some of the possible outcomes of the model will be foreclosed.

This is why the Cross Impact model is highly useful, even though the output is completely contained in the input. Once significant results are observed in the output, the forecaster can readily search the input to determine why they occurred.

The Cross Impact model also allows the forecaster to identify "critical" events. These are events whose occurrence or nonoccurrence makes a great deal of difference in the subsequent "history" of a model.

Another useful feature of the Cross Impact model is the ease with which the results of policy interventions can be tested. An intervention can be included as a specific event, and the model run with or without it; the differences will be due to the policy.

Finally, the Cross Impact model can be used to test the sensitivity of the outcomes to estimates of the timing and probability for events. If there is disagreement about the forecast for some event, the model can be run with each of the available forecasts to determine how much difference they make.

An important aspect of the Cross Impact model is that the output is guaranteed to be internally consistent, in the sense that two events that are contradictory or incompatible with one another cannot both occur. Moreover, output scenarios grow naturally from the input data; they are not imposed arbitrarily. Thus the forecaster who needs a set of alternative "contingency" scenarios can readily obtain them by adding or deleting events, or by altering probabilities of events. Thus alternative but internally consistent scenarios can be generated that span the range of possible outcomes inherent in a set of events and cross impacts.

The Cross Impact model is a systematic means for combining forecasts from a variety of sources. It provides explicit means for examining the interactions between pairs of events and it can be used to develop internally consistent scenarios. These in turn can be used for testing policies and identifying critical events. When the Cross Impact model is computerized, dozens or hundreds of events and cross impacts can be examined. Whether computerized or not, the Cross Impact model permits the forecaster to focus on events and their cross impacts rather than on the procedures for handling them, that is, on the forecasting aspects of the problem rather than on the data manipulation.

The Cross Impact model was originally developed to solve a problem that arose in the use of a Delphi. The panelists in some of Gordon and Helmer's original experiments stated that they had difficulty forecasting some events because these would be affected by other events they were also expected to forecast. The Cross Impact model was developed to allow these interactions to be taken into account. However, it now has little connection with Delphi; the forecasts used may come from any source or from several sources. Many variations have also been developed; some of the more important ones are discussed in the works in For Further Reference at the end of the chapter.

7. Summary

In many cases the forecaster has alternate ways of forecasting the same event or several forecasts of related events. A forecast can often be made more useful by combining several forecasts. The use of trend and growth curves together and the use of analogies with trend and growth curves can improve the forecast of a single technology. Where an aggregate and its components can be forecast separately, more information can often be gained by combining the separate forecasts. When a large number of related forecasts must be combined into a composite whole, either scenarios or a Cross Impact model may be used.

In all cases the forecaster should consider the possibility of combining two or more forecasts to offset the weaknesses of some with the strengths of others.

Reference

Kahn, Herman, and Anthony Wiener (1967). *The Year 2000* (New York: MacMillan).

For Further Reference

Chen, Kan, Kenan Jarboe, and Janet Wolfe (1981). "Long-Range Scenario Construction for Technology Assessment," *Technological Forecasting and Social Change* 20 (1), 27–40.

Ducot, C. (1980). "Futures Research Scenarios and Time Succession," *Technological Forecasting and Social Change* 17 (1), 51–59.

Duperrin, J. C., and M. Godet (1975). "SMIC-74—A Method for Constructing and Ranking Scenarios," *Futures* 7, 302–312.

Enzer, Selwyn (1980). "INTERAX—An Interactive Model for Studying Future Business Environments: Part I," *Technological Forecasting and Social Change* 17 (2), 141–160.

Enzer, Selwyn (1980). "INTERAX—An Interactive Model for Studying Future Business Environments: Part II," *Technological Forecasting and Social Change* 17 (3), 211–242.

Helmer, Olaf (1977). "Problems in Futures Research—Delphi and Causal Cross-Impact Analysis," *Futures* 9, 17–31.

Kaya, Y., M. Ishikawa, and S. More (1979). "A Revised Cross-Impact Method and Its Applications to the Forecast of Urban Transportation Technology," *Technological Forecasting and Social Change* 14 (3), 243–258.

McLean, M. (1976). "Does Cross-Impact Analysis Have a Future?," *Futures* 8, 345–349.

Turoff, Murray (1972). "Meeting of the Council on Cybernetic Stability: A Scenario," *Technological Forecasting and Social Change* 4 (2), 121–128.

Vanston, John H., W. Parker Frisbie, Sally Cook Lopreato, and Dudley J. Poston (1977). "Alternate Scenario Planning," *Technological Forecasting and Social Change* 10 (2), 159–180.

Problems

1. Discuss the advantages and disadvantages of combining separate forecasts.

2. A certain technology has shown progress, as given in the table below, over a period of roughly 50 years, and includes three successive technical approaches. The current year is 55 and you have been asked to prepare a ten-year projection of this technology. There appear to be no theoretical or practical limits to the continued growth of its functional capability. What method(s) will you use to prepare the forecast, and what is your forecast? (The years and values for the level of functional capability are arbitrary scales.)

Tech. app. A[a]		Tech. app. B[b]		Tech. app. C[c]	
Year	Functional capability	Year	Functional capability	Year	Functional capability
3	90	19	300	36	690
4	120	22	440	38	1100
6	210	23	660	39	1700
7	240	24	720	41	1900
9	310	26	900	42	2300
11	340	28	1000	43	2400
14	430	30	1250	45	2800
17	470	34	1350	47	3000
21	480	43	1550	49	3400
25	530			53	3600
				55	4000

[a] Theoretical upper limit 550.
[b] Theoretical upper limit 1600.
[c] Theoretical upper limit 4100.

3. You have been asked to prepare a 20-year forecast of a specific technology. The historical growth of the level of functional capability of this technology is shown in the table below (functional capability and years given in an arbitrary scale). This technology is widely used in the mining industry but has moderate use in other industries as well. You determine that the growth of the technology has been reasonably steady, except for the period following a major mineral strike in Alaska in the year 20. Last year (year 54) there was a major mineral strike in Canada. What is your forecast for the period through year 75? What levels of functional capability are to be expected in years 60, 65, 70, and 75?

Year	Functional capability
5	56
8	80
12	86
14	120
16	125
19	150
22	160
24	220
27	310
31	330
36	340
38	370
40	500
45	520
48	800
53	880

4. The director of a laboratory has compiled the following information on the number of scientific papers published by the scientists in the laboratory. You have been asked to estimate future publication growth. You are given the information that the size of the technical staff is expected to remain about the same as it was during the period covered by the data, that is, 200 people. What is your forecast of the future production of papers, and on what basis do you make this forecast?

Year	Number of papers
1960	9
1961	12
1962	11
1963	16
1964	14
1965	19
1966	17
1967	24
1968	22
1969	29
1970	28

[*Hint:* The absolute maximum number of papers produced by one scientist in a lifetime is about 900. Nobel Prize winners typically publish about 150 papers in their lifetime, and most of these before they win the prize. Only 1 scientist in 100 publishes more than 100 papers in a lifetime; only 1 scientist in 10 publishes more than 10 papers in a lifetime. The average scientist publishes four papers in a 40-year active lifetime.

5. Refer to Table A3.28, the U.S. total gross consumption of energy resources, in Appendix 3. Make an exponential fit to the historical consumption of each

Problems *(continued)*

energy source listed (i.e., anthracite, bituminous) and to the total gross energy input. Project each to 1980 (ignore the projections made by the Bureau of Mines in the last three rows). Does the total of your projections for the components add up to the projection of the total? Should the projection for the total be adjusted, or the projections of the components? If the latter, which components? What is the shift in the percentage of the total for each component between 1965 and 1980?

6. Generate ten simulations of the Cross Impact model in the chapter, obtaining random numbers either from the table in Appendix 2 or from a calculator. What is the frequency of occurrence of each event in your ten simulations? Are there any events for which the frequency of occurrence differs significantly from the original forecast?

7. Take one possible sequence of outcomes from the Cross Impact model in the chapter and write a scenario describing it. What additional information do you need to provide as background? What actors should be involved?

Chapter 11

Normative Methods

1. Introduction

The forecasting methods discussed in the previous chapters are "exploratory" methods; that is, they all start with past and present conditions and attempt to project these to estimate future conditions. They explore the possible futures implicit in past and present conditions.

A complementary approach is taken by "normative" methods. These methods have their foundations in the methods of systems analysis. They start with future needs and identify the technological performance required to meet those needs. In essence, they forecast the capabilities that will be available on the assumption that needs will be met.

Normative and exploratory methods are customarily used together. An exploratory forecast has implicit within it the idea that the capability will be desired when it becomes available. A normative forecast has implicit within it the idea that the required performance can be achieved by a reasonable extension of past technological progress. The complementary use of normative and exploratory forecasts will be taken up in later chapters on applications. This chapter will discuss the normative methods themselves.

Three common methods of normative forecasting will be described. These are relevance trees, morphological models, and mission flow diagrams.

2. Relevance Trees

Relevance trees are used to analyze situations in which distinct levels of complexity or hierarchy can be identified. Each successively lower level involves finer distinctions or subdivisions.

Figure 11.1 shows a relevance tree. At the top of the tree is an automobile. At the first level we have three elements of the automobile. One of these elements is further subdivided as the second and lower levels. The other two elements could be subdivided in the same way.

Each item on the tree is referred to as a "branch." The point from which several branches "depend" is a "node." Thus, except for those at the top and bottom of the tree, each branch depends from a node and each has a node from which several other branches depend. There is no requirement that each node have the same number of branches.

Figure 11.1 clearly shows the hierarchical structure of a relevance tree. A relevance tree also has several other characteristics that are not obvious from Figure 11.1 but which must be stated: First, the branches depending from a node must be a closed set. They must be an exhaustive listing of the possibilities at that node. In some cases this means simply listing all the members of a finite set; in most cases, however, the set is closed by the agreement that while there may be other items, they are not relevant or important by comparison with those listed. Second, the branches depending from a node must be mutually exclusive; they must have no overlap among them. Third, for a normative relevance tree the branches must be viewed as goals and subgoals. Each node is a goal for all branches depending from it; each goal is satisfied by the satisfaction of all the nodes below it, and, in turn, it derives its validity as a goal from the sequence of branches linking it to the top of the tree.

The branches of relevance trees used for normative forecasting will be

Figure 11.1. Relevance tree showing the major components of an automobile.

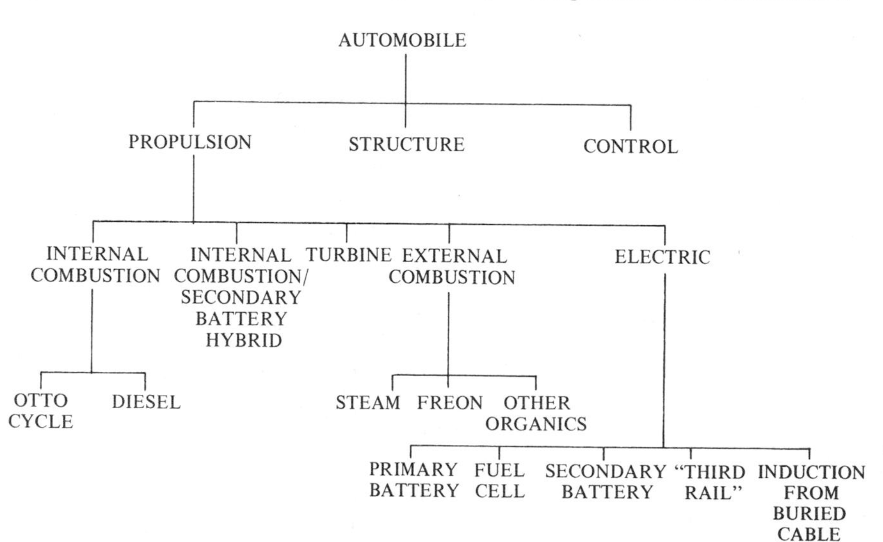

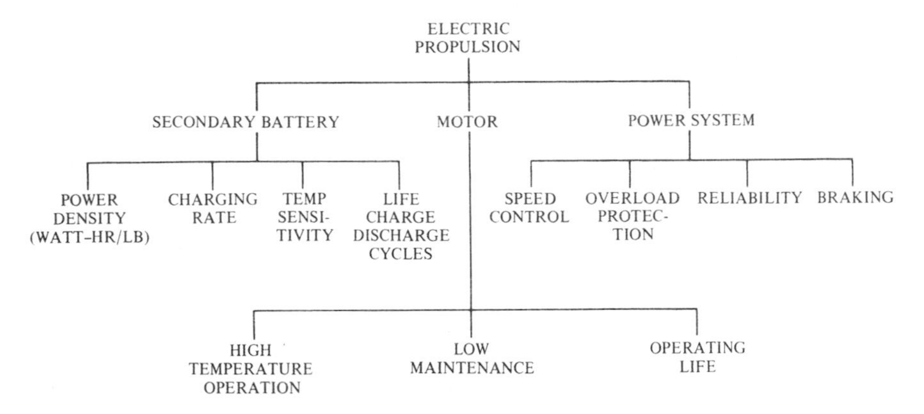

Figure 11.2. Problem tree for an electric automobile.

either problems or solutions. Thus two extreme or "pure" types of trees are the solution tree and the problem tree.

The solution tree has both "and" and "or" nodes. The "or" nodes are the most common. At each node the tree has two or more alternate solutions or answers to the question of how to achieve the solution at the next higher level; achievement of any single solution is sufficient. In some cases there will be "and" nodes; each of two or more partial solutions must be implemented if the higher-level solution is to be achieved.

Figure 11.1 is a solution tree. The branches at each node represent more refined or detailed means of achieving the solutions at the next higher level. At each node there are two or more alternative solutions.

A problem tree has only "and" nodes; at each node there are several problems that must be solved if the problem at the next higher level is to be solved. Figure 11.2 shows a problem tree. One segment of the relevance tree for an automobile is isolated here. At the top of the tree is the problem of electrical propulsion. At the first level there are three problems, each of which must be solved if electrical propulsion is to be achieved. At the next lower level there are several more problems, each of which must be solved if the problems at the first level are to be solved.

Normally the technological forecaster will use mixed trees, that is, trees containing both problems and solutions. The exact nature of the tree developed for a specific situation will depend upon the purposes it is to serve. There is no such thing as a universal or "all-purpose" relevence tree, even for a particular technology. The forecaster has considerable flexibility in designing or shaping a relevance tree. This flexibility should be utilized to assure that the tree satisfies the purpose for which it was prepared.

It is possible for a relevance tree to be internally inconsistent, in that branches in one portion of a tree are incompatible with branches in another portion. For instance, a branch included as a solution to a problem may be incompatible with a branch included as a solution to some other problem. The same may occur with problems; a problem in one portion of the tree may exist only in a particular environment or set of circumstances. The environment or circumstances may rule out a problem found in some other portion of the tree that can exist only in another environment or under other circumstances. This incompatibility is not really a matter for concern; it simply means that some combinations of solutions are impossible (for incompatible solutions) or unnecessary (for incompatible problems). For instance, one need not solve simultaneously the problems of operating a machine in a sandy desert and in a muddy swamp. After a tree is developed the forecaster may need to identify incompatibilities before using it to draw conclusions; however, the existence of incompatibilities does not undermine the utility of a tree.

Relevance trees can be used to identify problems and solutions and deduce the performance requirements of specific technologies. However, they can also be used to determine the relative importance of efforts to increase technological performance. Suppose some objective is to be achieved and suppose that there are three tasks that must be carried out to achieve this objective. We can then determine how much technological progress is needed in order to carry out each task. Assume that the first has twice as much need for progress as the second, which in turn has three times as much need for progress as the third. We can then assign these tasks numerical weights or "relevance numbers" of 0.6, 0.3, and 0.1. The relevance numbers must reflect the relative importance assigned to each of the tasks. In addition, the relevance numbers at a node must sum to unity. This requirement is referred to as "normalization" of the relevance numbers; it is an essential feature of the application of relevance numbers to relevance trees. Figure 11.3 shows a relevance tree with three tasks given the weights we have just assigned. Each of the tasks, in turn, may be achieved by two or more approaches, as shown. Relevance numbers have been assigned to each approach as well. The relevance numbers assigned to each task reflect the degree of improvement needed in the approach if the task is to be accomplished. Only the approaches for a particular node (task) are compared with one another. As before, the relevance numbers assigned are required to sum to unity at each node.

We can now determine the relative importance of the approaches with regard to the overall objective, not just with regard to the tasks to which they apply. For instance, A_{11} has a relevance of 0.7 to task T_1, which in turn has a relevance of 0.6 to the overall objective. Therefore the relevance of approach A_{11} to the overall objective is $0.6 \times 0.7 = 0.42$. The relevance of each of the approaches is shown at the bottom of Figure 11.3. The relative amount of effort to be allocated to each approach can now be determined from the relevance numbers. The advantage of this

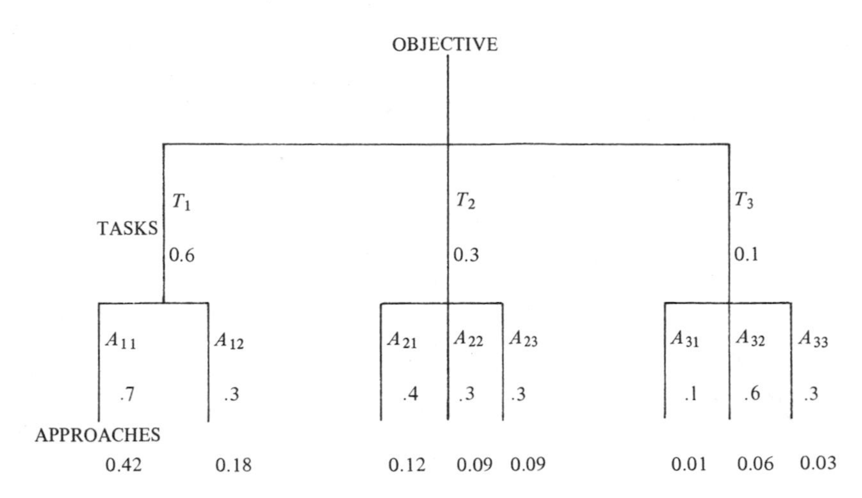

Figure 11.3 Relevance tree with relevance numbers.

technique is that approaches can be compared with one another even though they do not apply to the same task, since their relative importance to the overall objective can be determined.

A word of caution is in order. The relevance numbers assigned to individual branches are precise to only one significant figure. When multiplied together, however, they produce numbers of two or more significant figures. This represents a spurious precision. For instance, it is reasonable to say that A_{11} (with a relevance number of 0.42) is about twice as important as A_{12} (with a relevance number of 0.18). To argue that A_{12} is precisely 50% more important than A_{21} is reading more into the numbers than is really there.

This use of relevance numbers was originally given the name *PATTERN*. It was developed as a means of allocating R&D resources to the problems most needing solution.

Most relevance trees do not use relevance numbers. They are utilized to derive performance requirements for individual technologies in order to achieve some overall objective. Relevance trees can be extremely useful for this normative purpose alone. Their utility can be enhanced, however, by adding relevance numbers where this is appropriate. The use of relevance numbers increases the power of the relevance tree as a tool for identifying needed changes in technology.

3. Morphological Models

The word *morphology* is used in many sciences to describe the study of the form or structure of something (i.e., plants in biology, rocks in geology). The morphological model was invented by Fritz Zwicky (1957,

1969), who used it to analyze the structure of problems. The morphological model is a scheme for breaking a problem down into parallel parts, as distinguished from the hierarchical breakdown of the relevance tree. The parts are then treated independently. If one or more solutions can be found for each part, then the total number of solutions to the entire problem is the product of the numbers of solutions to individual parts. For instance, if a problem can be decomposed into 4 independent parts and there are 2 solutions to each, then there are a total of 16 solutions for the whole problem. Each solution for the whole problem is a combination of solutions for the individual parts. Some of these solutions will, of course, be impossible if a solution to one part is incompatible with a solution to another.

To make this concrete, Figure 11.4 shows a morphological model of automobile propulsion, which can be contrasted with the relevance tree of Figure 11.1. The model has six elements: the number of wheels, the number of driven wheels, the number of engines, the type of transmission, the engine type, and the power source. Each of these elements has two or more alternative components, as shown. This morphological model provides a total of $2 \times 4 \times 4 \times 3 \times 4 \times 5 = 1920$ distinct solutions to the problem of automobile propulsion. Some of the solutions are not really possible: For instance, the number of driven wheels obviously cannot exceed the number of wheels; some of the power sources shown are compatible with only one of the engine types shown; and so on. Once the impossible solutions are removed, however, there is still a significant number of different solutions to the problem of automobile propulsion.

Figure 11.4. Morphological model of automobile propulsion.

AUTOMOBILE PROPULSION MORPHOLOGY

P_1	WHEELS	3	4			
P_2	DRIVEN WHEELS	1	2	3	4	
P_3	ENGINES	1	2	3	4	
P_4	TRANSMISSION	NONE	MECHANICAL	FLUID		
P_5	ENGINE TYPE	INTERNAL COMBUSTION	EXTERNAL COMBUSTION	TURBINE	ELECTRIC	
P_6	POWER SOURCE	HYDROCARBON FUEL	PRIMARY BATTERY	SECONDARY BATTERY	FUEL CELL	THIRD RAIL

Once the impossible solutions are eliminated from consideration, the morphological model can be used to derive the performance requirements for each of the elements of the remaining solutions. These then become normative forecasts of the performance if the overall function is to be performed. In addition, the derived performance requirements can be used to estimate when the solution might be feasible if exploratory forecasts of the performance of each possible solution are prepared. Finally, if the required performance levels appear feasible by the time they will be required, they can be used as objectives for a R&D program.

In general, morphological models and relevance trees are suited to different types of problems; however, some problems can be treated by either method. Forecasters will ordinarily select the method that is easiest to tailor to their particular needs. A comparison of the two different methods as applied to the same problem is worthwhile. Comparing Figures 11.1 and 11.4, we see similarities and differences: Some items appear in both models of automobile propulsion. The various types of nonelectric engines, which show up as branches in the relevance tree, appear as alternate components under the "engine type" element of the morphological model. The sources of electrical energy that appear as branches in the relevance tree are also found as components under the "power source" element of the morphological model. This is actually a fairly general result. If the same system is modeled by a relevance tree and by a morphological model, the elements of the morphological model will correspond to major connected sections of the relevance tree, and the branches appearing at the bottom level of such sections in the relevance tree will be the components of the corresponding elements of the morphological model.

When a problem is obviously suited to one or the other approach, the forecaster has little difficulty in choosing between them. However, in those cases where either could be applied, the one that best suits the forecaster's purposes must be selected. This may require a careful study of the situation; there is no hard and fast rule that forces the selection of one over the other.

4. Mission Flow Diagrams

The mission flow diagram was originally devised by Harold Linstone as a means of analyzing military missions, which accounts for the name. However, it can be used to analyze any sequential process. It involves mapping all the alternative routes or sequences by which some task can be accomplished. All the significant steps on each route must be identified; the analyst can then determine the difficulties and costs associated with each route. In addition, it is possible to create new routes and identify the difficulties and costs associated with these. Once these difficulties and costs are identified, the performance requirements can be derived for the technologies involved and then used as normative forecasts.

Figure 11.5 shows a mission flow diagram for sending a message. It includes one path for which there is no physical movement at all, which illustrates that a mission flow diagram is not limited to cases involving physical movement. Figure 11.5 also includes cases of two paths merging into one, and of one path branching into two. This illustrates that paths may have common segments and that these may occur at the beginning, middle, or end of the paths.

Like the relevance tree and the morphological model, the mission flow diagram is of little value if it is restricted to simply describing some situation. The merit of the mission flow diagram is in allowing the user to identify problems and to derive the performance required of some technology to overcome the problems. Like the other two normative methods, the mission flow diagram permits the user to include new technologies and alternatives that do not yet exist.

Numerical weights can be placed on the alternative paths of a mission flow diagram, similar to the use of relevance numbers in relevance trees. These numbers could then be used to determine the relative importance of the technological solutions to each of the bottlenecks or difficulties on the paths. However, little work has been done in carrying out this approach. The greatest use of mission flow diagrams is to identify difficulties

Figure 11.5. Mission flow diagram for sending a message.

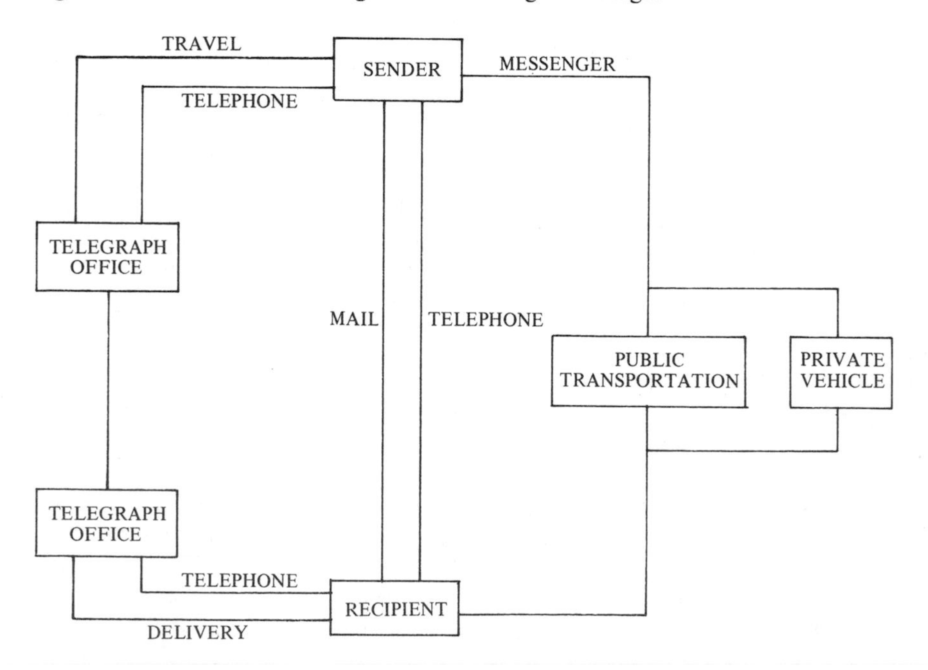

and to compare existing paths with alternatives that might be developed if desired.

5. Summary

In this chapter we have discussed normative, or goal-setting, methods of forecasting. Their application is to determine the level of functional capability that must be achieved if some problem is to be solved or some difficulty overcome. Relevance trees are used to carry out analyses of hierarchical structures; morphological models are used to analyze parallel structures; mission flow diagrams are used to analyze processes with steps in a spatial, temporal, or logical sequence. Often the same system can be analyzed by either a relevance tree or a morphological model. In such cases the forecaster should use whichever method is most convenient or appropriate for the problem in hand.

The strength of normative methods is that they tend to organize and structure the problem. They can help assure completeness, so that solutions holding some promise are not overlooked. By systematically laying out the structure of a problem, they can assist in the generation of new alternatives that may be superior to those currently in use. Even if all the alternatives uncovered with the use of normative methods prove inferior to those already known, this fact alone provides additional confidence in the choice of the superior method.

Normative methods also have disadvantages which they share with all systematic methods of problem-solving. One is that these methods may tend to impose rigidity on the solutions proposed. The problem itself may be distorted when it is fitted to a normative structure. Even if this does not happen, there may be a tendency to look only at the solutions that can be expressed easily within the formal structure of the model. Another difficulty arises when numerical weights are applied to the branches or elements of the model, as with PATTERN; it lies in the fact that the numbers tend to create their own validity. No matter how much subjectivity, and perhaps even sheer guesswork, went into the assignment of the numbers, they become very impressive once they have been written down. They can become even more impressive if they have been subjected to considerable manipulation in a computer. It often becomes very difficult to challenge the numbers under these circumstances, as though the computerization had somehow overcome any deficiencies in their origin. These disadvantages, however, are not sufficient to rule out the use of normative methods. With adequate care the disadvantages can be overcome completely, so that normative methods can, indeed, be very valuable. It is imperative, though, that the forecaster be aware of these difficulties.

Since one of the claimed advantages of normative methods is their completeness, it might be asked whether there is any guarantee that a normative model is complete. The answer, of course, is no. There is no formula or procedure that will guarantee that the constructor of a relevance tree or morphological model has not omitted something. Is this, then, not a weakness? Of course it is if the forecaster uses a normative model in such a way as to be vulnerable to surprise. For instance, the forecaster might assert that there is only one way in which a problem can be solved, since all the other paths on the mission flow diagram created are either infeasible or inferior to the path chosen. In such a situation the existence of an alternative not shown in the model can completely alter the forecaster's conclusions. If, however, the forecaster says only that there is *at least* one feasible solution (or however many are found) and that it places certain requirements on specific technologies, any alternatives that were missed will not overthrow the entire forecast. Finally, it should be noted that the systematic and orderly structure of any of the normative models we have discussed makes it easy to determine whether a specific solution has been omitted or not. It is not necessary to search through the set of all the configurations of a morphological model, for instance, to determine whether that set includes a particular well-defined solution; it is only necessary to check whether each of its components appears in the appropriate elements of the model. Hence even though there is no way to guarantee the completeness of a normative model, it is always possible to tell whether a specific solution has been missed. In addition, it is also usually possible to alter the structure easily to include solutions previously omitted.

In conclusion, it is essential that the normative methods be kept in proper perspective: They are in no sense a substitute for creativity or imagination; they merely provide a systematic means of examining technological requirements. When properly used, they provide a framework within which creativity and imagination can work effectively. However, it must be remembered that they are nothing but sophisticated ways of making lists of alternatives; and to revise the complaint of the author of Ecclesiastes, "of the making of lists there is no end." Nevertheless, even though normative methods are not in themselves a substitute for thought, they can be a very useful tool when applied properly.

Reference

Zwicky, Fritz (1957). *Morphological Astronomy* (Berlin: Springer-Verlag).

Zwicky, Fritz (1969). *Discovery, Invention, Research* (New York: Macmillan).

For Further Reference

Alderson, R. C., and W. C. Sproull (1972). "Requirement Analysis, Need Forecasting, and Technology Planning Using the Honeywell PATTERN Technique," *Technological Forecasting and Social Change* 3 (2), 255–265.

Blackman, A. Wade (1970). "Normex Forecasting of Jet Engine Characteristics," *Technological Forecasting and Social Change* 2 (1), 61–76.

Fischer, Manfred (1970). "Toward a Mathematical Theory of Relevance Trees," *Technological Forecasting and Social Change* 1 (4), 381–390.

Martin, William Tell, and James M. Sharp (1973). "Reverse Factor Analysis: A Modification of Relevance Tree Techniques," *Technological Forecasting and Social Change* 4 (4), 355–373.

Swager, William L. (1972). "Strategic Planning I: The Roles of Technological Forecasting," *Technological Forecasting and Social Change* 4 (1), 85–100.

Swager, William L. (1972). "Strategic Planning II: Policy Options," *Technological Forecasting and Social Change* 4 (2), 151–172.

Swager, William L. (1973). "Strategic Planning III: Objectives and Program Options," *Technological Forecasting and Social Change* 4 (3), 283–300.

Problems

1. What is the difference between normative and exploratory methods of forecasting? In what situations is each appropriate? How do they interact?

2. Construct the first two levels (below the complete automobile) of a relevance tree describing the materials used to fabricate an automobile.

3. Construct the first two levels of a relevance tree describing the habitability requirements of a moon base.

4. Take any of the second-level branches of your answer to Problem 3 and develop two levels of a problem-type relevance tree depending from that branch (i.e., the problems of providing for the habitability need).

5. Take any of the second-level branches of your answer to Problem 3 and develop two levels of a solution-type relevance tree depending from that branch.

6. Take any of the second-level branches of your answer to Problem 3 and develop four levels of a mixed problem–solution relevance tree; that is, the first level should consist of problems in meeting the requirement selected from the tree of Problem 3. Below these branches should go solutions to the problems; the next level should consist of the problems of implementing those solutions; and the last level should consist of solutions to the problems of implementation.

7. Construct a three-level relevance tree of improvements or changes needed in urban public transit. This may be a mixed-type tree if necessary. Place relevance numbers on each branch of the tree. (If this problem is worked on by a class or group, the Delphi procedure may be used to reach a consensus on the relevance numbers. In this case the panel estimate should be the mean instead of the median of the individual estimates, to preserve the property that the relevance numbers must sum to unity at each node.)

Problems *(continued)*

8. Develop a morphological model of the possible forms of wrenches.

9. Develop a morphological model corresponding to your relevance tree solution to Problem 5 above. Which of the two models appears better suited to this specific situation? Why?

10. Develop a morphological model of the problem of extracting food from the sea. For one configuration of your model prepare a normative forecast of the technologies involved. When might this configuration be needed? Will it be feasible by then?

11. Develop a mission flow diagram showing alternate paths for making the energy contained in unmined coal available in the home as electricity. For one path on your diagram that is not currently in use prepare a normative forecast of the technologies involved. What are the advantages of your chosen path over those currently in use? When might your path be feasible?

12. Devise an alternate path for Figure 11.5 that has the following properties: The recipient need not be in any specified physical location to receive the message, the message transmission must take essentially zero time, the method must permit a dialogue between the sender and the recipient, and a written text may be provided if desired. Develop a normative forecast of the technologies required if your alternative path is to be technically and economically feasible.

Chapter 12
Planning and Decision Making

1. Introduction

The preceding chapters have dealt with the methods of making technological forecasts. Their emphasis was largely methodological, and while some applications were mentioned, these were secondary to the discussion of methods.

The viewpoint taken in this book is that technological forecasting is an aid to decision making. Technological forecasts are not produced in a vacuum, nor are they produced for their own sake; if a forecast is not useful for making decisions, it is worthless. This is true regardless of the technical elegance of a forecast or the scope of the data behind it.

Each of the following four chapters will be devoted to a specific application of technological forecasting. The areas covered are those in which technological forecasting has been most widely used, although these are not the only areas in which technological forecasting is possible. The discussions will tend to focus on the unique problems of each application area. However, there is enough similarity among application areas that much of what is said can be applied, with the proper adjustments, to other areas as well.

The remainder of this chapter will discuss the features of planning and decision making common to all application areas. It thus provides an introduction to the chapters on specific application areas by placing them in a broader context. It also will provide a basis for extending applications of technological forecasting to application areas not specifically covered in the subsequent chapters.

2. The Purpose of Planning

One of the best-kept secrets of the planning profession is that planning has nothing to do with actions to be taken in the future. Instead, planning deals with actions to be taken in the present. Put another way, planning does not deal with future actions but, rather, with the futurity of present actions.

Nevertheless, the emphasis of planning is on the future. The entire purpose of planning is to achieve a preferred future, which may be a good future or simply the "least bad" future. In either case, planning is an attempt to apply rationality to the task of altering and shaping the future.

Shaping the future is something that is done in the present. By selecting one action over another, we alter the future to some degree. Shaping the future is thus inevitable. Even if we choose our actions at random, we cannot avoid shaping the future. We plan in order to select those actions in the present that will lead to a preferred future. Planning provides the link between a preferred future and a present action. Ultimately, it provides a reason for choosing one action over another: The chosen action leads to the preferred future.

How far in the future can this "preferred future" lie? Clearly, there is no fixed answer. One may engage in meal planning to the extent of buying a package of breakfast rolls today in order to eat them tomorrow morning, and in estate planning to the extent of establishing an annuity today that will be payable to one's children at a time perhaps half a century in the future. Clearly, planning may deal with almost any span of time.

Much discussion of planning involves "short-range" and "long-range" planning. These are often distinguished in terms of years. For instance, it might be asserted that short-range planning projects two years, five years, or some such number of years into the future, whereas long-range planning projects 20 years into the future. Distinguishing between short- and long-range planning on the basis of time span, however, is completely incorrect. The true distinction involves a combination of the lead time for actions to have their effects and the length of time until the "preferred future" comes into existence.

Consider a situation in which some preferred future is desired five years hence and suppose that the action that must be taken to achieve this future requires a lead time of five years to produce its results. The planner must take action right away if his desired future is to be achieved. Any delay in taking the action is tantamount to not choosing the "preferred" future. In this case planning for a time five years hence is a short-range plan.

Consider another situation in which some preferred future is desired three years hence and suppose that the action that must be taken to achieve this future requires a lead time of one year to produce its results.

This action, therefore, cannot even be taken for two years yet. In this case planning for a time three years hence is a long-range plan.

Is there any purpose to long-range planning? If action need not or cannot yet be taken to achieve the preferred future, why even be concerned with it? The reason is that when the time comes to take the necessary action, it may not be possible to do so; other events may have intruded to make the action impossible. Worse yet, the planner may have inadvertently taken some action, in carrying out a short-range plan, that precludes the action needed for the long-range plan.

Long-range planning, just as any other kind of planning, deals with actions to be taken in the present; it links these current actions with some preferred future. The essential feature of long-range planning is that it deals with present actions to *preserve* or *create* the *option* to take decisive, goal-oriented action at the time in the intermediate future when that action is necessary to achieve the preferred long-range future.

Another way to look at the distinction between short- and long-range planning is that in short-range planning the planner cannot wait for additional information; any decisions taken must be based on the situation as it then exists. Even though additional information might be available one-tenth of a "lead time" into the future, the planner cannot wait for it because the short-range plan requires action now. The long-range planner, in contrast, not only can but should wait for additional information. A good long-range plan is one that has some explicit built-in process for gathering the additional information that can be available before decisive action needs to be taken.

An implication of the possibility of additional information is that the planner in a long-range situation may not, and even need not, be completely certain of the specific action that will be taken when the time for it comes. There may be a need to create or preserve options for several alternate actions when the time for action arrives. The lead times for all these actions might not be the same or they may not yet be precisely known. Thus another aspect of a long-range plan is the need to gather information about the option that should be taken and to have this information available before the option with the longest lead time might have to be chosen.

There is, of course, no sharp boundary between short- and long-range planning. If the action needed to achieve some preferred future can be postponed by 24 hours, literally from today until tomorrow, this does not convert the situation from short- to long-range planning. From the perspective of the overall goal, if no new information can be obtained in the intervening 24 hours, the delay is simply procrastination. Thus a short-range plan need not be acted on right away; it is only necessary that no new information, change in resources, or change in external conditions be possible before the deadline for a decision imposed by the lead time for the action. Under such circumstances a plan is short range, and the decision is effectively made "right away." So long as something can change before the deadline for the decision, even if that deadline is only a few minutes off, the planning is effectively long range.

It is sometimes useful to talk of intermediate-range planning as well. Suppose that to achieve some preferred future we will need to exercise one of a limited set of options in ten years; however, to exercise these options, we will need some physical facilities that take five years to construct. Unfortunately, we do not yet know how to construct them. Thus our short-range plan may be to establish a R&D program to learn how to construct the facilities. The construction itself then becomes an intermediate-range plan. It is not really a long-range plan, because it does not of itself bring about the preferred future. It is instrumental in nature rather than being an end in itself; it derives its justification from the long-range plan, which is to lead to the preferred future. Not all situations require intermediate-range planning, but in some, however, the explicit recognition of plans that do not require immediate action, but which, on the other hand, do not lead directly to a preferred future, can be useful.

In summary, the purpose of planning is to choose actions in the present that will lead to a preferred future. A short-range plan is one that deals with actions which can no longer be delayed if they are to have their desired effect. A long-range plan is concerned with actions taken in the present to assure that at some later time it will be possible to choose a preferred future. This involves the preservation and creation of options for later choice. However, the distinction between short- and long-range planning does not alter the fact that both provide a link between actions taken now and some preferred future.

3. The Role of the Forecast

In the first chapter a number of purposes for forecasts were listed. These can be summarized in terms of taking actions to maximize the gain or minimize the loss from events beyond the control of the person using the forecast. This concept can now be placed within the context of planning.

A plan is analogous to a route chosen from a map. Where does the map come from? The forecast is analogous to the map; it describes the various possible events and the routes by which these events might be achieved. Thus the role of the forecast is to describe the alternatives open to the planner. The forecast informs the planner of the possible destinations, the routes there, and the relative distance or difficulty of each route. The forecast does not impose any specific choices on the planner; instead it merely defines them.

However, the analogy between a route and a plan and a forecast and a road map is not perfect. A traveller ordinarily expects to find towns and road intersections where a road map designates them, and to find that the distances shown on the map are accurate. Even the best of forecasts, however, may show towns and intersections where none exist and may fail to show others that do; in addition, the distances predicted by a forecast should not be expected to be as accurate as those on a roadmap.

Of course this has some correspondence to traveling as well: The experienced traveller will be aware that even the latest road map may fail to show all the roads under construction and all the closed bridges that will require detours. Nevertheless, we ordinarily expect a map to be much more accurate than a forecast.

Despite the potential inaccuracy, however, the role of a forecast is to define the possible alternatives. What possible destinations might one select? and what are the routes there? We expect a forecast to provide answers to these questions; that is, a forecast reduces the uncertainty about what lies ahead.

This, of course, is true of any forecast, be it economic, demographic, or technological. The important point about a technological forecast is that it provides information about technological alternatives, options, and consequences.

A technological forecast is important in the development of those plans that involve the development or deployment of new technology or the creation of technological options. It is also important in those plans that can be affected or altered by technological change. Plans can be affected by technological change in several ways:

1. A technological change may provide new ways of achieving objectives.
2. A technological change may render certain means of achieving objectives obsolete.
3. A technological change may render certain objectives obsolete.

The technological forecast then provides the background for selecting plans of any time horizon, short range or long range. It helps the planner identify those actions that must be taken in the short term either to achieve a particular technological capability by the time horizon of the plan or to create options which will be needed at future decision points.

4. Decision Making

Decision making can be defined as the act of choosing from among a set of feasible courses of action. This definition contains several elements, all of which are important.

First, decision making involves action. This action is either a change in the present situation or a deliberate choice to retain the present situation.

Second, the courses of action must be feasible; that is, decision maker must be able to carry them out. If there are no feasible courses of action, no decision is possible.

Third, there must be more than one course of action available. If there is only one course of action available, there is no decision to be made.

Fourth, decision making involves selecting one of several courses of action. This usually implies some limitation on the resources available. If all the available courses of action can be pursued simultaneously, no decision is needed. A decision becomes necessary only when more than one course of action exists and not all can be pursued simultaneously.

Finally, decision making is an act. This means that the choice among courses of action must not be passive or done by default. If the situation is allowed to drift until all courses of action but one are foreclosed, or until circumstances force the choice of a particular course, then there is no need for a decision maker.

The purpose of decision making is to change things or at least to make a deliberate choice not to change. But what is there to change? We can think of decision making in the context of an organization, although the same ideas can be applied to individual situations. A decision may change the objectives of an organization, its structure, the allocation of its resources, and the assignment of its personnel.

The objectives of an organization deal with the following questions: What function do we want our organization to perform in society? What kind of an organization do we want it to be? To the extent that we are able to shape society, what kind of a society do we want? Changing the objectives of an organization changes the answers to these questions.

The structure of an organization deals with questions such as these: Should we be more centralized or more decentralized? Should certain functions be more centralized while others are more decentralized? Should we integrate backward into producing some of the raw materials or supplies we use? Should we integrate forward into closer dealings with the final user? Should we integrate horizontally by expanding our present activities?

The allocation of resources within an organization involves questions such as the following: Are there elements of the organization that need a greater share of the available resources? Are there elements of the organization that are receiving a share of resources out of proportion to their contribution?. Are there elements of the organization that are bottlenecks, where additional resources would benefit the whole organization?

The assignment of personnel within an organization involves these questions: Do we need to move people with certain skills from one element of the organization to another? Do we need more people with a particular skill than we now have? Are we going to have more people with a particular skill than we can effectively use?

These questions always arise when an organization is faced with change. Our particular interest, however, is the way in which these questions arise in organizations faced with technological change. The role of technological forecasting is to allow the managers of such organizations to do a better job of answering these questions. Technological forecasts provide the decision maker with the information needed to make wise choices among the possible changes in an organization while there is still an opportunity to do so. Forecasts do not force particular choices.

5. Summary

The purpose of planning is to take into account the future consequences of present actions; in particular, it is to help select actions in the present in order to achieve a preferred future.

Forecasting provides the planner with an estimate of the kinds of futures possible and the specific actions that might lead to each of the alternatives. Technological forecasting in particular provides an estimate of the technologies that will become available so that the planner can take the greatest possible advantage of them and minimize their adverse impacts.

Any decision will change some aspect of the organization for which the decision is made. Decision making provides the ultimate justification for planning and forecasting, since the purpose of a plan is to lead to a set of good decisions.

Problems

1. What are the advantages of using a formal written forecast as the basis for a plan?

2. Consider a paper company that grows its own trees. These trees take 20 years to reach usable size. Does the company's planting of trees this year imply a firm commitment to use them for making paper in 20 years? Why or why not?

3. Discuss how the following features can be utilized to increase the flexibility of a plan:
   a. Inclusion of multiple options at future decision points.
   b. Inclusion of steps to collect additional information prior to needed future decisions.
   c. Inclusion of intermediate or lesser goals short of completion of the full plan.

   Is the inclusion of these features likely to make a plan more or less acceptable to a decision maker? Why?

Chapter 13

Technological Forecasting for Research and Development Planning

1. Introduction

The term *research and development* (R&D) is often used as though it described a homogeneous set of activities. This is not the case at all, and the differences among the activities lumped under this label are important. These activities do have some important similarities, however, and it is worth discussing them before taking up the differences.

All R&D activities share the characteristic of something being done for the first time. They involve pioneering in understanding, as well as experimentation and trial coupled with attempts to predict or explain the results on the basis of scientific laws. They often entail precise measurements and extensive calculations. In these characteristics they differ from most of the activities customarily excluded from the category of R&D.

However, one of the most significant common characteristics of R&D activities is uncertainty. The uncertainties of R&D programs can be classified into three categories: technical uncertainty, target uncertainty, and process uncertainty. Technical uncertainty involves the question of whether technical problems can be solved in order to achieve a desired outcome. There is no way to resolve this question except by carrying out the program. Target uncertainty involves the question of whether the technical characteristics specified for a R&D program are the correct ones; that is, even if a project meets its technical goals, will it perform satisfactorily in use? Process uncertainty involves the question of whether the human or organizational process of development will lead to a desired outcome. Even if technical goals are correctly chosen and are achievable, a project may fail for organizational or managerial reasons. Since the project is intended to do something that was never done before, there can be no certainty that the organization is correct.

The inherent uncertainty of R&D frequently gives rise to arguments that R&D is unplannable. However, this argument must be rejected. As with any other activity, present actions must be linked with preferred futures through planning. Nevertheless, one important point must be kept in mind: The specific outcomes of R&D projects cannot be prescribed in advance. The greater the technical uncertainty in a project, the less possibility there is of prescribing the outcome in advance; the greater the target uncertainty in a project, the less possibility there is of knowing what outcome to plan for; and the greater the process uncertainty, the less possibility there is of knowing how to obtain the planned outcome.

Technological forecasting can play two important roles in helping the R&D planner cope with these uncertainties. The first is to help set goals for R&D programs; the second is to identify opportunities that might be exploited. Before these roles can be discussed in detail, however, we must look at the differences among the various kinds of activities lumped together under the heading of R&D.

R&D can conveniently be divided into four categories: research, technology advancement, product development, and test and evaluation. The uncertainties in each of these are somewhat different, and their respective goals are quite different. Hence the role of technological forecasting can be discussed separately for each of the four categories. The remaining sections of this chapter will take up these categories separately.

2. Research

Research is a phenomenon-oriented activity; the researcher engages in it to gain new knowledge about some phenomenon in the universe. Research inherently involves a great deal of technical uncertainty. It is almost impossible to forecast the outcome of a research program except in general terms: "We will learn more about phenomenon *X*." Scientific breakthroughs are by definition unpredictable.

What, then, does technological forecasting have to offer? First, information about what phenomena are worthy of investigation, and second, information about what kinds of investigations will be possible. We will take these up in that order.

Phenomena will be investigated for a variety of reasons ranging from intellectual curiosity to the practical concerns of the organization sponsoring the research. It is primarily this latter group—the commercial and governmental organizations sponsoring research—that needs to know what phenomena should be investigated.

Such groups are involved with one or more technologies. In order to improve their functioning or to remain competitive in the marketplace, they must push their technologies to ever-higher levels of performance. This means that the devices embodying the technologies must face increasingly severe environments. These organizations should then be interested in knowing more about the phenomena that will be encountered in these environments. For instance, an aircraft company might be concerned with learning more about aerodynamics at higher speeds or altitudes. An automobile company might be interested in learning more about the behavior of fuels in internal combustion engines. A tire company might be interested in learning about the behavior of the materials it uses under conditions of high temperature and stress.

In all these cases the obvious motivation of the company is to learn how to build a product that will work in the expected environment. But the research *as such* will not lead directly to the desired product or product improvement. The output of the research effort is simply knowledge about phenomena. The company's designers can use the knowledge to design a better product, but the research effort itself may appear to have little relationship at all to specific products, especially to the company's present line of products.

If a company or some other organization is to remain competitive, it must have new knowledge to be used in the design of its products. But not just any kind of new knowledge will be helpful; the most useful new knowledge will be that about the phenomena a company's products will encounter or utilize as they deliver improved performance for the user. Here is an important role for technological forecasting in research planning. A forecast of the technology the organization employs or will employ can provide information about the knowledge that must be gained from research. A forecast of aircraft speeds and altitudes, for instance, can identify the range of phenomena that researchers for an aircraft company should be investigating. Similarly, a forecast of engine performance (e.g., horsepower-to-weight ratio, fuel consumption, etc.) can identify the range of phenomena that researchers for an automobile company should be investigating.

This use of technological forecasting applies to all organizations doing research that is intended to lead to improvements in their products or services. The technological forecasts cannot predict what will be learned by the research; however, they can identify research needs by identifying ranges of phenomena that will be encountered but for which knowledge is lacking. These knowledge "gaps" then provide the basis for research programs to make the improved products possible.

Some organizations, such as universities, really have no need to identify research "gaps" as the basis for planning research programs. Even so, however, they need to plan the research they will undertake in order to assure that they will be able to carry it out. At the very least they need to plan for the acquisition of new instruments and other research equipment. They may also need new buildings or other facilities for their research. If these are to be available when needed, planning must be done ahead of time. Furthermore, this argument applies to companies and government organizations as well. If they are to continue to do research, they must plan to have the equipment they will need when they will need it.

The technology of research instrumentation and equipment is just as susceptible to prediction as is any other technology. Forecasts of instrument technology, then, can be used by research planners to assure that their equipment will be adequate to serve the needs of their research staffs.

Figure 13.1 shows a plot of the accuracy of astronomical angular measure derived from the table in Appendix 3. The regression fit includes only

Figure 13.1. Accuracy of astronomical angular measure.

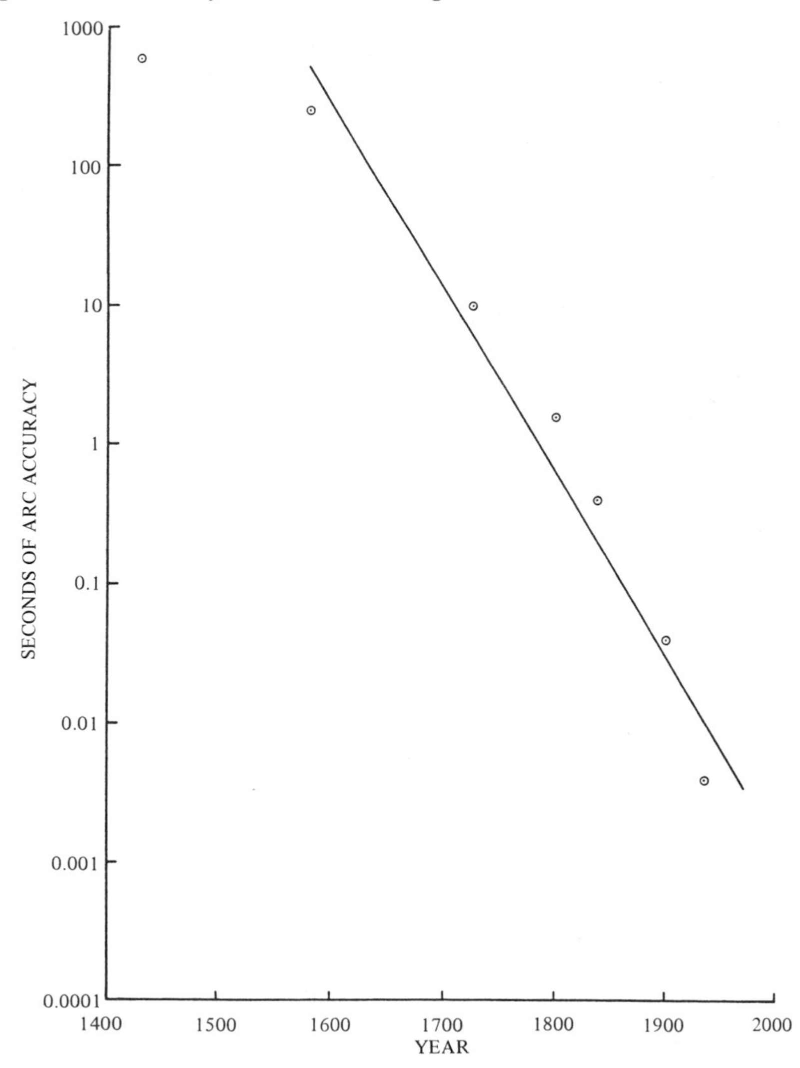

the data points starting with 1580; this is because the two earlier data points were made with simple quadrants and without the use of a clock. The data points from 1580 on represent the progress of modern astronomical instruments. As can be seen, the progress followed a regular pattern of exponential improvement from 1580 to 1935.

Appendix 3 contains data on the improvement of other scientific measurement technologies, each of which also shows considerable regularity in its improvement over time. This regularity allows the performance of the technology to be forecast with considerable confidence.

These examples are not merely isolated cases. As with most technologies, there are patterns of regularity in the improvement of measurement and instrumentation technologies. An important role of technological forecasting is to predict the future capability of instruments and equipment for scientific research. These forecasts can be used by research planners to determine the needs of their laboratories and to plan for the availability of equipment with the best available performance.

Research planners can utilize technological forecasting to help them link present actions with desired outcomes from their research programs. Technological forecasts can provide information about the specific phenomena that are worth investigating in order to provide knowledge needed for the improvement of particular technologies. In addition, forecasts of instrumentation technology can help research planners determine what kinds of research will be possible and assure that their laboratories are equipped to carry it out.

3. Technology Advancement

Unlike research, technology advancement is a problem-oriented activity. Research is concerned with learning more about a particular phenomenon; technology advancement, by contrast, is concerned with learning how to solve a particular problem or perform a particular function. Persons engaged in technology advancement are not limited to the use of only a particular phenomenon or to the use of only certain kinds of knowledge. Their concern is with solving a problem and they will draw upon any knowledge or phenomena that are useful.

Like research, however, technology advancement does not result in a specific product ready for use. The purpose of a technology-advancement project is to reduce technical uncertainty; it is intended to answer the question, Can it be done at all? This means that a significant portion of the time technology-advancement projects will reach negative answers. They may conclude that a particular problem cannot be solved at all or that at least it cannot be solved with the currently available techniques or knowledge. Such a result may not be popular with the sponsor, but it is a legitimate resolution to the technical uncertainty existing at the outset of a project.

Technology-advancement programs are not carried out for their own sake. The information about whether a problem can be solved by any of the available means is eventually intended to lead to a product for some user. There are two major issues facing the planner of a technology-advancement program: First, what level of performance will be required when the results of the program are ready to enter product development? And second, will the current technical approach be capable of delivering the level of performance? The first question can also be turned around to ask when will a specific level of performance be available for incorporation into a product? The second can be rephrased as, When will a new technical approach be required? Both these questions about the overall level of performance of a technology and the technical approach that will be utilized to achieve it can be answered by technological forecasts.

A forecast of the overall performance of some technology can help answer the first question. Here it is important to recognize that it is the overall technology, and not a specific technical approach, that is important. This forecast can be used either to identify the performance that can be expected by a particular time or to determine when a particular level of performance can be achieved.

Figure 13.2 illustrates the kind of information that can be provided by a forecast. It displays the ratio of energy output to energy input for nuclear fusion research projects. Since at least the mid-1970s this ratio has been growing exponentially. Figure 13.2 also shows a forecast of the possible range of performance of several new fusion devices in the early 1980s: These devices will "break even" before 1983, producing as much energy as they consume. An experimental device such as those shown in Figure 13.2 is still a long way from being a commercial fusion reactor. Nevertheless, the forecast provides the technology planner with an estimate of when the knowledge of how to achieve a certain level of functional capability will be available. Once that knowledge is available, the design of a commercial or operational device is possible.

Once the overall technology has been forecast, the technology planner can then look to forecasts of individual technical approaches. How long can the current technical approach keep up with the overall technology? If it will begin to reach its limits soon, is there a successor technical approach available? If so, when will its performance begin to compete with the current technical approach? If not, how much time is there to look for a successor?

Information about technical approaches is particularly important to technology planners. They need to know when current laboratory facilities will have to be updated or replaced, the kinds of skills that will be required in the laboratory staff, as well as the goals their specific projects should have so that they are neither striving after some level of functional capability that will be nearly impossible to achieve in the desired time,

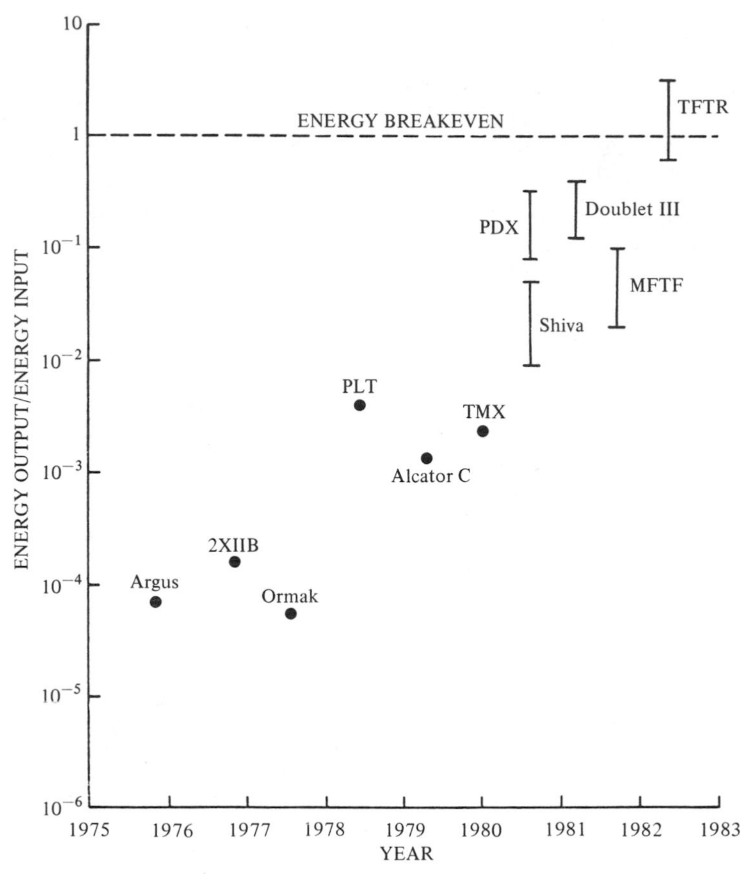

Figure 13.2. Energy characteristics of past and projected fusion power devices (courtesy of the Electric Power Research Institute).

nor pursuing something which is so easy that other competitive groups are likely to surpass it.

The technology planner needs to use forecasts of both the overall technology and of the current and potential successor technical approaches. These forecasts, both individually and in combination, indicate what performance to expect and whether it is time to switch to a successor technical approach. Once the appropriate technical approach is selected, the planner can determine what laboratory facilities and technical skills will be required to carry out a technology-advancement program.

4. Product Development

Like technology advancement, product development is a problem-oriented activity. However, its concern is not with learning how to solve a problem but, rather, with how to embody that knowledge in something a customer or user will find satisfactory. That is, technical uncertainty should already have been resolved by a technology-advancement program. Once the stage of product development is reached, there should be very little technical uncertainty left. The primary concern at this point is target uncertainty.

There are two aspects to target uncertainty: The first is what the customers or users will want; the second is what competitors will be offering for performing the same function. Technological forecasting can provide little or no information about customer or user desires. Its main function in product development is to provide estimates of the performance of competing devices or techniques that might be offered at the same time or shortly after a new product is to be marketed.

The selection of a design goal for a product-development project must represent a compromise between two conflicting objectives. On the one hand, the project should represent as small a technical challenge as possible. Aiming too high increases the cost of the project and increases the risk of technical failure. On the other hand, the project should be as ambitious as possible. Aiming too low increases the risk that a competitor will bring out a superior product shortly after the project is completed. The goal set for the project must therefore balance the risks of technical failure of the project and early obsolescence of the product.

A measure of technology obtained by regressing time on the technical performance of a series of devices was presented in Chapter 6. This technology measure can be used to estimate the likelihood that a particular level of functional capability will be reached by a particular time.

Use of the method can be presented most easily if the technology is described by only a single parameter. Ordinarily a regression fit of the data for that single parameter is made using the following equation:

$$Y_i = A + BT_i + e_i,$$

where Y is functional capability, T is time, e is a random error, and the coefficients A and B are obtained from the fitting procedure. Confidence intervals can be placed on the fitted equation, giving the range of Y that might be achieved at a specific time T. The technology measure reverses this process, using instead the equation

$$T_i = A' + B'Y_i + e_i,$$

where all the terms are as above except that A' and B' are the coefficients obtained by regressing T on Y. Placing confidence intervals on this equation gives the range of T values by which a given level Y might be achieved.

Figure 13.3 illustrates such a "reverse" regression: T has been regressed on Y. However, Figure 13.3 is plotted in the conventional form, with T as the abscissa and Y as the ordinate. Confidence intervals for 50 and 90% have been plotted; however, they must now be read horizontally instead of vertically. The 50% confidence intervals, for instance, give the

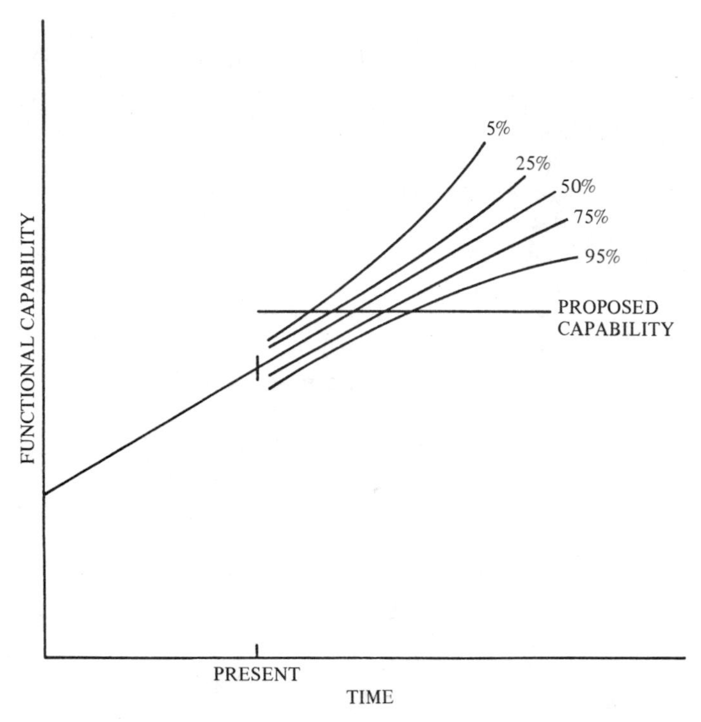

Figure 13.3. "Reverse regression" of time on capability to show the confidence limits on the time of achievement.

range of T within which it is 50% likely that some proposed level of capability Y_p will be achieved. Since the probability distribution is assumed to be symmetrical about the fitted line, the early 90% probability contour gives the time by which it is 5% likely that the value of Y will have been achieved. The late 90% probability contour gives the time by which it is 95% likely that the value of Y will have been achieved. The 50% probability contours likewise give the times by which it is 25 and 75% likely that the value of Y will have been achieved. Other likelihood values can be plotted if desired.

Note that this reverse regression can be carried out by regressing time on the logarithm of Y if functional capability has been growing exponentially. If the functional capability has been following a growth curve, the value can be linearized as described in Chapter 4, and time regressed on the linearized value. Therefore this approach is not limited to technologies that grow linearly.

When functional capability involves two or more technical parameters, the same procedure can be applied. It cannot be illustrated as easily as for the case of only one technical parameter; however, the standard error of the forecast, $\overline{S}(T \cdot Y_1, Y_2)$, can be computed as described in Appendix

1, and confidence bounds for T can be computed. These give the probability that a given value of the measure of technology will be achieved by a particular date; put another way, they give the probability that a particular tradeoff surface for the technical parameters will have been reached by a particular date.

The forecast of the likelihood of a particular level of functional capability being achieved by a particular date gives the risk that some competitor will have a product with that performance on the market. Once this estimate of the risk is obtained from the forecast, it can be balanced against the risk of technical failure of the development project, and a goal that properly balances these risks can be chosen. As usual, the forecast provides information on which to base a decision, but it does not force a particular decision. The R&D planner is still free to select the goal on the basis of estimates of the cost of failure and the cost of early obsolescence.

5. Testing and Evaluation

Testing and evaluation are also problem-oriented activities. They are concerned with determining whether some new device will actually perform as it should in the environment in which customers or users will employ it. There are three different groups that might be concerned with testing and evaluation: The developer of a device is interested in validating a design and determining whether changes are needed before the device is marketed or deployed for operation; the customer or user is interested in determining whether a device meets its performance specifications and whether it is suitable for use; and finally, a regulatory agency may be interested in determining the effectiveness or potential hazards of some newly developed device, material, or technique. Each of these three groups may then subject the item to testing and evaluation.

Testing and evaluation may be carried out by either of two means or by a combination of both. First, the new item may be tested in the actual environment in which it is to be used. Second, the new item may be tested in a simulated environment.

A simulated environment may be something like a wind tunnel, an altitude chamber, or a water bath. It is intended to create an artificial environment that matches the one in which the item will actually be used. A simulated environment may be cheaper to achieve than the real thing, especially if the environment is an "exotic" one such as outer space. A simulated environment is usually easier to instrument for measurements than is a real one. Finally, a simulated environment makes it possible to repeat tests under identical conditions. However, if an environment is to be simulated, it is necessary to specify it in detail, usually in terms of chemical and physical properties. Even then, there may be some question about the extent to which the "test chamber" actually matches the environment in which the item will be used.

Despite the problems of instrumentation and the difficulty of precisely repeating sets of conditions, testing in an actual rather than a simulated environment is done frequently. The major advantage of this is that it is not necessary to specify the environment in detail. In fact, if the environment cannot be specified in detail, there is no substitute for testing in the actual environment.

In dealing with testing and evaluation the R&D planner is faced with two different problems: The first is determining what will need to be tested; the second is determining what testing will be possible. It is necessary to know what will have to be tested in order to plan for adequate testing capability. It is necessary to know what kinds of instrumentation or other test equipment will be available in order to determine how the required testing can best be carried out.

A forecast of the technology to be tested can provide much useful information to the planner responsible for testing and evaluation. It can be used to determine by how much an existing test facility must be upgraded or extended, or even whether an existing facility can be upgraded to the degree required. If a new test facility is to be constructed, such a forecast can help avoid building one that will be outgrown before it is physically worn out. A forecast can also help avoid making a major investment in a facility to test an obsolescing technical approach and direct investment toward a facility that will handle a successor technical approach.

A forecast of testing technology, including both simulation technology and instrument technology, can help the planner determine whether testing should be done in a real or simulated environment and how it can best be done in either environment. For instance, it may be desirable to conduct certain tests in a simulated environment because repeatability is needed or because the risk of catastrophic failure of the test item is intolerable in the real environment. A forecast of simulation technology will allow the planner to determine whether it will be possible to conduct the testing in a simulated environment and, if so, the range of environments that may be adequately simulated. If testing is desired in the actual environment, forecasts of instrument technology will allow the planner to determine the extent and nature of the tests possible.

In some cases forecasts of both simulation technology and instrument technology will aid in selecting between the two or in selecting the proper mix of the two. For instance, if advances in simulation technology make it possible to reproduce accurately and cheaply an environment that can be reached only with difficulty in nature, a switch from testing in the real environment to testing in a test chamber might be appropriate. Conversely, if a forecast of instrument technology shows that instruments will be lighter or more rugged or have remote sensing capability, it may be possible to switch from simulation testing to testing in the actual environment.

Whether the testing is done by simulation or in the real environment, there may be several suitable alternative instrument technologies. A forecast of these technologies will allow the planner to maintain the option of using whichever turns out to be the best at the time; otherwise, certain actions might inadvertently be taken that preclude the use of one or more technologies.

Testing and evaluation is an important "last step" in the R&D process. Forecasts of the technology to be tested can be used to assure that the testing and evaluation capability will be available when needed. Forecasts of the technology available for test and evaluation, including simulation technology and instrument technology, can be used to assure that the testing and evaluation activities are carried out in a technically effective manner, at the lowest cost.

6. Summary

The R&D planner is faced with planning for something that is to be done for the first time, with no certainty that it can be done at all. In fact, the reason for attempting to do so is to determine whether it can be done. The R&D planner, like all other planners, must decide what to do in the present in order to enhance the likelihood of the future success of the R&D program.

Technological forecasts can be helpful to the R&D planner by identifying reasonable goals for projects, likely levels of the performance of products offered by competitors, and the capabilities of instruments used in R&D. Technological forecasts can also be helpful in identifying when specific technical approaches may have to be abandoned in favor of successor approaches.

The role of the technological forecast is to provide information for decision making. The forecasts do not force particular choices or courses of action on the R&D planner; however, the information they do provide can be combined with other information about the costs and benefits of various outcomes to determine what actions should be chosen in the present, and what options preserved for possible future choice.

For Further Reference

Beckerhoff, Dirk (1970). "Goal Systems for Government R&D Planning," *Technological Forecasting and Social Change* 1 (4), 363–370.

Glushkov, V. M., G. M. Dobrov, Y. V. Yershov, and V. I. Maksimenko (1978). "CMEA Experience in Multinational Forecasting of Science and Technology Advance," *Technological Forecasting and Social Change* 12 (2, 3), 111–124.

Martino, Joseph P. (1971). "Examples of Technological Trend Forecasting for Research and Development Planning," *Technological Forecasting and Social Change* 2 (3, 4), 247–260.

Sheppard, William J. (1970). "Relevance Analysis in Research Planning," *Technological Forecasting and Social Change* 1 (4), 371–380.

Problems

1. Describe the differences between R&D and the following activities: mass production, construction of a one-of-a-kind item such as a large dam, and testing the composition of some manufactured product.

2. The staff at the Lawrence Radiation Laboratory, University of California at Berkeley, pioneered in the development of liquid hydrogen bubble chambers for research in particle physics. The sizes of the bubble chambers built by this group are given in the following table [adapted from: Shutt, R. P., *Bubble and Spark Chambers* (New York: Academic Press), 1967].

Year	Size	Remarks
1954	$1\frac{1}{2}$ in. diam, $\frac{1}{2}$ in. deep	First liquid hydrogen chamber to show tracks.
1954	$2\frac{1}{2}$ in. diam, $1\frac{1}{2}$ in. deep	First metal-wall chamber
1955	4 in. diam, 2 in. deep	
1956	10 in. diam, $6\frac{1}{2}$ in. deep	
1957	15 in. diam	Replaced 10-in. chamber
1959	$72 \times 20 \times 15$ in.	Held 300 liters of liquid hydrogen

The functional capability of these bubble chambers is determined by the area (of the circular face, for the cylindrical chambers; by length times width, for the rectangular chamber). Calculate the area for each of the above chambers and plot them on an appropriate scale. Should the first chamber, which was built solely to demonstrate the feasibility of liquid hydrogen bubble chambers, be included in the determination of a growth law for bubble chamber area? What growth law seems to describe bubble chamber area? Is there any imminent upper limit to the size of bubble chambers?

3. For this problem use the data on the characteristics of integrated-circuit gates from Table A3.10 in Appendix 3. The speed–power product is a measure of the technology of these gates. Based on historical changes in performance, what would you recommend as the performance goal for a technology-advancement project that was to be completed and ready for product development by 1980?

4. Use data from the table in Appendix 3 giving the performance of civilian passenger transport aircraft (i.e., delete data on military aircraft, those designated by the letter *C* followed by a number). Use only those civilian aircraft that had the highest performance in terms of passenger-miles per hour at the time they were introduced. Regress time on the logarithm of aircraft performance. Interpret your results for a level of aircraft productivity of 3,300,000 passenger-miles per hour.

5. Using the table of electrical transmission voltages from Appendix 3, prepare a forecast of the range of transmission voltages likely during the period 1975–1985. What voltage range would you recommend for testing facilities to be constructed by electrical equipment manufacturers, with construction starting in the period 1970–1975? In retrospect, how well would your forecast have turned out? What were the actual voltages used?

Chapter 14

Technological Forecasting in Business Decision

1. Introduction

This chapter deals with the use of technological forecasting by business firms, that is, by organizations that are privately owned, produce goods or services for sale in a market, and are in competition with other firms providing similar goods or services, as well as with other firms providing alternative goods or services. The essential characteristic of a firm is that it must obtain income from the sale of goods and services in excess of the cost of providing them. Moreover, it must do this in the face of competition from other firms that are also trying to attract customers.

From the standpoint of society, an important aspect of the business firm is its expendability. If a business firm fails to satisfy its customers, it goes out of business. This is in contrast with organizations such as a post office, a municipal hospital, or a state university, which remain in business no matter how poorly they satisfy their customers. Thus the set of business firms in existence at any time consists primarily of those that are doing a good job of satisfying their customers.

From the standpoint of the individual firm, this expendability is viewed with much less equanimity. The failure of a firm results in direct losses to the owners and workers. Losses are also experienced by the suppliers who must find new customers, and by the customers who were still satisfied and who must now find new suppliers. Thus the failure of a firm is something the people involved with the firm wish to avoid.

If a firm is to stay in business, it must continue to satisfy its customers. It must anticipate changes in their wants and needs, as well as changes in the ways these wants and needs can be satisfied. There are a great many aspects to this anticipation of change. Technological forecasting is one tool the business manager can use to help anticipate such changes.

2. What Business Are We In?

An important element of business success is to identify just what business one's firm is in. All too often firms identify the nature of their business only in the most superficial terms: "We manufacture product *X*," or "We provide service *Y*." This identification is often too narrow, leaving the firm vulnerable to shifts in technology (as well as in other things). A firm's vulnerability can often be reduced by recognizing that the nature of its business can be identified more broadly than by simply a good or a service.

How can the business of a firm be identified? A positive answer to one of the following questions will identify the nature of the firm's business:

Do we perform a specific function?

Do we make a specific product or provide a specific service?

Do we utilize a specific process?

Do we utilize a specific distribution system?

Do we possess a particular set of skills?

Do we use a specific raw material?

At any point in time, of course, a firm makes one or more specific products, uses a particular distribution system, and so forth. The important issue is which of these items characterizes the firm? Which is the last item the firm would give up if it had to change? Which is the one that gives the firm whatever unique advantages it has in satisfying customers? The single feature that most characterizes the firm identifies the firm's business.

Some firms perform a specific function. Banks, for instance, are in the business of borrowing money from depositors and lending it to borrowers. They may adapt to technological and social change by using automated teller machines, providing bank credit cards, accepting electronic deposits from employers and electronic withdrawals from depositors, and so on. The exact form of the business may change, but the function remains the same.

Some firms make a specific product. Steel mills, for instance, sell steel to a variety of customers and over the past century have made steel by a succession of different technologies. A steel company may change its technology, its sources of ore, or the way in which it ships products. The product itself, however, remains the same.

Some firms utilize a specific process. Telegraph companies, for instance, send messages by using electrical signals that encode the information presented by the sender. Originally these companies transmitted text messages between individuals; later the bulk of their business was data of various kinds, such as stock prices and weather reports. Now the bulk of their business is becoming the direct transmission of data between computers instead of between people. While the nature of the data and the type of customer served have changed, the process has remained the same.

Some firms are identified with specific distribution systems. Mail-order firms such as Sears are perhaps the prime example of this type of firm. A firm identified with a specific distribution system may change the nature of the products it sells; it may change the type of customer it seeks; and it may change the geographical area it covers. The thing that remains constant is the use of a particular distribution system.

A firm may be identified with a particular set of skills. Firms specializing in law, engineering, architecture, accounting, and computer programming are examples of such firms. They are not identified with a particular raw material or a process; they may change their customers and the way in which they deliver services. The constant factor is the set of skills they offer.

A firm may be identified with the use of a specific raw material. Here we are not concerned with firms that sell a specific raw material to others, such as steel firms which sell steel bars or plates to automobile or refrigerator makers; these steel firms can be considered as identified with a product. Instead we are concerned with firms that utilize a specific raw material themselves. Wood products firms are an example of this identification with a raw material: They may switch their output from plywood to boards to chipboard as the market demand shifts; however, they are always concerned with finding something that can be made from wood. It is wood that is the identifying mark of the business, not the things into which the wood is converted.

Only the managers of a firm can decide what business the firm is in. Outsiders can never be certain what decisions will be made in the face of change: to stick with a product but make it from a different raw material or by a different process; to stick with a set of skills and seek different customers; or whatever. However, the managers of a firm do not have unlimited freedom to choose the nature of their business. This is constrained by past history and present circumstances. Often it is a matter of discovering what business the firm is in, rather than one of deciding. The managers must determine the aspect of the business that is the hardest to change, the one which would take the longest to change, or the one least profitable to change. Whether by decision or discovery, however, identifying the nature of the business in which a firm is engaged is one of the major responsibilities of its managers.

Once the managers of a company decide what business they are in, they are in a better position to evaluate the consequences of technological change. In some cases technological change can bring an end to the very business they are in. A frequently cited example is the destruction of the buggy-whip business by the automobile. A less drastic change may leave the basic business in existence but change the way in which it is conducted. For instance, steel companies have adopted a succession of steel-making techniques, each of which reduced the cost of production. Petroleum refineries have adopted a succession of new processes for extracting gasoline from crude oil, each of which reduced cost, produced less pollutants, "cracked" more crude into gasoline, or provided some other advantage over the older processes.

It is therefore extremely important for managers to be alert to technological changes that can destroy the business their firm is in. However, such dramatic changes are fairly rare. Lesser changes that alter the way in which the fundamental business of a firm is conducted are more common. These changes, if not anticipated, may leave the firm unable to compete effectively with other firms in the same business that did anticipate the changes properly. Hence such changes can be just as fatal as a change that eliminates the business itself. Thus the managers of a firm must also be alert to technological changes that leave the basic nature of the business intact but which alter some important aspect of the way in which it is conducted.

3. Business Planning for Technological Change

Once a firm's business has been determined, threats and opportunities from technological change can be more clearly identified. With regard to the business of the firm, a fundamental question is, Will this business continue to exist? If it will, and the firm wishes to remain in that business, the next question is, How will this business be conducted?

Technological change can alter the fundamental nature of the firm's business; more frequently, it can alter the way in which that business is conducted. To illustrate these possible impacts we will review some historical examples of technological change that altered important aspects of certain businesses.

Function. A firm may perform some function directly or it may sell a product that performs some function. In each case the firm is vulnerable either to a technological change that makes the function unnecessary or to one which performs the function in some other manner. At the end of the 19th century, for instance, the most common means of lighting homes in large cities was the gas lamp. By 1910, however, the gas light was on its way out. The thing that destroyed the gas companies, however, was not a cheaper way of producing gas or a more efficient way of distributing it; gas lights were replaced by a superior technology, electric lights. The function of lighting was performed in an entirely different manner. Gas, of course, is still used for heating. However, electricity is no longer used just for lighting but also for heating and for operating electrical and electronic appliances of all kinds.

Product. A product is sold to a set of customers who utilize it for some purpose. A technological change may allow some different product to be utilized for the same purpose. Transistors, for instance, have replaced vacuum tubes except in very specialized applications. The consumer who buys a radio really does not care whether it contains vacuum tubes or transistors, just so that it performs satisfactorily. The advantage of transistors was that they allowed radios to perform the same old functions, but more cheaply and in a smaller, lighter package. From the standpoint of the user, a transistor radio is a better radio, but still a radio. It is worth noting, however, that none of the firms that were leading manufacturers of vacuum tubes are now leading manufacturers of transistors. Moreover, most of the leading manufacturers of transistors did not produce vacuum tubes at all. The advent of transistors gave new firms an opportunity to take the market away from those firms that served the same function.

Process. In many cases a new process for manufacturing some product or providing some service can almost directly replace an old process. The advent of a sequence of new processes for making steel and for refining petroleum has already been mentioned. When these new processes were innovated, the firms that did not adopt them lost a share of the market or even went out of business because they could not compete with firms which did adopt the new processes.

These new steel-making and petroleum-refining processes replaced only one or a few steps in the total manufacturing process. The raw material remained the same, and most of the steps in the manufacture remained the same. In some cases, however, a technological change may affect the total process of manufacture. At one time in the electronics industry, for instance, one could distinguish among firms that made electronic materials, firms that made components out of these materials, and firms that assembled final products out of these components. A single firm might be involved in each of these steps, but the steps were still distinct and often carried out in different plants. With the advent of integrated circuits, however, the situation has changed. No longer are discrete components manufactured out of special materials and then assembled into products. The entire product may be formed as an integrated circuit by "depositing" layers of materials on a "substrate." By putting the layers down in a specific order and by masking and screening certain portions of the circuit between layers, thousands of individual components may be formed on a single circuit less than 1 in. in diameter. Producing a portable radio now may mean nothing more than attaching a speaker, a battery, and a tuning circuit to an integrated circuit that contains everything else needed for the radio. The sequence of steps of preparing specialized materials, making components out of them, and assembling the components has been replaced by a single process. As with transistors and vacuum tubes, most of the firms that are leading producers of integrated circuits simply did not exist before, and few of the firms which were once leading manufacturers of components are now active in producing integrated circuits. A technological change in a process allowed new firms to displace those that formerly dominated the market for electronic components.

Distribution. Technological change has altered the way in which firms distribute their products. The use of air freight has meant that high-value products can be delivered to a customer across the country on an overnight basis. While air freight costs more than other transportation on a ton-mile basis, its use may save enough to offset its cost. If regional warehouses near customers can be replaced by a central warehouse plus air freight, the size of the inventory needed can be reduced. This reduces storage costs, insurance costs, and the cost of capital tied up in the inventory. Firms failing to make use of distribution innovations such as air freight may find themselves at a cost disadvantage relative to competitors who do make use of them. Thus technological innovations that impact on distribution systems may affect a firm's competitiveness.

Skills. Technological change may alter the skills required by the workers in a firm. There are at least two ways in which this can occur: If the process for making a given product changes, the workers may need new skills for the new process; if the nature of the product itself changes, the skills needed to make it may change as well. An example of the first situation is the replacement of manually controlled machine tools with computer-controlled ones. With the former tools the skill of the operator was the critical element in making a part to specifications: The operator directly controlled the tool, measured the part during machining, and was responsible for the quality of the finished part. An automated machine tool performs these operations by itself. The operator, however, must be able to set it up, insert the proper program, and verify that the tool is working properly. The operator may actually need as much skill as the old-time machinist, but the specific skills needed are different. An example of the second situation is the replacement of mechanical cash registers with electronic ones. The firms that formerly made mechanical cash registers had large work forces of machinists, tool and die makers, and similar workers. When mechanical cash registers were replaced by electronic ones, a completely different set of skills was required in the work force. The skills that had provided the basis for the superiority of the leading firms were no longer of any use; electronic assembly skills were required. The firms involved had to alter not only their designs but their work forces as well.

Raw Materials. Technological change can cause a change in the raw materials used by a firm to carry out some process or make some product. Frequently this comes about through the replacement of a natural material by a synthetic one that is cheaper or better. The replacement of silk by nylon and the replacement of cotton by a whole host of synthetic fibers are examples. In other cases the change may not involve the raw material directly but the product itself, causing an indirect change in raw materials. As aircraft technology improved, wood-and-fabric construction was replaced by aluminum. Later, aluminum was replaced by titanium and by stainless steel for applications requiring temperature resistance.

Historically each of the major aspects of a business has been impacted by technological change. However, a firm may be impacted in two other aspects as well: management itself and support functions within the firm. Impacts on these are reviewed below.

Management. Technological change may impact on the way in which a firm is managed. The impact may arise directly from new management technology or it may arise from technology that allows more efficient management. New management technology has had a major impact on firms within the past several decades. The introduction of management science and operations research has reduced to routine calculations many of the decisions that formerly required management judgment. Examples include such things as setting inventory levels and reorder points in stockrooms and warehouses, selecting warehouse locations to minimize shipping costs, and selecting the numbers of clerks, service booths, and so forth to balance the cost of service and loss from waiting time. The introduction of a new technology, especially communications technology, may alter the way in which management operates. For instance, the introduction of the telegraph in 1851 allowed the centralized dispatching of trains. Previously, when a train was delayed, all the others that were to meet it or use the same track were held up as well. Only local control was possible. The local dispatcher would not dispatch a train until he was certain the track would be clear, through arrival of the delayed train. In general, improved communications allow a more central control of stocks of goods, of shipments from different locations, and of the movement of vehicles. This centralized control often (but not always) allows the total resources of a firm to be used more efficiently, since the need for safety stocks at many locations, or vehicles held in readiness at many points, for example, is reduced or eliminated.

Support. This, in a sense, is a catchall term for the activities a firm must carry out simply to operate at all. There have been many technological changes that impact the support functions of a firm. Changes in behavioral technology, such as aptitude tests, have had an impact on hiring practices. Computers have affected support functions such as computing payrolls, writing paychecks, storing and maintaining records, and the production of routine reports from stored data. The development of copying machines has impacted the handling and filing of correspondence, the maintenance of records, and the distribution of written communications within a firm.

These examples illustrate that technological change can impact all the aspects of a firm's business, including its management and support activities. An aggregate look at the impact of technological change on firms is also of interest. Consider the 100 largest firms in the United States in 1917. Fewer than half of these are still among the top 100 firms, and over a quarter of them are no longer in business. Central Leather was in 1917 the 24th largest firm in the United States, with profits larger than those of Sears Roebuck, and a net worth greater than that of General Motors. It failed to respond quickly enough to changes in shoe-making technology that required different kinds of leather: It was undercut by other companies that responded more quickly and is no longer in business. American Woolen did not respond quickly enough to the consumer demand for synthetic fibers; it is no longer in business. Executives of Baldwin Locomotive publicly and frequently expressed their doubts of the diesel locomotives replacing the steam locomotive. They were wrong, and Baldwin Locomotive is no longer in business. In 1940 Curtiss-Wright was the 28th largest company in the United States; it was larger than either Boeing or Lockheed and was a major producer of high-performance aircraft during World War II. Its managers decided that the jet aircraft had no future as a civilian vehicle; they were wrong. The company is still in business, but not only is it not among the top 100 firms, it is not even among the top 500.

A major problem that often arises in business forecasting is the core assumptions often established by a company (perhaps implicitly), biasing their forecasts. Core assumptions will be discussed in detail in Chapter 17. The problem is mentioned here as one that technological forecasters in business organizations should be aware of.

4. Secondary Impacts

The previous sections have focused on the direct impacts of technological change on a business firm. A technological change may alter the function performed by the firm or provide an alternate way to produce a product, and so on. However, each of these direct impacts may have secondary or tertiary impacts, which must also be taken into account.

Some of these have already been mentioned in passing. For instance, a technological change that affects the process by which a product is made may also require changes in the skills of the workers. However, the change in worker skills is a secondary impact, resulting from the prior change in process.

While these impacts are secondary, they can still be important to a firm. It is therefore just as important to anticipate them as it is to anticipate the direct impacts of technological change. This simply means tracing out the consequences of changes on a firm. If a technological change will alter some major aspect of a firm, what other changes will be necessary as a result? What additional changes will be necessary as a result of these other changes? And so on. The impacts of a major change in one aspect of a firm can ripple throughout the entire firm.

The task of a firm manager is therefore not finished when the initial impact of some anticipated technological change has been identified. The subsequent changes must be identified as well. These changes should be systematically traced out to be certain that as many as possible have been anticipated. By identifying the changes induced by the initial impact of technological change, a manager can reduce the degree of turbulence and confusion that would otherwise accompany a major change in some important aspect of a firm. In the long run, this may be just as important as keeping the firm in business by anticipating the initial impact of the technological change.

5. Summary

Failure to adapt to technological change can destroy a firm, or at best significantly reduce its size. The demise of a firm that fails to adapt to technological change has some social value, of course: It transfers resources out of the control of those who have demonstrated incompetence, into the control of those who have proved to be competent. Nevertheless, the demise of a firm can bring hardship to a great many people.

The role of technological forecasting in business planning is to prevent damage to firms by allowing them to anticipate technological change while they can still adapt easily. By forecasting technological threats and opportunities related to the major aspects of its business, management, and support activities, a firm can counter threats and capitalize on opportunities. Its managers will thereby demonstrate their competence and justify their continued use of resources; in addition, they will protect the interests of their stockholders, workers, and suppliers, all of whom might otherwise be hurt by technological change.

For Further Reference

Leporelly, C., and M. Lucertini (1980). "Substitution Models for Technology-Fostered New Production Inputs," *Technological Forecasting and Social Change* 16 (2), 119–142.

Patterson, William C. (1980). "Technological Trends in the Automobile Industry and Their Impact on Aluminum Usage," *Technological Forecasting and Social Change* 18 (3), 205–216.

Tsuto Kono, Tsutomu (1980). "Economic and Management Implications of Technological Advances in the World Steel Industry," *Technological Forecasting and Social Change* 16 (1), 33–46.

Utterback, James M., and Elmer H. Burack (1975). "Identification of Technological Threats and Opportunities by Firms," *Technological Forecasting and Social Change* 8 (1), 7–21.

Problems

For Problems 1 through 5, consider direct impacts from technology only. Do not consider the indirect impacts resulting from the direct impacts.

1. If controlled thermonuclear power were developed to the stage of practicality, which of the activities of an electric power company would be affected?

2. If the direct broadcasting of television programs from satellites to home receivers became practical, which of the activities of a television network would be directly affected?

3. If low-cost, transparent, weatherproof domes were developed that could be placed over existing cities, which of the activities of a clothing manufacturer would be directly affected?

4. If strains of bacteria suitable for the low-cost, large-scale extraction of specific minerals from ore were developed, which of the activities of a mining company would be directly affected?

5. If a practical and low-cost fuel cell were developed that could use fuels which are cheap and widely available, which of the activities of an automobile manufacturer would be directly affected?

For Problems 6 through 10, consider the indirect impacts resulting from the technological change referred to.

6. Identify the indirect impacts on a power company resulting from the direct impacts you identified in Problem 1.

7. Identify the indirect impacts on a television network resulting from the direct impacts you identified in Problem 2.

8. Identify the indirect impacts on a clothing manufacturer resulting from the direct impacts you identified in Problem 3.

9. Identify the indirect impacts on a mining company resulting from the direct impacts you identified in Problem 4.

10. Identify the indirect impacts on an automobile manufacturer resulting from the direct impacts you identified in Problem 5.

Chapter 15

Technological Forecasting in Government Planning

1. Introduction

Some activities of governments are very similar to the activities of other organizations. R&D planning in a government organization is hardly, if at all, different from R&D planning in other organizations. Similarly, when a government operates a business-type activity such as a post office, the planning required is very much like that in any other business. Hence there is no need to discuss those activities of governments that are similar to the activities treated in the preceding three chapters.

This chapter will consider several types of activity that are unique to governments and which therefore present planning problems different from those of other types of organizations. The role of technological forecasting in these activities is therefore unique and deserves specific treatment. The activities to be treated in this chapter are internal operations of government, the regulation of economic and social activity, the provision of public goods, and changes in the form of government.

2. Internal Operations of Government

Most government decision makers find that the objectives of their organization are established by law and are therefore beyond their control. Their budgets are determined by the legislature, and even the sizes of their work forces are fixed by factors over which they have little control. In many ways, therefore, their problems are different from those of private decision makers.

However, government decision makers are still faced with technological changes that may alter the internal operations of their organization. Thus even though their problems differ from those of private decision makers, there is still a need to utilize technological forecasts to anticipate these changes. We will consider two issues in which technological change may impact the internal operations of a government organization: centralization versus decentralization, and skills of the work force.

All government organizations are faced with the issue of the proper degree of centralization or decentralization. The advantages of decentralization are flexibility of operation, responsiveness to local conditions, and ease of access by the citizens who must deal with the organization. The advantages of centralization are reduced operating costs, more uniform policies, ease of access to higher-level decision makers, and ease of access to central files. Any organizational structure represents a balance between centralization and decentralization. In a well-run organization this balance reflects the relative costs of centralization and decentralization.

Technological change can alter the relative costs and thereby allow a shift in the balance between centralization and decentralization. Communications, for instance, can make it possible for remote offices to refer the "hard cases" to higher-level decision makers. This allows some of the benefits of decentralization while still retaining some of the benefits of centralization. Copying machines allow the replication of central files in remote locations, with frequent updates. This allows more decentralization, while retaining one of the benefits of centralization. A central computer can significantly reduce operating costs, thereby shifting the desired balance toward centralization. Remote access to that computer, however, can shift the balance back toward decentralization.

Planning for the future of an organization, in terms of locations, equipment needed, and so on, should be based on the technology to be expected during the period being planned for. To make commitments to specific locations and select personnel on the assumption of a static technology simply means that the organization will not have the best balance between decentralized and centralized operations. The agency will operate at costs higher than necessary and will not meet its legally required objectives as well as possible.

Decision makers in government organizations thus should use forecasts of the technology that will affect their operations, such as communications, recordkeeping, filing, document transmitting, and copying. These forecasts can help determine the proper balance between centralization and decentralization. If new technology permits a shift in the balance, planning for that shift ahead of time can allow the organization to carry out its legal responsibilities at minimum cost.

Regardless of the effect on the balance between centralization and decentralization, a change in technology may require different skills on the part of the work force. Some instances are obvious: The adoption of new equipment requires that users or operators have the skills required to utilize the equipment; however, more subtle effects are also possible. Changes in equipment in one part of the organization may require changes in skill in other parts as well. The installation of a computer, for instance, requires the addition of programmers, computer operators, and other workers with skills directly related to making the computer function. However, people in other parts of the organization may also have to acquire new skills. Their previous decision-making procedures may have been shaped by the kind of information that was available in the agency's files and the ways in which that information could be extracted. The availability of a computer means that information can be extracted in ways that were not possible before. Both the depth and breadth of the information potentially available may increase; new skills will then be required if the decision maker is to take advantage of this information.

The government agency planner is faced with making some combination of three choices when technological change occurs: hire people with the newly required skills, train existing workers to have the newly required skills, and find other uses for workers whose skills are no longer required. The planner is often severely constrained in his choice by laws and regulations intended to protect the agency worker against arbitrary action. A forecast of technology that will require the availability of new skills within the agency can allow the change to be made on an orderly basis, within the legal protection provided to workers, while maintaining both effectiveness and efficiency.

3. Regulatory Agencies

There are two different aspects to regulation by government agencies. One type of regulation is focused on a specific industry; it concerns itself with economics, that is, the rate of return to companies in the industry, prices charged by the industry, entry into the industry, and operations of the industry as they affect services provided to the customers. The other type of regulation is concerned with the safety of products, the safety of working conditions, and the safety of third parties not involved in the industry. This type may or may not be limited to a specific industry. The Food and Drug Administration is concerned with safety, rather than rate of return, but is restricted to a particular industry. The Occupational Safety and Health Administration, on the other hand, is concerned with a single subject, workplace safety, but in all industries. Planning for each of these types of regulation requires technological forecasting; however, since the intents of the two are different, they must be treated separately.

A major problem in the economic regulation of industry has been the failure to adapt to technological change. Examples are abundant: The Interstate Commerce Commission for years hindered the growth of the "piggy-back" transportation of truck trailers on trains because of the potential threat to over-the-road trucking. This arrangement would have saved energy, reduced the wear on the highways, and provided lower costs to customers. These advantages were not considered important, however, by comparison with the need perceived by the regulators to keep the trucking and railroad industries distinct. The Federal Communications Commission restricted the growth of cable television for years because of a perceived threat to commercial television networks. Many similar examples could be cited.

The central feature in these examples is the attempt made by regulatory agencies to freeze industries into a pattern based on a particular set of technologies. This attempt is aggravated by the concept that competition must mean competition between firms using different technologies; that is, trucking companies, railroads, and barge companies must compete against one another. The concept of a "transportation company" that could provide a mixture of services and which would compete against other companies providing a mixture of services is for some reason not acceptable.

Because of this insistence on forcing regulated firms to restrict themselves to single technologies, regulatory agencies have a serious problem. They must take technological change into account to prevent the extinction of firms that have been trapped into obsolete technologies by agency regulations.

Technological forecasts, then, must be used to anticipate technological changes that will alter the competitive posture of firms which are required to utilize only a single technology. Currently, when a technological change threatens to alter the competitive situation, regulators tend to prohibit it or restrict its introduction and diffusion, as in the examples cited above. It would be far more beneficial to society at large if these technological changes could be adopted rapidly, and this would require anticipating the changes and their impacts so the affected firms can adapt rapidly. The regulatory agency may even need to consider allowing a restructuring of the regulated industry instead of trying to force the new technology into a pattern based on earlier technologies, benefiting the public as well as the firms involved. Here again, technological forecasts could be of significant use in early planning for restructuring. Those identifying the changes to be expected can provide the basis for an orderly change instead of a crisis.

Health and safety regulations involve two types of problems: The first is anticipating threats to health and safety; the second is measuring them. Technological changes may alter the nature of these threats, as well as improve people's ability to measure and identify them.

Despite much of the concern voiced about threats to health and safety from "new" technology, the fact remains that most of the threats today arise from the massive use of comparatively old technologies. The combustion of coal and oil, for instance, contribute far more to air pollution than any new technology. Nevertheless, it makes sense to anticipate threats to health and safety when possible. In particular, given the limited resources available to agencies responsible for health and safety regulations, it may make more sense for them to get a head start on threats posed by new technologies than to expend their resources on the threats from old or obsolescing technologies. For instance, suppose a new industrial process is introduced that is expected to replace an existing process through a normal substitution. This new process may expose workers to chemicals not currently present in the workplace. A regulatory agency that has a forecast of the substitution might begin investigating the health hazards posed by the new process. This would have two advantages: It might make it possible to alter the process, at low cost, to eliminate or reduce the hazard; second, if the process cannot be altered, precautions can be taken against worker exposure before any harm is done. It may well make more sense to get a head start on the new problem, and eliminate or control it before it becomes serious, than to spend resources protecting workers against the problems of the declining technology and ignore the new technology until serious damage is done. By failing to take advantage of technological forecasts of potential new hazards, the regulatory agency is always spending its efforts dealing with large but declining problems, allowing small problems to grow into large ones owing to lack of advanced planning.

Forecasts of the technology for measuring and identifying threats can also be of value to agencies regulating health and safety. As an extreme example, consider a regulation or law that prohibits the presence of a dangerous substance at any level whatsoever (the Delaney clause, which bans food or drugs with any detectable contamination by known carcinogens, is a real-life example). Below some value the level of contamination may be irrelevant, in the sense that the additional risk it poses is insignificant. However, as the technology of detection improves, even these inconsequential levels of contamination will eventually be detected, and a "zero-contamination" law or regulation will require the banning of a product that is actually harmless. Regulations on the levels of allowable contamination should be based on known hazard levels and the capabilities of both contemporary and future detection technology.

But aside from the problem of "zero-level" regulations, there is still that of tailoring the regulations of exposure and contamination levels to detection capabilities. Forecasts of detection technology can be of great use in determining what is worth regulating and how. For instance, if a contaminant cannot be measured accurately even at a level that is unsafe, elaborate and expensive precautions against contamination may be the only possible measure of protection. Once detection technology is adequate, however, emphasis can be shifted from prevention, which might approach "overkill," to actual measurement, with the screening out of those items actually contaminated. Thus while present regulations may emphasize manufacturing procedures, technological forecasts may be used to determine when the emphasis should start shifting to actual measurement.

Whether regulatory agencies are concerned with economics or health and safety, the nature of problems they face will be altered by technological change. Forecasts of technological change can make it easier to be "on top of" upcoming problems before they become serious.

4. Public Goods

One function of governments is to provide so-called "public goods." These are goods (or services) that by their very nature cannot be made available only to specific individuals but, if made available at all, must be available to everyone. There are three general categories of such public goods: protection such as law enforcement or national defense, infrastructure such as highways, and safety facilities such as lighthouses and air traffic control. We will briefly discuss the use of technological forecasting in each category.

The applications of technological forecasting in national defense are already well known and need not be described here. However, law enforcement is an important area of public protection in which technological forecasting is not as widely used as is desirable. There are two ways in which technological change affects law enforcement: The first is the creation of new types of crime; the second is the creation of new ways of preventing or detecting crime.

Many kinds of technological change increase the criminal's scope of activity. The mobility provided by the automobile certainly aided the bank robber during the 1920s and 1930s. The adoption of telecommunications by the police eventually offset this mobility to some extent. However, the computer has created new opportunities for the bank robber, some involving the direct alteration of bank records, others involving the penetration of data links between banks and the insertion of false information. The importance of computer crime is revealed by the fact that the average nonelectronic embezzelement amounts to about $23,000, while the average computer fraud amounts to over $400,000.

It may be difficult to forecast just how criminals will exploit a new technology. If the manner of exploitation were obvious, it would often be designed out of the technology in the first place. The vulnerability of computers in banks is largely due to the failure to consider the possibility of computer theft. However, forecasts of technological change can often be the first step in identifying the potential for new crimes.

A forecast of the use of computers in banks, for instance, might be the first step in predicting the scope of computer crime and in planning measures against it. Thus, while a technological forecast may not provide much information about the exact nature of future crime, it can provide information about what new technology might be vulnerable to criminal exploitation or available to aid criminals.

Much of the thought given to technological measures against crime focuses on so-called "scientific crime detection." This generally involves technical means for connecting pieces of physical evidence with specific individuals. Fingerprints, blood typing, paint matching, and so forth are now widely used for such purposes. Technological change can still improve the capabilities of scientific crime detection. The increased sensitivity and accuracy of measuring instruments will continue to contribute to this important aspect of law enforcement. Forecasts of detection and measurement technology can help police officials keep their crime laboratories up-to-date.

Other technological changes also aid law enforcement, however. The use of telecommunication has already been mentioned. Computer files of criminal characteristics and behavior are becoming widespread. The West German police, for instance, have combined a central person index, a fingerprint-identification system, a criminal-activities data base, a documentation directory, and a "persons–institutions–objects–things" data base into a computerized person-tracking system. Burglar alarms of varying degrees of sophistication are widely available. The use of radars to catch speeders is now commonplace. Forecasts of these other technologies that aid law enforcement can be used to plan the acquisition of necessary equipment, identify new skills in which law enforcement personnel must be trained, and identify necessary alterations of laws to allow new types of evidence or to protect the public from the unwarranted use of a new technology.

Governments often provide social infrastructure such as highways, flood-control dams, bridges, and tunnels. They also take responsibility for dredging rivers and harbors, digging canals, and maintaining waterways. This infrastructure is presumed to provide benefits to the public at large, as well as to the direct users. Land near highways, for instance, often increases in value because of the availability of better transportation.

As with any investment, the planners and builders of infrastructure may make either of two mistakes: They may overbuild or they may underbuild. Either error results in unnecessary costs to the public as well as to direct users. In a competitive business errors committed by a single company are not particularly serious: If a company underinvests, some other company may take up the slack; if a company overinvests, the owners lose their own money, not anyone else's. With publicly funded infrastructure, however, the public at large pays for the mistakes of the planners. Thus it is extremely important that the planners responsible for selecting the level of investment in infrastructure take technological change into account. Such change may alter the level or type of construction needed.

For over a decade there has been discussion of an alleged tradeoff between transportation and communication. This has been expressed in terms such as, How much would videophone rates have to be reduced in order for the videophone to substitute for (some or all) business travel? In the past this question has not been particularly meaningful. Travel itself was so cheap and provided so many advantages in terms of "body language" and informality that even free videophone might not have been an acceptable substitute; however, the cost of travel is increasing as the real cost of energy rises. With the advent of satellites the cost of video conferencing is decreasing, and the quality increasing. For the first time the question of a tradeoff may be meaningful. What does this mean in terms of infrastructure? The public authorities responsible for highways and airports might be well advised to incorporate forecasts of telecommunication technology into their plans. The level of construction required might well be considerably less than that which would be appropriate were there no tradeoff between communications and transportation. Conversely, a forecast of some technological change that reduced the cost of transportation might serve as the basis for increasing the construction of highways and airports to reduce the cost of congestion.

Some technological changes alter not only the level of infrastructure needed, but also the type. For instance, extensive use of ground-effect machines for carrying cargo on inland waterways (lakes and rivers) would alter the type of infrastructure required (see Problems 2–5 at the end of this chapter). Cargo carriers would no longer unload at docks at the water's edge; instead they would require ramps that would allow them to leave the river or lake and travel to a warehouse. A forecast of the use of these carriers would be useful in selecting the proper mix of docks and ramps to be built.

Lighthouses are a classic example of safety facilities that must be built by governments, it being allegedly impossible to collect fees from the ships that utilize the light from the lighthouses. This is an issue we need not become involved in. It is sufficient to note that governments do build and maintain lighthouses, air traffic control systems, harbor traffic control systems, and so on. The planners for these facilities are faced with two issues: What functions must these facilities serve and what technology will be available to serve them: A look at the history of aerial navigation will serve to illustrate these two issues.

Initially, commercial aircraft flew only in the daytime because of the difficulty of navigating at night. The first nighttime navigation aids were beacon lights similar to the lighthouses long used at sea. Chains of these beacons were established along routes between major cities; however, they were ultimately unsatisfactory because they could not be seen by aircraft flying above clouds or in bad weather. That is, when they were most needed, they were not usable. The next major step was the "radio beam," an arrangement under which an aircraft could follow a specific track in the sky by properly matching signals from several transmitters. These beams in effect laid out aerial routes that could be followed despite bad weather or clouds. The problem with these routes was that they were narrow, leaving most of the sky unused and eventually severely restricting the traffic capacity of the system. Two later developments, radar and directional beacons, eliminated the need to have all aircraft follow the same narrow tracks.

In this example we see that the function to be served changed over the years. The air traffic control system first had to meet the needs of individual aircraft flying on clear nights (they could not fly at night in bad weather anyway). Later it had to meet the needs of aircraft flying in bad weather, and, ultimately, those of large numbers of aircraft flying in clouds or bad weather. The system was originally intended to allow aircraft to find their destinations; later its function was expanded to keep aircraft from colliding with each other. Thus the function to be served became more difficult and broader in scope.

Similarly, the technology changed over the years. Visual means of navigation were abandoned for electronic means, and the electronic means, in turn, were refined and extended in performance. This refinement is still going on, with computer tracking replacing the manual tracking of aircraft on their assigned routes.

The planners of a network of safety equipment such as an air traffic control system need forecasts of the technology available to the users of their system. For instance, forecasts of aircraft speeds, altitudes, and numbers define the performance that will be required of an air traffic control system. Forecasts of the technology that will be available for the system, on the other hand, define the system's possible performance. Technological forecasts can thus be used to avoid building systems that will not meet the needs of the users (e.g., chains of lighted beacons which are invisible from above the clouds) and to start the installation of systems that will (e.g., when will the system need to be usable above a certain altitude?). Similarly, forecasts of capability will allow planners to avoid excessive investment in obsolete equipment when something better will be available.

5. Changes in the Form of Government

The previous sections have dealt with technological changes that bring about changes within a fixed form of government. Here we consider technological changes that make changes in the form of government possible, desirable, or necessary. Two examples will illustrate the ways in which technology has brought about changes in the form of government.

In large cities people (at least for the most part) used to live, work, shop, and so on in a single district; the interests of individuals were identified with the interests of the districts in which they lived. The appropriate form of city government then involved the election of representatives from these districts. Aldermen were supposed to represent the interests of all those living in their ward, involving not only issues relating to their constituents as householders, but as shopkeepers, employees, students in public schools, users of municipal services, and so on. In general, it was the same set of people who appeared in all these roles within a district.

The structure of cities has been changed since then by improvements in communications and transportation. Now individuals may live in one district, work in a second, shop in a third, while their children attend school in a fourth, and so on. It is no longer the case that a given law or regulation will have uniform effects on the residents of a district; a measure may help some people as householders but hurt them as workers. These changes mean that a city government based on the representation of geographic districts no longer meets the needs of the citizens. In many cities, therefore, power has shifted from the city council to the mayor. This shift is a reality even though it may not be reflected in the city charter. Only the mayor really represents the interests of the entire city, and only the mayor, then, fully represents the people whose interests are distributed throughout the city.

The second example involves advances in managerial technology. Techniques of operations research, systems analysis, computer modeling, and so on offer the possibility of reducing many political issues to technical ones. In the past different political groups may have proposed different courses of action with regard to some issue. The problem was to determine whether they differed on the desired outcome or merely the means with which to achieve it. The advent of computer models of the economy, for instance, appears to separate the two issues. Congress can determine as a policy matter what the desired outcome should be. An agency using a computer model can then be assigned the task of achieving that outcome in the most efficient fashion. This is an example of the kind of change in the form of government that can take place when technology makes it possible to separate the technical issues from the political ones.

However, a warning is in order here. In the late 1800s public utilities were often regulated directly by the city council or state legislature. Even as mundane a matter as changing the rates a utility could charge required a new ordinance or law. Many people of that time, under the concept of "good government," argued for the regulation of utilities by appointed commissions of experts rather than directly by legislators. They pointed to examples of predatory behavior exhibited by legislators, bribery on the part of utility companies, and the mixing of political and technical issues. The "good government" people won the day and obtained regulation by commissions rather than by legislatures. It is not clear whether the results have been completely satisfactory. Many genuinely political issues (such as having one class of users subsidize another) have been decided by utility commissions as though they were merely technical issues.

Despite this warning, it is clear that changes in social technology do permit and encourage changes in the form of government. Forecasts of the appropriate technologies can allow such changes to be made in a more considered manner, rather than as a hasty response to onrushing events. In the same way, forecasts of other technologies can be used to identify in advance possible impacts on the form of government. Possible changes in the form of government can be considered carefully while there is ample time, rather than undertaken in haste.

6. Summary

Changes in technology can lead to changes in the operations and form of a government. The types of decisions undertaken by government planners in response to changes in technology are somewhat different from those facing decision makers in nongovernment organizations. Despite this difference, the use of forecasts of technological change can provide the same benefits as in other types of organizations. When changes are identified in advance, it is possible to adapt to them smoothly and routinely. Thus planners and decision makers in government should make use of the forecasts of those technologies that will affect either their activities or the manner in which they are to carry them out.

The problem of core assumptions is also worth mentioning. (These will be discussed in detail in Chapter 17.) Technological forecasters in government organizations should be aware of the possibility that the core assumptions made by their organization, often implicitly, can seriously bias forecasts. It is important to identify and validate the core assumptions that are often tacitly accepted throughout an organization.

For Further Reference

Anderson, Mark W. (1978). "The Institutionalization of Futures Research in the U.S. Congress," *Technological Forecasting and Social Change* 11 (4), 287–296.

Crickman, Robin, and Manfred Kochen (1979). "Citizen Participation Through Computer Conferencing," *Technological Forecasting and Social Change* 14 (1), 47–74.

Singh, Manohar (1978). "Technological Forecasting Activities in Selected U.S. Federal Agencies," *Technological Forecasting and Social Change* 11 (4), 297–302.

Problems

1. In the past city governments have been organized on the basis of functional departments, such as a fire department, a department of building inspection, a department of sanitation, and so on. Each department kept records on those aspects of the city and its residents that pertained to its operations or functions. Thus the records pertaining to any one person could be scattered through many departments. With the use of computers, however, it is now possible to consolidate records so that all those pertaining to a specific person are combined. What might be the long-term impact of this capability on the structure and organization of city government?

The four problems given below are based on hypothetical forecasts contained in the following table. This table describes four stages in the development of ground-effect machines (GEMs). These devices obtain lift by accelerating a stream of air downward, much like a helicopter; however, this lift is increased by allowing the air stream to interact with the surface of the earth (''ground effect,'' hence the name). In effect, they can ''hover'' a few feet off the earth while moving forward; they do not need to travel over a prepared surface. Any reasonably smooth surface (even boulders smaller than 1 ft in diameter) makes a satisfactory ''roadway.'' They can also travel over mud, swamp, ice, snow, or open water with equal ease. Because of lack of friction with the surface, speeds of 50 mph are reasonable for current normal use, and speeds of up to 100 or 150 mph will be technically feasible in the future.

Stage	Length (ft)	Gross weight (tons)	Cargo (tons)	Unrefueled range (miles)	Cost per ton-mile
1	100	80	20	200	Competitive with motor trucks
2	220	400	100	500	Competitive with railroads
3	780	5,000	1,000	4,000	Competitive with cargo freighters
4	2,500	50,000	10,000	10,000	Competitive with medium tankers

2. When the development of GEMs reaches stage 1, what problems will be presented to road-building agencies? Agencies regulating commerce and transportation companies? Agencies regulating water-borne traffic? What problems will require the joint action by two or more of these agencies?

3. When the development of GEMs reaches stage 2, what additional problems will be created for the agencies listed in Problem 2? How does the magnitude of these problems depend on the decisions made at stage 1? How could these problems be reduced by taking into account, at stage 1, a forecast of the fact that GEMs will eventually reach stage 2? What problems will be presented

Problems *(continued)*

to agencies responsible for international commerce and for the merchant marine, the U.S. Department of Transportation, and the Internal Revenue Service?

4. When the development of GEMs reaches stage 3, what new problems will be presented to the agencies discussed in Problems 2 and 3?

5. When the development of GEMs reaches stage 4, what new problems will be presented to the agencies discussed in the preceding problems?

Chapter 16
Technology Assessment

1. Introduction

There have been numerous definitions of the term *technology assessment* (often abbreviated TA). One of the best is the following definition by Joseph F. Coates:

> [TA is] a class of policy studies which systematically examine the effects on society that may occur when a technology is introduced, extended or modified. It emphasizes those consequences that are unintended, indirect, or delayed (Coates, 1974).

The idea behind TA is that technology does affect society and its effects should be examined before it is deployed. Karl Marx was one of the first to express the notion that technology determines society; however, his notion was quite deterministic. "The hand-mill gives you society with the feudal lord; the steam-mill gives you society with the capitalist." In this, as in virtually everything else he wrote, Marx was wrong. The English census of 1086 AD counted a total of over 5000 water-driven mills in England alone. At that time England was still a feudal society. On the other hand, many mills owned by capitalists at the beginning of the industrial revolution were driven by hand or by wind power before steam engines were adopted; moreover, the hand mill itself can be traced back thousands of years to ancient Sumer, a society that can best be described as socialist. Thus there is no rigorous deterministic link between the nature of a technology and that of the society using it.

This does not mean that technology has no effect on society. On the contrary, the whole purpose of technological change is to produce social change, lifting burdens, lightening loads, easing work, and making life more pleasant. The impetus for TA arises, then, not from the fact that technology has social consequences, but because some of these consequences are unforeseen and unintentional.

2. Some Historical Cases

Even though there is no deterministic link between technology and social change, new technology has often brought about social changes that were unanticipated and unintended by the sponsors of the technology. Here we will review the consequences of the automobile, the railroad, and the mechanization of agriculture. To provide a framework for analysis we will utilize the dimensions of the environment introduced in Chapter 3. Not all the dimensions will be relevant in each case, but using them as a checklist provides a means of assuring completeness in the search for consequences, as well as an analytical scheme that allows for comparisons between different technologies.

The Automobile. The history of the modern automobile can be traced to the German inventors Carl Benz and Gottlieb Daimler, who independently devised vehicles powered by gasoline engines. The basic design of the modern automobile was essentially fixed in 1891. In that year E. C. Levassor, of the French firm of Panhard–Levassor, designed an automobile using Daimler's patents.

Early inventors of the automobile, such as Benz, visualized it as completely replacing the horse and revolutionizing transportation. However, for the first decade and a half of their existence, automobiles were primarily toys for the wealthy. In 1908 Henry Ford brought out the Model T, which was the first successful mass-owned automobile. While Ford is often praised for introducing mass production to the automobile industry, this was not an end in itself. Ford's basic goal was an automobile that would be cheap enough to be owned by the average person, rugged enough to operate successfully on the primitive roads of the day, simple enough for anyone to operate, and inexpensive to maintain. So Ford set out to realize the original dreams of Daimler and Benz on the automobile; that is, he deliberately set out to produce social change.

The automobile did in fact achieve the social change Ford and the earlier inventors hoped for. It replaced the horse on a worldwide basis. However, it had many consequences beyond this first-order, intentional one. The secondary consequences, particularly the unintentional and unanticipated ones, are the concern of TA; hence a few are worth examining.

In the technological dimension it might be expected that widespread use of the automobile would lead to inventions directly related to its manufacture and use. However, as shown in Table 16.1, an enormous increase in the rate of inventions in the petroleum industry can be attributed directly to the need for new technology in producing gasoline from

Table 16.1. Petroleum-Refining Patents Issued

Year	Number of patents	Year	Number of patents
1860	10	1915	91
1870	24	1920	248
1880	13	1925	831
1890	20	1930	646
1900	18	1940	461
1910	38	1950	237

crude oil. The economic dimension is shown in Table 16.2, which summarizes the size of the automobile industry in terms of establishments, sales, payrolls, and employees. The social dimension of the impact of the automobile can be seen from the Table A3.23 which shows the number of automobile registrations per year. By 1979 there was one passenger automobile in the United States for every two persons. The political dimension of the automobile's impact can be seen in part from Table 16.3, which shows state and local highway debt. This debt averaged between one-tenth and one-fifth of the total state and local debt for the period shown. The ecological dimension of the automobile's impact is illustrated by the fact that by the late 1960s, before pollution reduction was emphasized, 60% of the air pollution was emitted by automobiles. These dimensions could be examined in more detail, and information from other dimensions added, but all these examples are sufficient to illustrate that the automobile had impacts on society far beyond those intended by its inventors. These unintended impacts are the concern of TA.

Railroads. The intent of the inventors of the railroad, and of the people who built them, was to provide cheap and reliable transportation. This they certainly achieved. However, the railroads also had consequences that were unforeseen and unintended by their inventors and sponsors. We will briefly examine some of these.

Table 16.2. Economic Impact of Motor Vehicles—1977[a]

	Establishments (thousands)	Sales (millions of dollars)	Payroll (millions of dollars)	Employees (thousands)
Auto dealers	139	149,952	10,654	1,115
Gas stations	177	56,468	3,830	673
Auto wholesale dealers	39	147,112	5,148	423
Repair, parking, rent	200	21,576	4,456	483
Manufacturing	4.2	36,880	15,996	873

[a] *Source*: *Statistical Abstract of the United States*, U.S. Bureau of the Census, Washington, DC, 1978.

Table 16.3. State and Local Public Highway Debt

Year	Debt (millions of dollars)
1950	4,436
1955	9,658
1960	13,166
1965	15,316
1970	19,008
1975	23,801
1980	25,117

The technological dimension of the impact of the railroad is indicated by Table 16.4, which shows the number of railroad patents issued from 1837 to 1950. Railroads drew the attention of a significant fraction of the nation's inventive talent. The economic dimension is shown in Table 16.5, which shows the gross capital formation for railroads and for all U.S. industry. During the era of railroad construction the railroads drew on a significant fraction of the total U.S. investment capital. The social dimension of the impact of the railroad may have been its most important impact. In a very real sense, railroads provided the glue that held the United States together. Even in George Washington's time there were separatist movements in the trans-Allegheny West. For the nation to survive it had to be possible for trade and travel to be conducted over long distances, at tolerable cost. When agricultural or mineral commodities were hauled by wagon, the transportation costs ate up the profits for distances over about 150 miles. First the canals and then the railroads made it possible for the United States to exist as an economic and political entity. Without railraods, population and commerce would have been constricted to narrow ribbons along suitable waterways; separatist move-

Table 16.4. Railroad Patents Issued

Year	Track	Nontrack	Total
1837	4	11	15
1840	4	21	25
1850	3	27	30
1860	11	107	118
1870	48	278	326
1880	45	531	576
1890	242	1526	1768
1900	133	954	1087
1910	453	1438	1891
1920	284	1147	1431
1930	192	1038	1230
1940	53	602	655
1950	39	363	402

Table 16.5. Average Gross Capital Formation (Billions of Dollars)

Year	Total U.S. industry	U.S. railroads	Percentage of total (%)
1873	1.47	0.26	17.7
1878	1.72	0.15	8.7
1883	2.18	0.59	27.1
1888	2.56	0.24	9.4
1893	3.06	0.27	8.8
1898	3.48	0.11	3.2
1903	4.91	0.24	4.9
1908	6.42	0.59	9.2
1913	6.94	0.54	7.8
1918	16.7	0.50	3.0
1923	16.2	0.80	4.9
1928	19.2	0.82	4.3
1933	6.68	0.20	3.0
1938	15.9	0.38	2.4
1943	28.7	0.58	2.0
1948	47.9	1.05	2.2

ments, centered around major river basins, would have been inevitable. But the railroad could go anywhere and haul goods more cheaply than competing means. Thus the existence of a transcontinental United States may be one of the most important social impacts of the railroad. The cultural dimension of the railroad's impact was also important: During the era of railroad construction newspapers and magazines were filled with accounts of railroad activities: projects, speed records of trains, accidents, and the profits of the companies. Railroad lore appeared in songs, poems, and speeches. The ''spirit of the times'' was embodied in the image of the railroad—the Age of Steam.

The Mechanization of Agriculture. The intent of the inventors and manufacturers of agricultural machinery was to reduce the amount of human labor involved in producing farm products. They succeeded in their objective, as shown in Table 16.6. The amount of labor required to

Table 16.6. Productivity of Farm Labor

Year	Wheat[a]	Corn[a]	Cotton[b]
1800	373	344	601
1840	233	276	439
1880	152	180	318
1900	108	147	280
1920	87	113	269
1940	47	83	191
1950	28	39	126

[a] Person-hours per 100 bu.
[b] Person-hours per bale.

Table 16.7. Value of Farm Implements

Year	Value (millions of dollars)	Farms (thousands of dollars)	Average ($)
1850	152	1,449	104
1860	246	2,044	120
1870	271	2,660	102
1880	406	4,009	95
1890	494	4,565	108
1900	750	5,737	131
1910	1,265	6,406	197
1920	3,595	6,518	550
1930	3,302	6,546	506
1940	3,060	6,350	482
1950	11,216	5,648	2,000

produce certain crops has been reduced by a factor of 10 or more. However, there were other consequences of the mechanization of agriculture, some unintentional.

The economic dimension is illustrated in Table 16.7. As mechanization gained momentum in the late 1800s, the investment per farm increased dramatically; from 1850 to 1950 it grew by a factor of 20. But the social dimension of the impact is perhaps the strongest of the unintended consequences. Table 16.8 shows the migration to and from farms between 1921 and 1957. The mechanization of agriculture resulted in enormous numbers of people moving off farms and into towns and cities. The census data show that these migrants improved their condition, as compared with conditions prior to migration; nevertheless, this migration represents a major social upheaval.

The brief review of these three technological changes showed that technology does affect society; moreover, many of the effects are unanticipated and unintentional. This is not to say that all unanticipated changes are bad. It does point out, however, that careful prior assessment of the changes brought about by new technology may help alleviate the undesirable impacts and gain more benefit from the desirable ones. Min-

Table 16.8. Migration of Farm Population (Thousands of Migrants per Year)

Year	To farms	From farms	Net loss
1921	560	896	336
1930	1604	2081	477
1940	819	1402	483
1950	995	2309	1314
1957	459	2695	2236

imizing the unfavorable consequences of technological change is the ultimate objective of TA.

3. The Role of Technological Forecasting

There are many good books available on TA. This chapter cannot cover the subject in such detail as these books; instead, its purpose is to present the role of technological forecasting in TA. The other aspects of TA, important though they are, will not be treated here.

The examples in the previous section described the consequences of several past technological changes. TA, however, is intended to identify in advance the effects of future technological change. Thus technological forecasts play a key role in the conduct of a TA. The would-be technology assessor needs estimates of the technology's performance, the rapidity with which it will be adopted, and its ultimate scope of deployment. Without these estimates there is no way to identify the consequences of the technology.

Forecasts of the performance of a technology are required to estimate its direct technical effects, such as its efficiency, the nature of its energy requirements and other inputs, the nature of its waste products, and other characteristics of its devices. Information of this type is precisely what technological forecasts can provide. Hence before a TA can be initiated, the assessor must obtain forecasts of the technical characteristics of the new technology.

The rapidity with which the new technology will be adopted determines the nature of some of its consequences. If the technology is likely to be adopted slowly, for instance, work-force adjustments can be made through normal retirements; however, users will not get the benefits quickly. If adoption is rapid, users and customers benefit quickly, but work-force adjustment may be more difficult. Forecasts of substitution, using the appropriate methods described in earlier chapters, are important inputs to the TA process.

The ultimate scope of deployment of the new technology determines the scope of its consequences. Even a technology that requires significant shifts in the relevant work force or which substitutes one type of raw material for another may have little overall effect if the total scope of deployment is small. Technological forecasts can be useful in estimating the scope of deployment by identifying those applications in which the new technology is likely to be superior to both improved versions of the current technology and other new technologies.

Thus, while technological forecasts do not address, for example, the social and economic consequences of a new technology, they provide the basis from which these consequences can be estimated. The role of technological forecasting in TA is to provide estimates of the extent and nature of the technological changes that will have social consequences.

4. An Example

Let us briefly examine some of the consequences of a new technology. In Chapter 8 a simulation model was presented for the growth of the use of solar energy in homes for space heating and hot water. The model focused only on the growth of the technology itself. Let us now expand that model to include some variables related to broader consequences of the technology. Let us add the following variables to the original model:

1. Level of atmospheric pollution. The current value is taken as 0.5 of its maximum and is given the plotting symbol *A*.
2. Fossil fuel use. The current value is 0.7 of its maximum and is given the plotting symbol *U*.
3. Number of coal miners. The current value is 0.8 of its maximum and is given the plotting symbol *M*.
4. Number of solar energy technicians. This is taken as being 0.1 of its maximum value and is given the plotting symbol *T*.

Next we consider the impacts on each of these variables. Because some variables have been added, some of the earlier impacts given in Chapter 8 have changed. Since the use of fossil fuel is now explicitly included, its price should no longer directly affect itself. Price should affect use, which in turn should affect price.

The proportion of homes using solar energy is affected positively by the price of fossil fuel, positively by the performance of solar devices, and positively by the availability of solar energy technicians.

The price of fossil fuel is affected negatively by the proportion of homes using solar energy, positively by the level of atmospheric pollution (pollution-control devices raise the cost of using fossil fuel), positively by fossil fuel use, and positively by the outside world (price will rise with increasing scarcity, even without the demand for home heating).

The performance of solar energy devices is affected positively by increased use and R&D.

R&D on solar energy devices is affected positively by the fraction of homes with solar energy, and positively by the level of atmospheric pollution.

Atmospheric pollution is affected negatively by the fraction of homes with solar energy, and positively by fossil fuel use.

Fossil fuel use is affected negatively by the fraction of homes with solar energy, negatively by the price of fossil fuel, negatively by atmospheric pollution, and positively by the outside world (it is used for other things besides home heating).

The number of coal miners is affected positively by fossil fuel use.

The number of solar energy technicians is affected positively by the fraction of homes with solar energy, and positively by R&D on solar energy.

The impacts among the variables are shown in Table 16.9, which should be compared with Table 8.4 to see the changes that have been made with the addition of the four extra variables.

The additional values can be inserted in the DATA statements of the KSIM program in Appendix 4. When the program is run, the results are as shown in Figure 16.1, which should be compared with Figure 8.5 to

Figure 16.1. KSIM run of the TA problem in text.

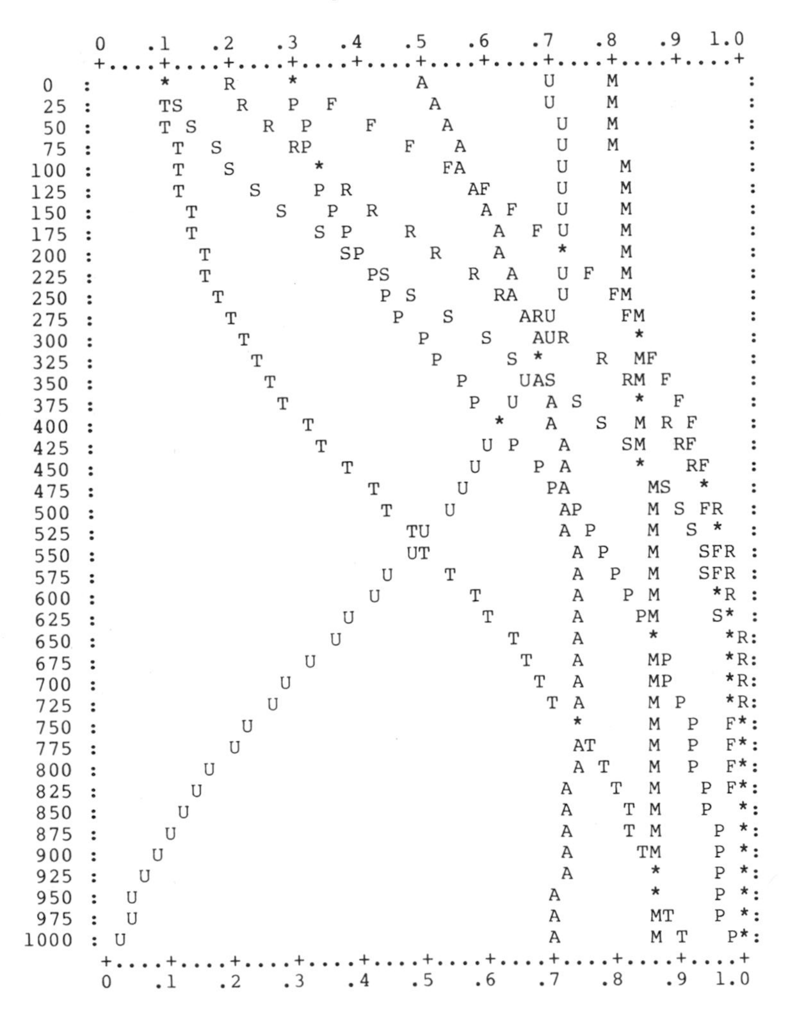

Table 16.9

	1	2	3	4	5	6	7	8	9[a]	Initial value	Symbol
1	0	5	4	0	0	0	0	3	0	0.1	S
2	−2	0	0	0	3	2	0	0	4	0.3	F
3	3	0	0	2	0	0	0	0	0	0.3	P
4	8	0	0	0	5	0	0	0	0	0.2	R
5	−1	0	0	0	0	3	0	0	0	0.5	A
6	−3	−2	0	0	−5	0	0	0	5	0.7	U
7	0	0	0	0	0	1	0	0	0	0.8	M
8	3	0	0	1	0	0	0	0	0	0.1	T

[a] Outside world

see the differences of the simpler model and the behavior of the additional variables.

The model here, of course, is not prescriptive in any sense. It does not tell the decision maker what should be done. However, it shows some of the consequences of the deployment of solar energy technology in terms of atmospheric pollution, the employment of coal miners, and the employment of solar energy technicians.

Not all TAs are centered on computer models; in fact, most are not. Assessments do, however, attempt to take into account the full range of consequences of the introduction of a new technology. As shown in the model, these consequences include not only the physical effects on the environment, but impacts on poeple, such as changes in employment. Further impacts, such as illness and death from pollution, might also be included.

Whether a TA is centered on a computer model or utilizes some other means of examining the consequences of technological change, the emphasis is on what happens to society, and not simply on the technology itself. However, the basis for forecasting the effects on society is a forecast of the technological change itself.

5. Summary

Technology produces changes in society; indeed, the primary purpose of technology is social change. It is introduced to perform some function deemed useful by at least part of society. However, technology often has consequences beyond those intended by its inventors and sponsors, and it is these unintended and unanticipated consequences that are the concern of TA. If these consequences are to be anticipated, it is necessary to have estimates of the performance, the rate of adoption, and the scope of deployment of the new technology. The technological forecaster, then, plays an important role in the team carrying out a TA. The forecaster's task is to provide a description of the technology the team is trying to assess.

For Further Reference

Coates, Joseph F. (1976). "The Role of Formal Models in Technology Assessment," *Technological Forecasting and Social Change* 9 (1, 2), 139–190.

Coates, Joseph F. (1974). "Some Methods and Techniques for Comprehensive Impact Assessment," *Technological Forecasting and Social Change* 6 (4), 341–358.

Lee, Alfred M., and Philip Bereano (1981). "Developing Technology Assessment Methodology," *Technological Forecasting and Social Change* 19 (1), 15–32.

Martino, Joseph P., and Ralph C. Lenz, Jr. (1977). "Barriers to Use of Policy-Relevant Information by Decision Makers," *Technological Forecasting and Social Change* 10 (4), 381–390.

Porter, Alan A., and Frederick A. Rossini (1977). "Evaluation Designs for Technology Assessments and Forecasts," *Technological Forecasting and Social Change* 10 (4), 369–380.

Skolimowski, Henryk (1976). "Technology Assessment in a Sharp Social Focus," *Technological Forecasting and Social Change* 8 (4), 421–438.

Sullivan, James B. (1976). "A Public Interest Laundry List for Technology Assessment," *Technological Forecasting and Social Change* 8 (4), 439–440.

White, Lynn, Jr. (1974). "Technology Assessment From the Stance of a Medieval Historian," *Technological Forecasting and Social Change* 6 (4), 359–370.

Problems

1. In the model presented in the chapter atmospheric pollution is affected only slightly by the use of solar energy. Increase the size of this effect from −1 to −5 and run the revised model.

2. In the model presented in the chapter the number of solar energy technicians lags behind the growth of the use of solar energy. Is this reasonable? Increase the impact of solar energy use on the number of solar energy technicians from +3 to +8 and run the revised model. Are there any factors in the "outside world" that might affect the number of solar energy technicians (either positively or negatively), in addition to the use of solar energy itself? If so, make the appropriate change in the model and run it.

3. Add an "illness and death" variable to the model in the chapter. How is this variable affected by atmospheric pollution? How is it affected by the outside world? Is it affected by any other variables in the model? Does it affect any of the variables in the model? Make the necessary changes in the table of cross impacts and run the revised model.

Chapter 17

Some Common Forecasting Mistakes

1. Introduction

It has been frequently suggested that the art of forecasting can be improved by the study of past forecasts and determining why they went wrong. As several chapters in this book have indicated, whether or not a forecast comes true is hardly the proper criterion by which it should be judged. Nevertheless, there certainly is some appeal inherent in this criterion. First, it appears unambiguous. Second, if the forecast itself is not of a self-defeating nature, then its utility for decision making is probably highly correlated with its eventually coming true. Third, in retrospect there is probably no other measure of the utility of a past forecast, since in general we do not have access to the decisions that were based on the forecast. Hence, despite the admitted shortcomings of "coming true" as a measure of the utility of past forecasts, until better records of decision making are kept and made available to scholars, it may be the best we can manage.

It is interesting to observe that in some cases it is hardly possible to tell whether a past forecast has come true or not. For instance, the forecasts contained in the original Gordon–Helmer Delphi study were examined six years later in an attempt to assess their validity. It was found that despite all the attempts during the original study to make statements of events as unambiguous as possible, in many cases it was impossible to tell whether an event in question had occurred yet or not. From the hindsight vantage point of six years later, it was clear that many of the statements were a great deal more ambiguous than they were originally thought to be. Hence it may not always be possible to tell whether or not a forecast has come true.

Despite this problem, and that of self-defeating forecasts, however, there is much to be learned by examining past forecasts, particularly those that "went wrong" in some spectacular fashion. To the extent that the sources of error can be specifically identified, the present-day forecaster can learn to avoid them. We will look at some typical examples of forecasts that turned out to be dramatically wrong and examine some of their identifiable sources of error.

One example of an erroneous forecast is given by Lord Bowden (1970) in a description of some of his early experiences with Ferranti, the British computer firm:

> I went to see Professor Douglas Hartree, who had built the first differential analyzers in England and had more experience in using these very specialized computers than anyone else. He told me that, in his opinion, all the calculations that would ever be needed in this country (i.e., Great Britain) could be done on the three digital computers which were then being built—one in Cambridge, one in Teddington, and one in Manchester. No one else, he said, would ever need machines of their own, or would be able to afford to buy them. He added that the machines were exceedingly difficult to use, and could not be trusted to anyone who was not a professional mathematician, and he advised Ferranti to get out of the business and abandon the idea of selling any more of them. (Bowden, 1970)

George Chatham (1970) quotes a forecast regarding the future of the turbine engine for aircraft:

> In its present state, and even considering the improvements possible when adopting the higher temperatures proposed for the immediate future, the gas turbine could hardly be considered a feasible application to airplanes mainly because of the difficulty in complying with the stringent weight requirements imposed by aeronautics. The present internal-combustion-engine equipment used in airplanes weighs about 1.1 pounds per horsepower, and to approach such a figure with a gas turbine seems *beyond the realm of possibility with existing materials*. The minimum weight for gas turbines even when taking advantage of higher temperatures appears to be approximately 13 to 15 pounds per horsepower. [italics in original] (Chatham, 1970) (Note: This forecast was prepared by The Committee on Gas Turbines, of the National Academy of Sciences, and published on June 10, 1940.)

Several examples of erroneous forecasts are given in Jantsch (1967). He quotes an IBM estimate made in 1955 that by 1965 there would be 4000 computers in use in the United States, whereas the actual figure was 20,000. He also quotes the chairman of the board of United Airlines as stating that he had "never visualized the speed and comfort of today's jet travel, the volume of air traffic, or the role of air transportation in American national life."

In 1935 C. C. Furnas wrote a book entitled *The Next Hundred Years* in which he made a sizable number of forecasts of the technological advances to be expected by the year 2035. In the mid-1960s he reviewed some of his own forecasts as being spectacularly off the mark (1967). Some of the forecasts he selected as being in serious error were the following:

> It seems obvious that one task confronting aviation is to improve design of planes so that they can improve performance at the higher altitudes. It certainly can be done. . . . Three hundred fifty miles per hour for transport planes at the higher altitudes is not an unreasonable figure to expect. That brings San Francisco within 8 hours of New York and London within 10 hours. . . .
>
> If man could only induce matter, any kind of matter, to explode into energy then he would have a power house that really is something. Half a pound of matter, says Professor Aston, would contain enough energy to drive a 30,000 ton vessel from New York to Southampton and back at 25 knots. Can the atomic physicists show us how to tap that reservoir profitably? The last word has not been said yet but do not buy any stock in an Atomic Energy Development Company. You will certainly lose. . . .
>
> It is about time that some of the glory of taking lunar trips be robbed from the pages of Jules Verne. Silly chatter? Perhaps. Perhaps not. There is nothing fundamentally erroneous in the idea. I do not doubt that some day a great crowd will gather at some aviation field to watch a man be locked into a peculiar bullet-like machine on the first bonafide trip to the moon. I hardly expect it to be in my day. . . . (Furnas, 1967)

Some of the most extensive studies of past forecasts have been carried out by James R. Bright. One forecast on which he has reported (1969) is a study sponsored by the U.S. Government Department of the Interior in 1937. According to Bright, about 60% of the individual forecasts in this study have actually come to pass; nevertheless, the authors of the study missed radar, the jet engine, nuclear power, antibiotics, nylon, and a number of other economically significant technological advances of the decade or two following their study.

These examples of the errors of past forecasts have all been instances of underestimating possible progress. It should not be thought, however, that this is the only possible direction in which to err; errors of overestimation are made too. For instance, the authors of the 1937 forecast mentioned above anticipated a rapid growth in private flying in the 1940s and 1950s, which simply did not take place. They also forecast for the immediate future the facsimile printing of newspapers in the home.

Thomas J. O'Connor (1969) made a study of the so-called von Karman report, prepared in 1945 by a team of some of the most outstanding scientists in the United States, under the leadership of Theodore von Karman. This report was prepared at the request of General H. H. Arnold, Commander of the U.S. Army Air Forces, and was intended as a guide for Air Force research and development in the post-World War II era. The authors of this report forecast the following: the use of Flying Wing aircraft, which in fact turned out to be inferior to other types; the use of nuclear power for the propulsion of aircraft, which is still far from technical feasibility; the supersonic propeller, which was bypassed by the jet engine; and the widespread use of "supersonic pilotless aircraft" (later to be known as "cruise missiles"), which were bypassed by the ballistic missile. In all of these cases the authors of the report overestimated the technical capabilities of the devices they forecast.

These instances of errors contained in forecasts of the past are not intended in any way to denigrate the competence of their forecasters. In general, the forecasts quoted above were made by persons of outstanding stature. Especially in the cases of reports requested or commissioned by government organizations, the forecasters were among the most competent scientists in the country. Our objective in studying incorrect forecasts is to recognize that even competent people can make errors. Our main concern is to identify the causes of errors and avoid repeating them.

In the remainder of this chapter three major types of causes of error will be discussed: environmental factors, personal factors, and core assumptions. Procedures that can help reduce the likelihood of errors from these sources will also be discussed. Even if these errors of the past are avoided, however, this will not guarantee perfect forecasts. As Winston Churchill is reputed to have once remarked, "These mistakes will not be repeated. We will make enough of our own." Nevertheless, eliminating the known causes of error to the greatest extent possible cannot help but improve the utility of forecasts to decision makers.

2. Environmental Factors That Affect Forecasts

Many of the studies of past forecasts have concluded that a significant cause of error is the underestimation or omission of certain factors in the environment that could have an impact on the technology being forecast. That is, the forecaster takes a narrow look at only the technology to be forecast. As a result, the forecast may be rendered invalid by some factor that was either completely ignored or whose impact was grossly underestimated. The possibility of this type of mistake can be reduced by systematically examining environmental factors that can have an impact on the technology being forecast. We will look at some of these factors, utilizing as a framework the dimensions of the environment first introduced in Chapter 3. Each dimension will be examined separately, and some factors whose impact had led to the failure of past forecasts will be identified.

Technological. A common mistake of forecasters is to fail to look at all of the stages of innovation, particularly the early ones. Most of the items missed in the 1937 forecast discussed in the previous section, such as penicillin, were actually in existence in one or the other of the early stages of innovation. Had the forecasters looked closely at items that were then perhaps at the stage of scientific findings or laboratory feasibility, they might not have missed taking them into account. Hence the forecaster should deliberately search for such items in the early stages of innovation as might impact on the forecast.

Another common failing of forecasters is to ignore developments in other fields, or other countries, that may supersede the technology to be forecast. For instance, someone trying to forecast the future of vaccum tubes in 1945 would have missed the work on semiconductors that led to the transistor had only the field of vacuum tubes been investigated. In an actual case a 1950 forecast of materials for the United States failed to identify the impact the basic oxygen process would have on steel making within the subsequent decade, even though this process was already being adopted on a large scale in Austria. Thus the forecaster should look at other technologies, industries, and countries for developments that might supplant or supersede the technology to be forecast. Not only the later stages of innovation but also the earlier ones should be examined.

Another common failing of forecasters is to ignore the impingement of one technology on another. The widespread growth of the computer, for instance, might not have been possible without the transistor or something that shared its properties of low cost, high reliability, and low power consumption. Hence forecasters should examine not only the technology with which they are concerned, but also its supporting and complementary technologies; the interactions between them may be such that a predictable improvement in a supporting or complementary technology may greatly alter the growth rate of the technology being forecast.

Finally, a common failing of forecasters can be classified as "scientific error" or "incorrect calculation." This may take the form of an assumption that the present state of scientific knowledge represents some final or ultimate level of knowledge; it may also take the form of a calculation that is formally correct but which uses unduly pessimistic or optimistic values for certain factors that have not yet been evaluated; or it may take the form of an assumption that something will be done "the hard way" or "by brute force." Such an error may be particularly significant when it occurs in the calculation of an upper limit to a growth curve. The forecaster must then make sure that the assumptions and calculations of the forecast do not treat the present level of scientific knowledge as the last word, that the values assumed for some unknown factors are all biased one way (a good practice is to carry out the calculations with both an optimistic and a pessimistic bias and see what the differences are), and that the forecast takes into account the cleverness people have shown in the past in overcoming supposedly "ultimate" barriers.

Economic. Typical failures here arise from an overoptimistic estimate of the acceptance of some innovation, or an overpessimistic estimate of the problems of introducing it. For instance, the forecast of facsimile newspapers in the home, cited earlier, failed to take into account the costs to the consumer. This failure often amounts to mistaking the "operational prototype" stage of innovation with the "commercial introduction" stage. Technical feasibility does not necessarily imply commercial success. On the other hand, innovations in financing a new device may overcome the cost barrier. For instance, had IBM insisted on selling its computers, or Xerox on selling its copying machines, probably neither would have been a commercial success. By leasing their products to the user, both IBM and Xerox overcame the barrier presented by the high cost of manufacturing the devices. Hence, if a forecaster is required to predict the "commercial introduction" stage of innovation or some later stage, both overoptimism and overpessimism in the evaluation of economic factors should be avoided.

Managerial. The primary problem here seems to have been the inability to see the impact of advances in managerial technology. New management techniques may make a much more rapid advancement of technology possible. For instance, the managerial technique known as PERT, which was devised for the Polaris program, had a major impact on making that program manageable. Managerial technology has not reached its limit any more than have many other technologies. Hence the forecaster should be alert to the possibility that the advancement of a technology may be speeded up significantly as a result of better management techniques.

Political. Changes in the political environment can have a significant impact on technological progress. This may arise from the creation of new agencies charged with achieving specific goals (e.g., National Aeronautics and Space Administration, Atomic Energy Commission, Federal Water Pollution Control Agency). It may arise from the replacement of one official having one set of views by another having different views. Such a replacement may take place through death, retirement, transfer, or the loss of an election. Changes may also arise from efforts to foster a specific technology such as agriculture, to control the detrimental effects of another such as pollution from automobiles, or to restrict the growth or spread of yet another, as the U.S. Atomic Energy Commission attempted to restrict the development of centrifuge technology for separating uranium isotopes on the grounds that it would foster a proliferation of nuclear weapons. If some event of this nature has already taken place (i.e., the death of an individual, the creation of a new agency), the forecaster should take into account its impact on the technology being forecast. If some change is possible (e.g., an officeholder is reaching retirement age, or the creation of a new agency has been proposed), the possible impact should be looked into. If necessary, the forecaster may attempt to obtain a political forecast from an appropriate source. In any case, the impact of possible political changes on the technology being forecast must be taken into account.

Social. A major source of error in past forecasts has arisen from the failure to take into account changes within society, particularly population growth, changes in the age distribution, and the growth of affluence. In addition, the impact of special-interest groups within society, such as labor unions, conservation groups, consumer organizations, and so on, must be taken into account. Many forecasts made during the 1930s were seriously in error because they grossly underestimated the population growth in the United States and thereby underestimated the demand for certain technologies and devices such as the automobile. Present population forecasts calling for significant growth in the U.S. population by the year 2000 may turn out to be equally wide of the mark. Current-day forecasts based on the assumption of a population growth continuing at the rate of the last decade or so may be as overoptimistic as those of the 1930s were overpessimistic. Thus the forecaster must realize that population can have an impact on a forecast and should therefore be consciously aware of the population growth assumptions (or forecasts) being built into it. The same holds true with assumptions about per-capita income. The forecaster must examine the possible impact of actions by various groups within society. Even similar kinds of groups may have different attitudes. The United Mine Workers Union, for instance, has encouraged the use of new technology in coal mines. The International Typographers Union, however, has been very slow to permit the utilization of new technology in the printing industry. Thus the forecaster must identify the groups that may have an impact on the technology being forecast, determine what their attitudes are, and take these into account. Resistance from some group may succeed in slowing the rate of progress well below that which would be feasible from the technical standpoint.

Cultural. Changes in the values subscribed to by society can have an impact on a forecast. One example of a radical shift in the values of American society is the change in attitudes toward education in science and mathematics that resulted from the launch of the first successful artificial satellite, the *Sputnik*, by the Soviet Union. Some commentators on the American scene have claimed to see a shift toward a "postindustrial society" that will downgrade the values of the preceding industrial society, values which were responsible for the rapid economic and technological growth of the past generation. Such a possibility would certainly have an impact on any technological forecasts intended to cover the next two or three decades. These shifts in values may be set off by some external event such as the *Sputnik*; they may also be set off by the actions of some individual or by some book, article, or play that brings some problem to sudden public attention. Ralph Nader's book *Unsafe at Any Speed* brought the problem of automobile safety to public attention and triggered a shift in values toward this subject. On the other hand, values may shift slowly and cumulatively, as a result of factors either internal or external to the society, with little notice being taken of the shift until it has reached significant proportions. For instance, the decade 1960–1970 saw a slow but steady shift in the values of youth in the United States until a significant proportion of the young people no longer subscribed to the value systems of their parents. In retrospect, this shift seems obvious, but the rate at which it crept up on society was so slow that it was hardly noticed until it had become quite large. The technological forecaster must take into account any recent shifts or current trends in the cultural values subscribed to by society, and should also be aware that a forecast may be invalidated by a subsequent shift in such values. A technology may have its rate of growth slowed down simply because people no longer care.

Intellectual. Here again we are concerned with values, those of the intellectual leadership of society. These values are of concern to the technological forecaster from two standpoints: First, the values of the leadership may differ from those of society. In the short term the values of the intellectual leadership may be more important than those of society at large, since the former are likely to have more influence on the course of the development of technology. Conversely, over the long term the attitudes of society at large may have more influence than those of the intellectual leaders, who cannot remain leaders indefinitely if they have no followers. If the values of society at large remain unchanged, new intellectual leaders will arise who reflect those values. Second, the values of the intellectual leaders may be precursors of the values of society at large. The technological forecaster may be able to identify forthcoming shifts in the values of society at large by determining what values are currently subscribed to by the intellectual leaders. In either case, whether the values of the intellectual leadership shift to match those of society or vice versa, the forecaster should not ignore the impact of this dimension on technology; instead an attempt must be made to determine what the impact will be, and it should be included in the forecast.

Religious–Ethical. At the present time it seems unlikely that any religious group could have a large-scale impact on technological advance, at least on doctrinal grounds. However, it is quite possible that at some time in the future religious, ethical, and professional groups, together or separately, may come to have an impact on technological advance by raising questions about the proper goals and objectives for humankind. The "we can, therefore we should" attitude regarding technology is already under attack, and the strength of this attack may increase. In addition, these groups may take stands regarding the personal responsibility of the technologist for any dangers or unwanted side effects arising from a new design. There have already been suggestions to the effect that a proper code of ethics for engineers would include such personal responsibility. The first of these attitudes may slow the rate of growth of technology generally; the second may give rise to increased empiricism and incrementalism in design, making engineers hesitant to depart too far from something that is known to work safely and reliably, although at a comparatively low level of performance. If either of these changes took place, they would have a significant impact on technology, and therefore they are important to the forecaster. In addition, it is possible that other similar changes will have an impact on technology and will therefore also be of concern to the forecaster. Hence the religious–ethical dimension must be considered in preparing forecasts.

Ecological. This dimension is likely to have a growing impact on the course of the development of technology and is therefore of considerable importance to the technological forecaster. For instance, the 1937 forecast mentioned above presented the following forecast regarding waste disposal: "There is no cheaper or better way of disposing of sewage than by dilution with water. . . . Rivers . . . must carry at least all of the soluble wastes of life and industry" (Bright, 1969). The authors of this forecast clearly did not anticipate the impact of this attitude on the environment when practiced by a much larger and more affluent population. Since society is becoming more conscious of the necessity to maintain a satisfactory environment, and may well be less likely to tolerate technology that damages the environment, the forecaster must take this dimension into account. In particular, it must be recognized that if a certain scale of use of a particular technology produces an intolerable level of damage to the environment, that technology may not be allowed to reach so large a scale or may be supplanted by a less damaging technology.

Each of these dimensions of the environment represents a possible source of impact on technology. Therefore, in order to prepare a useful forecast of a technology, it is essential to look at each dimension to attempt to identify the impact. The impact may be an alternative technology, an acceleration or deceleration of the rate of progress because of a change in the level of interest, demand, or support, or a consequence arising from an interaction of several technologies. Whatever the impact, it must be identified and incorporated into the forecast, which is otherwise unlikely to be of any great utility for decision-making purposes. It may even mislead the decision maker into neglecting the impact of some important developments.

3. Personal Factors That Affect Forecasts

In the previous section we have examined some environmental factors that could have an impact on technology and which, if ignored, could render a forecast worthless. These were derived from studies of the failures and errors of past forecasters. In this section we will examine some factors internal to the forecaster that have likewise been found to invalidate forecasts. These may be harder to identify in specific cases, because the forecaster may not want to admit to any shortcomings; however, if aware of the possibility of their existence, he or she has a better opportunity to search for them. To the extent that a forecaster can identify and make allowances for deficiencies, a better forecast will be made that will certainly be more useful and, if not self-defeating, more likely to come true.

Vested Interest. This problem arises when forecasters have a personal interest in an organization or a particular way of doing things that might be threatened by a change in technology. They may have spent their professional lifetimes with a particular organization or industry; they may have many favorable or pleasant personal associations or experiences with an organization or a group of people; or they may have a feeling of personal commitment to an organization or to a way of life. In any of these cases they may tend to suppress a forecast that appears to threaten any of these interests. Bright (1964) cites two statements made by executives of locomotive-manufacturing companies that illustrate this point. Robert S. Binkerd, vice-president of the Baldwin Locomotive Works, said in a 1935 speech to the New York Railroad Club,

> Today, we are having quite a ballyhoo about stream-lined, light-weight trains and Diesel locomotives, and it is no wonder if the public feels that the steam locomotive is about to lay down and play dead. Yet over the years certain simple fundamental principles continue to operate. Sometime in the future, when all this is reviewed, we will not find our railroads any more dieselized than they are electrified. (Bright, 1964)

W. C. Dickerman, president of the American Locomotive Company, said in a 1938 speech to the Western Railway Club in Chicago, "For a century, as you know, steam has been the principal railroad motive power. It still is and, in my view, will continue to be." However, as we saw in Chapter 8, from Mansfield's work on the rate of the diffusion of innovations in various industries, the diesel locomotive followed the typical growth curve quite closely. In fact, Mansfield's data showed that by 1935, 50% of the railroads in the United States had already bought at least one diesel locomotive. Diesels accounted for far less than 50% of all locomotives at that time, but the trend was clearly there (or at least it is clear now in hindsight; it obviously was not clear to everyone in 1935). At any rate, here we have a clear-cut example of two forecasters who had a commitment to a particular technical approach, so much so that their companies failed to anticipate the transition to a new technical approach despite the available evidence. Thus forecasters must be aware that a vested interest in some organization or way of doing things may obscure the true implications of the evidence they are examining. They may consciously or unconsciously refuse to forecast a technological development that would threaten their vested interest. Note, however, that this would be a disfavor to the forecaster or to the decision maker who utilizes the forecast. The impact, when it comes, will be all the more severe for being unexpected. Thus forecasters must do everything possible to make sure they are not being biased by vested interests.

Narrow Focus on a Single Technology or Technical Approach. This error is committed, when the forecaster looks at only one specific technology or technical approach. It differs from the previous one in that the narrow focus arises from the forecaster's experience or training, rather than from a vested interest; because of experience or training, the forecaster is conditioned to think only in terms of one type of solution. This may lead making a "window-blind forecast" of the type described in Chapter 1, and a technical approach that could replace or supplant the one being forecast may fail to be identified. A major recent example of this error is the behavior of firms making mechanical calculators and cash registers. The major innovations in their industries involved electronics and computer technology, which were introduced by firms outside or peripheral to the calculator or cash register industries. The companies that led these industries 20 years ago either dropped out altogether or had to scramble to catch up to newcomers to the industry. Their major problem was a focus on a particular mechanical technology, which led them to overlook the electronic technology that took over the industry.

Commitment to a Previous Position. This situation arises when a forecaster has prepared a forecast on a particular topic at some previous time and is asked to take another look at the same topic or a related one. As a result of the earlier forecast, the forecaster may be unwilling to make any alterations, even in the face of new information. This may result from a false personal commitment to one's previous work, leading to an unwillingness to recognize that recent information may require one's inferences to be changed; or it may result from a feeling that if one's previous position has changed, one's credibility or stature will be lost in the eyes of clients or the public. Thus even if it is recognized that more recent information has significant implications for the field in question, the forecaster may be unwilling to make changes in the previous forecast. It should be apparent that no matter what the reason for reluctance to change a previous stand, the situation can be dangerous. The forecaster's reputation is likely to suffer much more if clients recognize an unwillingness to take new information into account, than it is if he or she admits that an earlier position was based on incomplete information and no longer appears as soundly based as it did at the time of the previous forecast. Forecasters must examine their own attitudes carefully to be sure they are not inadvertently committing this mistake.

Overcompensation. This situation arises when the forecaster bends over backward to prevent a personal commitment to a particular view, a vested interest in some technology, or a strong desire to see a particular result from distorting a forecast. As a result, the forecast is distorted in the opposite direction. This distortion is just as bad as the one arising from a commitment to a particular technology, organization, or viewpoint. It is harder to detect, however, by both the forecaster and clients. Hence forecasters must be just as aware of this possibility as they are of the more obvious forms of bias.

Giving Excessive Weight to Recent Evidence. It is only natural that recent events will loom large in the mind of the forecaster, larger than events of the more distant past, even when the latter may contradict the recent events. If some activity has been following a long-term rising trend, for instance, but has recently had a downward fluctuation similar to many minor downward fluctuations in the past, the forecaster may give more weight to the recent downward fluctuation than to the long-term trend. While this may be natural, it is certainly not conducive to effective and useful forecasting. The forecaster must look for long-term trends and make a forecast that is contrary to these trends only when there is good reason to believe that the phenomenon is in fact departing from the trend. To ignore the trend and base a forecast solely on the latest fluctuation, either up or down, is asking for trouble. If the long-term trend is still continuing, there will inevitably be a counterbalancing fluctuation to offset the one on which the forecast is based. It is essential, then, that the forecaster give the proper weight to all relevant evidence, including that from the bulk of the past. If one of the rational and explicit means of forecasting described earlier in this book is used to prepare the forecast, the danger from this problem is somewhat reduced. In such a case it is obvious to both the forecaster and the forecast user which data have been used and which omitted. The reasons given by the forecaster for omitting specific items of data are more readily analyzed. However, even in this case there is still the possibility that the forecaster will fail to use earlier data, not because of any rational conviction that it is no longer relevant, but simply because of a subjective feeling that "it doesn't fit with more recent events." Hence there is no sure cure for this problem, and the forecaster must take pains to avoid making such a mistake.

Excessive Emphasis on the Troubles of the Recent Past. This is a more vicious form of the preceding deficiency. Serious troubles are likely to make a deep impression, predisposing the forecaster toward a particular outlook. Troubles may come to be viewed as permanent instead of a temporary phase that may but need not recur. Schoeffler (1955) gives an extensive description of this problem, in terms of economic forecasts prepared in the early 1940s, with regard to the reconversion of the U.S. economy from wartime to peacetime production. Virtually all the forecasts made at that time foresaw extensive unemployment, slow growth in the level of production, and general economic stagnation. Schoeffler identified some specific errors made in these forecasts, a few of which are itemized below:

1. Industrial reconversion tempo underestimated.
2. Construction revival misjudged.
3. Quick retirement of emergency workers after V–J day.
4. Underestimate of effective foreign demand.
5. Rush to build up inventories underestimated.

There were a considerable number of these specific errors, which Schoeffler has summarized thus:

> Continued reflection upon the etiology of the forecasting errors leads the writer to the conclusion that the underlying cause of the errors was *psychological*—not theoretical, statistical or methodological.
>
> *The originating cause of most of the reconversion forecasting errors was the prevailing psychological 'mind-set toward depression' among the forecasters.* Most of the forecasters had an unmistakable predilection to look at the economy through the dark glasses of the 1930s. This predilection had profound consequences for economic forecasting of postwar conditions. Almost inevitably, it led to an acceptance of the stagnation of the 1930s as the economic norm for the American economy of the twentieth century. Given this bias, it was natural to regard *re*-conversion as a return to basic prewar conditions and to minimize the continuing economic consequences of the war. (Schoeffler, 1955) (italics in original)

Nor were the effects of the depression limited to the years immediately following it. Ikle (1967), writing in 1966, stated,

> We make the mistake of focussing our predictions where our shoe hurts. If we have been hurt particularly badly, our predictions will look backwards

> to our old pain for a long time. The Great Depression of the 1930s had a depressing effect on all social predictions; this cast a longer shadow into the future than the real dislocations of the Depression itself. On a number of issues, our predictions are *only now* recovering from the Great Depression. Thus the volume *Recent Social Trends,* written just before the worst years of the Depression, takes a view of the future in many ways closer to our current agenda for predictions than the planning and forecasting done for some twenty years following the Depression. (Ikle, 1967) (italics added)

Since this problem is closely related to the one described in the preceding section, the procedure for handling it is much the same. The forecaster must make a deliberate effort to look for long-term trends and behavior, instead of concentrating on recent events. This will be especially difficult, since as both Schoeffler and Ikle indicate, serious troubles may easily be interpreted as representing basic structural changes in the situation that render older data irrelevant. The awareness that this can happen, however, can be helpful in counteracting the tendency to assume a structural change, since rendering this assumption explicit makes it easier to examine and either verify or invalidate.

Unpleasant Course of Action. Sometimes the course of action that seems to be made necessary by a forecast will appear so unpleasant that the forecaster shrinks from making the forecast itself in the hope of thereby avoiding the necessity for the action. Ikle (1967) made the following comments on this problem: "Some predictions imply such a horrible course of action that we experience a failure of nerve and find ourselves unable to go through with the indicated choice. This might be called the 'non-Freudian Oedipus effect.'" The reference here is to the parents of Oedipus, who lacked the courage to make sure their son was killed. Our concern here is with the forecaster who lacks the courage to take the necessary action and therefore refuses to make the forecast that would require the action. This attitude can only be described as "head in the sand." The forecaster whose nerve fails in an adverse situation is helpful to no one. It is admittedly not very helpful to give advice such as "one must have the courage to face the consequences of one's own forecasts, or else stay out of the forecasting business." While the advice is sound, the forecaster who needs it is probably least likely to heed it. However, if one is aware of the possibility of stumbling into this pitfall, one has a better chance of avoiding it.

Dislike of the Source of an Innovation. There are cases where an innovation that already exists should be taken into account as having an impact on the technology to be forecast. However, the forecaster may for some reason have acquired a dislike for the source of the innovation. This source may be an organization or an individual. In either case such a dislike of the source may cause the forecaster to ignore the innovation and its potential consequences. This attitude is not rational, nor is it conducive to making an effective and useful forecast. It is, however, human. If one is aware that the possibility of this attitude exists and honest about whether one suffers from it, one has a much better chance of avoiding this mistake.

Systematic Optimism–Pessimism. It has been observed, through studies of many forecasts, that there is a systematic shift from optimism to pessimism as the time length of a forecast increases. In the short run many forecasters tend to be optimistic, especially about work they are responsible for; they feel confident that the immediate obstacles can be overcome, and usually underestimate the effort required to overcome them. In the long run, however, they tend to be pessimistic. They see many difficulties and barriers that they do not at the moment envisage overcoming. Thus these difficulties loom large and make the forecaster forget that difficulties of similar magnitude have been overcome in the past. In effect, they make an implicit forecast that the historical rate of innovation will slow down or come to a complete halt. A useful rule of thumb is that the shift from optimism to pessimism comes at a point about five years in the future. While this may not be absolutely accurate, it can be helpful in warning the forecaster of where to expect one psychological attitude to be replaced by another. In any case the forecaster should know that this tendency to shift from short-term optimism to long-term pessimism is a natural one, and that by being aware of it he or she can minimize its effect on forecasts.

It should not be thought that every forecaster is inevitably afflicted with each one of these deficiencies; many will be afflicted with at most one or two, and some fortunate souls will escape them entirely. However, forecasters should take into account the fact that they are human beings and are therefore afflicted with human frailties. Some of the frailties that have in the past contributed to reducing the usefulness of a forecast have been mentioned above. If one takes into account the possibility of being afflicted by them; one has a much better chance of avoiding them, thereby increasing the usefulness of one's forecasts.

4. Core Assumptions

William Ascher (1978) has shown that one of the most serious sources of error in forecasts is what he has called "core assumptions." These are the underlying assumptions made by the forecaster, regarding the subject area to be forecast. As Ascher puts it, once the core assumptions are made, selection of the forecasting method is usually either obvious or trivial.

An example drawn from the recent history of the beverage-can industry will illustrate the importance of choosing core assumptions correctly. [This material is based on the research of John Machnic of the University of Dayton Research Institute (personal communication).]

Since 1960 there have been three technical approaches to the function of providing a metal can to contain beverages. They are the following, in order of development:

1. Three-piece steel. The can is a cylinder of tinned steel, with tinned steel top and bottom soldered to the cylinder.
2. Two-piece aluminum. The can is a drawn, cup-shaped aluminum piece to which an aluminum top is crimped.
3. Two-piece steel. The can is the same as the two-piece aluminum, except that it is made of steel.

The two newer technical approaches are supplanting the older three-piece steel can. Decision makers in the can industry and in the steel and aluminum industries would be concerned about which of the new technical approaches is going to be dominant. There are (at least) three possible interpretations of the situation:

1. A multilevel substitution is taking place. The two-piece aluminum can is substituting for the three-piece steel can, while the two-piece steel can is substituting for the two-piece aluminum can. Ultimately the two-piece steel can will dominate the market. This multilevel substitution is plotted in Figure 17.1.
2. The three-piece can is being replaced by the two-piece can. Ultimately the market will be dominated by two-piece cans, with price determining whether steel or aluminum is used as the raw material (the latest can-making machines can use either steel or aluminum). Substitution of the two-piece can for the three-piece can is shown in Figure 17.2.
3. The aluminum can is substituting for the steel can, with the two-piece steel can simply being a tactical action on the part of the steel industry to slow their decline in market share. The substitution of aluminum for steel is shown in Figure 17.3.

Each of the substitutions, shown as Fisher–Pry plots, appears plausible; yet they lead to dramatically different outcomes: One leads to the dominance of steel, another to the dominance of aluminum, and the third to price competition between aluminum and steel. How do we select among these predictions? We can make a statistical analysis of each plot.

If the first interpretation (multilevel substitution) were correct, we would expect the rate of decline of the three-piece steel can to remain unchanged after introduction of the two-piece steel can, but the rate of growth of the aluminum can should slow down after introduction of the two-piece steel can. Regressing the logarithms of the Fisher–Pry trans-

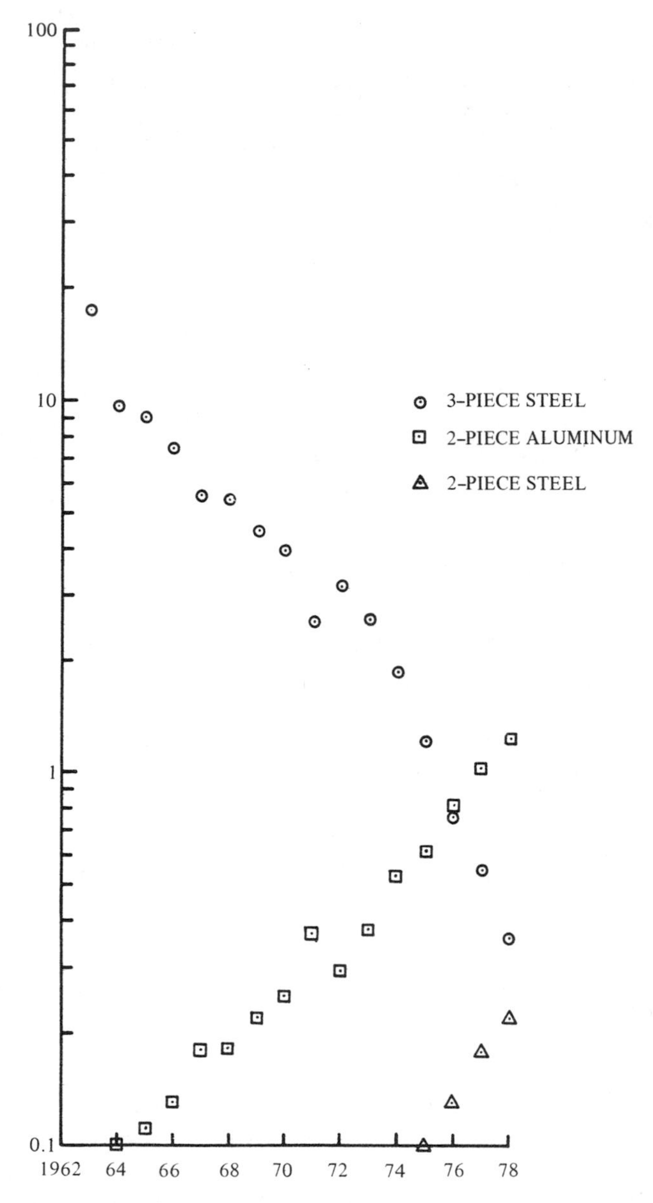

Figure 17.1. Fisher–Pry plot of the multilevel substitution in beverage cans.

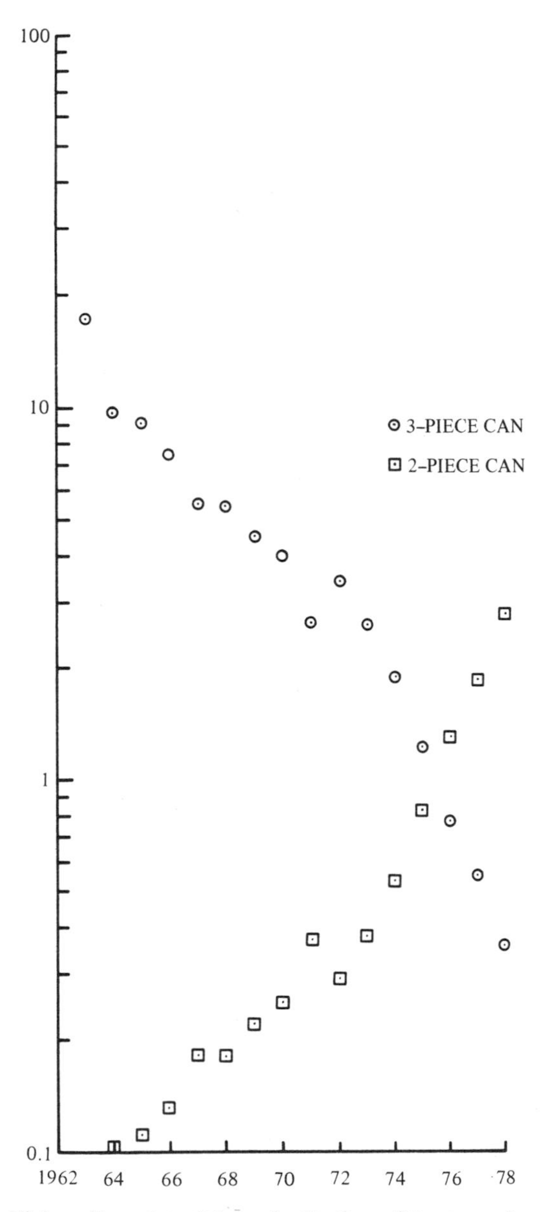

Figure 17.2. Fisher–Pry plot of the substitution of the two-piece can for the three-piece can.

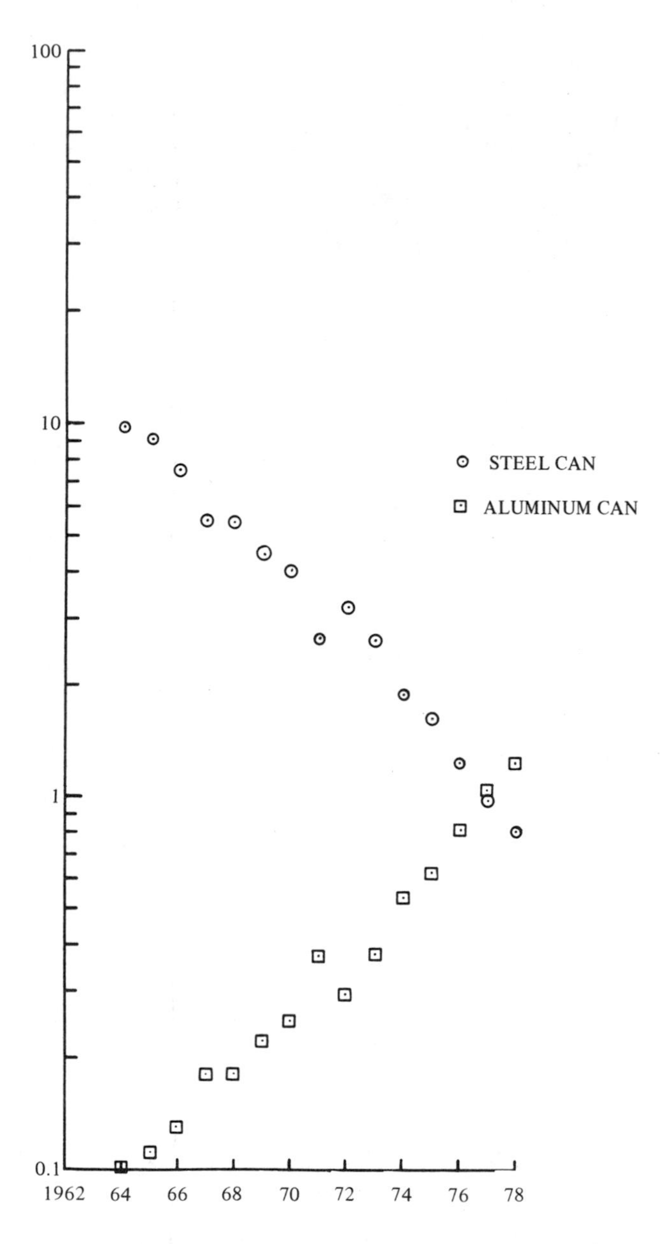

Figure 17.3. Fisher–Pry plot of the substitution of the aluminum can for the steel can.

formed market shares on time, we obtain the following results: The rate of decline of the three-piece steel can increases from −0.222643 (0.0140794) in 1963–1978 to −0.397327 (0.0150179) in 1975–1978 (the numbers in parentheses are standard errors). This difference is about 14 standard errors and is highly significant. The growth rate of the two-piece aluminum can *increases* from 0.171952 (0.0118235) to 0.230999 (0.0113858). This difference is about 4.9 standard errors and is also highly significant. Thus we can reject the explanation that a multilevel substitution is taking place.

If the second interpretation (the two-piece can substituting for the three-piece can) were correct, we would expect the rate of decline of the three-piece steel can to remain unchanged after introduction of the two-piece steel can. We would also expect the combined rate of growth of the two-piece steel can and the two-piece aluminum can to be the same as the growth rate of the latter by itself prior to introduction of the former. But we have already seen that the rate of decline of the three-piece steel can increased after 1975, and the rate of growth of the two-piece aluminum can alone was greater in 1975–1978 than it had been in 1963–1974. Thus the growth rate of all two-piece cans has to be greater after 1975 than it was previously.

If the third interpretation (aluminum substituting for steel) were correct, we would expect the combined rate of decline of the two-piece and three-piece steel cans to be the same, after 1975, as the rate of decline of the three-piece steel can prior to 1975. We would also expect the growth rate of aluminum cans to be the same before and after 1975. The rate of decline of all steel cans for 1975–1978 is −0.228809. The difference between this and the rate of decline of the three-piece steel can in 1963–1974 is not statistically significant, being much less than one standard error. However, we have already seen that the growth rate of aluminum cans did not remain constant after 1975. On the other hand, the actual Fisher–Pry values for aluminum cans for the period 1975–1978 are above the 50% confidence limit but inside the 90% confidence limit for a projection based on 1963–1974. Similarly, the values for all steel cans are within the 90% confidence limits for a projection based on the steel market share for 1963–1974.

What can we conclude from this? The data are not consistent with either the first or second interpretation. The data for steel cans are completely consistent with the third interpretation, whereas the data for aluminum cans are consistent with it on one criterion but not on another.

The important point here is that no matter how precisely we fit curves to the data, the forecast is going to be dominated by our core assumption about what is actually taking place. If we assume that aluminum is being substituted for steel, then our forecast will show an ultimate dominance of aluminum, and the fitted curve will only tell us how rapidly this will occur. However, if we assume that a multilevel substitution is taking place, with a succession of technical approaches, then our forecast will show the ultimate dominance of the two-piece steel can, and the fitted curve will only tell us how rapidly this will occur. No amount of precision in curve fitting can save the forecast if the forecaster has *chosen the wrong interpretation* for what is taking place.

Although the data are most consistent with the third interpretation, this does not constitute sufficient grounds for assuming that this interpretation is correct. The forecaster should attempt to verify this core assumption by consulting with experts in the technology. The technological forecaster should not make the mistake of assuming that a knowledge of forecasting techniques can be substituted for an understanding of the technology itself, any more than an understanding of the technology can be substituted for knowledge of forecasting techniques.

More generally, forecasters should recognize that their core assumptions will dominate the outcome of a forecast. They should therefore attempt to identify the assumptions being made and verify them from appropriate sources such as experts in the technology. Unfortunately, there are no clear-cut rules or procedures for identifying core assumptions. All too often these are assumptions of which forecasters are unaware. Nevertheless, if forecasters recognize the problem they are more likely to avoid it.

5. Summary

This chapter has covered some common mistakes made in forecasting. These have been identified by studies and reviews of past forecasts and can be grouped into three categories: environmental factors, personal factors, and core assumptions.

Environmental factors represent influences originating in the environment of the technology being forecast. If the forecaster concentrates solely on the technology concerned, these factors are likely to be missed, thereby producing a forecast that is less useful than it might have been. By systematically examining the environment, it is possible for the forecaster to reduce the likelihood of omitting or overlooking some significant factor and thus invalidating the forecast.

Personal factors represent psychological and other factors influencing the forecaster that may cause the inadvertent or unconscious distortion of a forecast. This chapter has provided a checklist of such factors to help the forecaster avoid them.

Core assumptions represent those assumptions that underly the forecast and which are often made implicitly by the forecaster. These assumptions dominate the forecast and may have more influence than the actual methodology on the outcome. The forecaster should attempt to identify the core assumptions behind a forecast and verify their validity.

There can be no absolute assurance that the forecaster is avoiding the mistakes described in this chapter; however, by being aware of them and making a conscious effort to avoid them, the forecaster has a much better chance of preparing an error-free forecast. The forecaster should always keep in mind that mistakes have been made by some of the most eminent scientists and engineers of the past, that no one is immune to them, but by a conscious and systematic effort, they can certainly be minimized.

References

Ascher, William (1978). *Forecasting—An Appraisal for Policy-Makers and Planners* (Baltimore, MD: Johns Hopkins University Press).

Bowden, V. (Lord) (1970). "The Language of Computers," *American Scientist* 58 (1), 43–53.

Bright, J. R. (1964). *Research, Development and Technological Innovation* (Homewood, IL: Richard D. Irwin).

Bright, J. R. (1969). "Some Insights from the Analysis of Past Forecasts," *Technological Forecasting Conference, Industrial Management Center.*

Calder, N. (ed.) (1965). *The World in 1984* (Baltimore, MD: Penguin Books).

Chatham, G. N. (1970). "Open Season on Aeronautics," *Astronautics and Aeronautics* 22.

Furnas, C. C. (1967). "The Next 100 Years—Three Decades Later." (Paper given to the Thursday Club, April 13, 1967.)

Gamarra, Nancy T. (1967). "Erroneous Predictions and Negative Comments Concerning Exploration, Territorial Expansion, Scientific and Technological Development," Washington, DC, Library of Congress, Legislative Reference Service.

Ikle, F. C. (1967). "Can Social Predictions be Evaluated?," *Daedalus* (Summer) pp. 733–758.

Jantsch, E. (1967). *Technological Forecasting in Perspective* (Paris: Organization for Economic Cooperation and Development).

O'Connor, T. J. (1969). "A Study of Some Effects of the Scientific Advisory Group's Recommendations on Air Force Research and Development for the Period 1945–1950," unpublished Master's thesis, University of Denver, Denver, Co.

Paterson, Robert W. (1966). *Forecasting Techniques for Determining the Potential Demand for Highways* (Columbia, MS: University of Missouri).

Schoeffler, S. (1955). *The Failures of Economics: A Diagnostic Study* (Cambridge, MA: Harvard University Press).

Whelpton, P. K. (1936). "An Empirical Method of Calculating Future Population," *Journal of the American Statistical Association* 31, 195.

Problems

1. What benefits can forecasters derive from studying past forecasts produced by others in their company, industry, government agency, and so forth?

2. What benefits can forecasters derive from studying their own past forecasts?

Problems *(continued)*

3. In which dimensions is the opinion of a specialist likely to be of help in identifying factors that might affect a forecast? In which of these dimensions is it likely to be essential?

4. Many professional forecasters are employed to help in the planning process of large organizations. Much of their work involves forecasting those technologies with which their organization is involved. What can they do to help overcome the problem of vested interest?

5. What were the errors made by the authors of the following forecasts?
   a. Roger Revelle, in Calder's *The World in 1984*, written in 1964, forecast that a worldwide network of meteorological satellites coordinated with buoys in the open ocean, carrying meteorological instruments, and reporting surface conditions by radio would be established by 1970.
   b. In 1936 P. K. Whelpton prepared a forecast of the U.S. population through 1980. Most of his work involved some very elaborate statistical methods for estimating unregistered births, so as to obtain an accurate estimate of the population in each age bracket in 1936. He then assumed three different fertility rates and mortality rates, which were as follows:
   High fertility. The historic decline in fertility would halt and fertility would continue at the 1930–1934 rate, 2177 births per 1000 women.
   Medium fertility. The historic decline in fertility would slow down, reaching a level of 1900 births for 1000 women living through the childbearing period in 1980.
   Low fertility. The historic decline in fertility would slow down only slightly dropping to 1500 births per 1000 women.
   Low mortality. Life expectancy at birth would rise from 62.4 years in 1934 to 73 in 1980.
   Medium mortality. Life expectancy at birth would rise to 70 in 1980.
   High mortality. Life expectancy at birth would rise to 67 in 1980.

Whelpton combined these possible birth and death rates into six different series. The highest and lowest series as percentages of the actual population in various years are shown in the following table.

Year	Low series (%)	High series (%)
1940	99.2	100.3
1945	95.8	994.
1950	89.7	96.3
1955	82.9	92.4
1960	76.3	88.7
1965	70.1	85.6

[Whelpton's forecast can be found in his 1936 article. The material given here is taken from the discussion in Paterson's *Forecasting Techniques for Determining the Potential Demand for Highways (1966)*.]

c. After the Wright brothers' successful flights in December 1903, there was considerable popular interest in aviation. Octave Chanute, the aviation pioneer, whose work with gliders antedated the Wrights' experiments, made this forecast of the future utility of aircraft in March 1904:

> The machines will eventually be fast, they will be used in sport, but they are not to be thought of as commercial carriers. To say nothing of the danger, the sizes must remain small and the passengers few, because the weight will, for the same design, increase as the cube of the dimensions, while the supporting surfaces will only increase as the square. It is true that when higher speeds become safe it will require fewer square feet of surface to carry a man, and that dimensions will actually decrease, but this will not be enough to carry much greater extraneous loads, such as a store of explosives or big guns to shoot them. The power required will always be great, say, something like one horse power to every hundred pounds of weight, and hence fuel can not be carried for long single journeys. [Quoted in Gamarra (1967).]

Chapter 18

Evaluating Forecasts as Decision Information

1. Introduction

How does a decision maker know whether a forecast is a good one? The obvious measure of the goodness of a forecast is whether or not it comes true; however, this measure cannot be used by the decision maker, since at the time a forecast is needed the outcome is not known. A forecast can be compared against the actual outcome only after the fact. Our concern here is with evaluating forecasts before the fact, when they are needed.

The purpose of a forecast is to help the decision maker reach a better decision than would have been possible without the forecast. The criterion for judging a forecast, then, must be based on its utility for decision-making purposes. Even here, we must be careful to distinguish evaluation before the fact from evaluation after the fact. A decision that looked entirely reasonable in the light of the information available when it was made may turn out to be worse than some other alternative that looked unreasonable before the outcome was known. However, the most we can demand of a decision is that it look reasonable when made. To demand that it appear right after the outcome is known creates the same problem as demanding that a forecast come true. It leaves us without any means of judging the quality of a decision at the time we need to do so. Thus we must demand that a decision take into account the best information available when it is made; we can demand no more than this. To judge a forecast's utility for decision-making purposes, then, we must determine whether the estimate of the future it provides is based on the best information available.

A forecast is a statement about the future, based on data about the past. But past data, in and of itself, never implies anything about the future. The forecaster must use some law, pattern, or logical relationship to link data about the past with statements about the future. Therefore any judgment about the quality of a forecast must consider not only the data utilized, but the pattern linking that data with the future.

This chapter presents a forecast-evaluation procedure that focuses on how well a forecast makes use of information about the past and about the nature of change in the subject area of interest. It is used to determine whether the forecast provides the best possible estimate of the future that can be obtained on the basis of available information.

2. The Interrogation Model

The evaluation procedure is called an interrogation model because it consists of a series of questions to be asked about the forecast. The order in which the questions are asked is chosen deliberately to narrow down the evaluation as quickly as possible. Early in the procedure items that cannot be evaluated are identified and eliminated from further consideration; only those that can be evaluated further are allowed past each screening. The interrogation model thus allows an efficient search for items that are important in each specific case.

The interrogation model involves four steps: (1) interrogation for need; (2) interrogation for underlying cause; (3) interrogation for relevance; and (4) interrogation for reliability. The first two steps address the problem area; the third and fourth steps address the forecast to be evaluated. These steps are described below.

Interrogation for Need. This step focuses on the decision maker and addresses the issue of why the forecast is needed. The following specific questions help elicit the specific need; their order is important, since the answer to each question determines the subject of the next question:

1. Who needs this forecast? That is, what specific person, agency, business firm, or other actor is to make the decision for which the forecast is needed? (In what follows, this specific actor will be referred to as "we.")
2. What do we need to know about the future? This depends on both the identity of the actor, and the decision to be made. The initial statement of need should be as broad as possible to avoid the risk of excluding elements that might later turn out to be important.
3. How badly do we need to know the future? This depends upon the decision to be made. What is at stake? What is the cost of not knowing the future? How much is knowing it worth?
4. Through what future time range must our forecast apply? What are the lags, delays, or rates of change involved in our decision? How much time is involved between making the decision and obtaining the payoff?

Interrogation for Underlying Cause. This step focuses on the subject area of the forecast. It attempts to identify the nature of changes that can occur in the subject area during the time range found to be of interest in the preceding step. The interrogation is carried out iteratively; that is, each time a possible source of change is identified, that source is addressed in turn to determine what sources of change affect it. The result of this iteration is a relevance tree of the causes of change. The following specific questions can be used at each step of the iteration:

1. In what ways can the subject area change? Can there be changes in magnitude (size, numbers, performance, capability)? Can there be changes in composition (increase or decrease in components, addition or deletion of components)? Can there be changes in character (organization or interrelation of components)? Once the modes of change have been identified, the next question is asked.
2. What can cause these changes? At this step each mode of change is examined to determine possible causes of change. Note that there may be more than one possible cause of change at a given step. A possible cause may not be sufficient or even necessary for change; it may be only contributory. Nevertheless, all these possible sources of change must be identified at this step. The sectors of the environment that have been used before can serve as a checklist for searching out possible causes of change. Appropriate questions in each sector are given below.

Technological. Can a competing technology cause change? Is a change in a supporting or complementary technology possible? What effect would that have?

Economic. Will costs of production, use, or operation change? Will the general economic climate change? Will financing be available on the scale required? Will there be a change in the potential market?

Managerial. Would a possible technological change require a scale of enterprise beyond the capabilities of present managerial technique? Will there be a change in managerial technique that would make a larger-scale enterprise feasible?

Political. Who benefits by change? Who suffers from it? What are these parties' relative political strengths? Will there be changes in the missions or responsibilities of existing institutions? Will new institutions be established? Will there be changes in political leadership? Are vested interests involved? Would certain political objectives unrelated to the technology in question be hurt or helped by change? Would the acceptance of change from a particular source help or hurt some political objective?

Social. Will there be a change in population size? In age distribution? In the geographic distribution of the population? Will there be changes in major institutions such as the family, schools, and business? Will there be changes in society's "ideal image" of itself?

Cultural. Will there be large-scale changes in the values held by society? Will there be changes in the values held by major groups within society? Will demographic or geographic changes shift the relative importance of groups with different values?

Intellectual. Will there be changes in the values held by society's intellectual leaders? Will there be changes in the makeup of society's intellectual leadership through death, retirement, and so forth? Will the values of the intellectual leaders be subject to fads or fashions?

Religious–Ethical. Is a change possible in the religious or ethical attitudes that have supported the current situation? What is the likely response of each major religious or ethical group to a possible change?

Ecological. Is the ecological impact of current practice likely to cause change? What would be the ecological impact of a specific change? Are there alternatives with less ecological impact?

At this point we have identified all the possible sources of change of the subject area concerned. In effect, this gives us the first level of a relevance tree, the top of which was our original subject area.

3. We iterate the process for each of the factors identified as a source of change, identifying the causes of change and again using the sectors of the environment as a checklist. This iteration continues until a set of "fundamental" causes of change is obtained. This set consists of causes of change that are not worth further investigation (e.g., it is probably not worth investigating whether the sun will continue to shine during the period of interest). Once these fundamental causes have been identified, the relevance tree of causation is complete.
4. What are the key elements of the subject area? Once the iteration of searching for causes of change has terminated in a set of fundamental causes, it will normally be the case that these causes appear several times at the bottom of the relevance tree. These repetitive causes are the key elements of the subject area. Of course, a factor

that appears only once may be a key element if its influence is strong enough. The influence of each fundamental cause must be determined and the important causes of change identified.

Interrogation for Relevance. This step focuses on the forecast to be evaluated. The information generated in the two preceding steps is utilized to determine the relevance of the forecast as a whole and in parts. This is done by comparing the forecast with the key elements identified in the preceding step.

In general, there will be three categories of key elements: (1) relevant and included, that is, found to be key elements and included in the forecast; (2) relevant but excluded, that is, found to be key elements but not included in the forecast; and (3) irrelevant but included, that is, not relevant to our needs but included in the forecast. The presence of irrelevant material does not necessarily indicate a fault in the forecast; the forecast may have originally been prepared for some other purpose or for a subject area that only partially overlaps the area of concern.

The elements that are irrelevant can be ignored; those which are relevant but not included are items that must still be sought out, for without them, the forecast does not provide complete information and is to this extent inadequate as a basis for decision.

Interrogation for Reliability. The information that was both relevant and included in the forecast must next be evaluated for reliability. This process involves examining the past data, the assumptions, and the methods connecting past data with estimates of the future. These three features are inevitably intertwined, but we will consider them separately.

1. *The methods.* The methods used to produce the forecast must be identified. In some cases they will be described clearly; this is usually the case when models or explicit procedures are used. In other cases the methods will not be described at all and only the results will be given. For each result the method must either be identified, or categorized as not given. When the method is not given, the forecast cannot be evaluated for reliability. Hence some results may have to be classified as having unverifiable reliability. When the methods are given, two further questions are possible.
   a. *Are the methods replicable?* A replicable method is one that can be used by someone other than the original forecaster to obtain the same result. If the methods are judgmental, the results will not necessarily be wrong; it simply means that the reliability of the methods cannot be verified.
   b. *Are the methods formally consistent?* Once a method has been determined to be replicable, it must be examined to determine

whether it is logically consistent. Does the conclusion follow from the premises? Have any logical errors or fallacies been committed? A logical fallacy will make any further conclusions unreliable.

These questions group conclusions into three categories: verified as arising from reliable methods; verified as unreliable; and of unverifiable reliability. This last category indicates an area where further information must be sought. The results of the second category must be rejected. Results from the first category can be examined further.

2. *The assumptions.* Assumptions are used in place of facts when the facts are unknown or are inherently unknowable. The assumptions, then, are critical to the forecast. The following questions should be asked, in order.
   a. *What are the assumptions?* In many cases the assumptions are not clearly listed; some may be hidden. In some cases the forecaster might not have been consciously aware of all the assumptions; hence ferreting them out may require a careful examination of the forecast. Some of the kinds of assumption found in forecasts are the following: that a particular method is appropriate in a specific case; that a given group will behave in a particular way; and that data from a sample are representative of the entire population.
   b. *Are the assumptions adequately defined?* Before an assumption can be tested for validity, it must be adequately defined. The definition must specify the conditions to which the assumption does and does not apply and the evidence that is necessary for affirmation or denial. The task of defining assumptions can be made easier by the following questions.
      i. *Is this a static or dynamic assumption?* Static assumptions refer to states or conditions; they must specify the time at which they are valid. Dynamic assumptions involve a rate, an acceleration, a sequence, a delay, or a causal linkage; they must specify a dimension of change and an interval of time during which the change takes place. A static assumption, valid for one point in time, may not be reliable if applied to some other point. A dynamic assumption, valid for one time interval, may not be reliable if applied to some other interval.
      ii. *Is this a realistic or a humanistic assumption?* Realistic assumptions concern phenomena essentially independent of human opinion. These may include facts about the physical universe or the past actions of humans. Humanistic assumptions refer to the values, perceptions, future actions, intentions, goals, and so on, of human beings. A humanistic assumption must specify the group or individual to which it applies, since it may not be true of other groups or individuals.

Assumptions of worth, cost, or risk must specify to whom they apply.

c. *How are the assumptions supported?* An assumption for which no support is offered is of unverifiable reliability; if support is offered, reliability may be determined by the following steps:
   i. *Is the assumption a necessary consequence of some law, principle, or axiom?* If so, how is this law, principle, or axiom supported? Is the law so well accepted that no support is needed? If not, is evidence or data offered in support?
   ii. *Is the assumption supported by the opinion of an expert in the field?* If so, is evidence offered in support of the expert's credibility?
   iii. *Is evidence offered in direct support of the assumption?*
d. *Is the evidence relevant to the law, expert, or assumption it is to support?* That is, is the evidence of the type demanded by the nature of the assumption? Is it an adequate test of the assumption? The following criteria can be used to evaluate the relevance of the evidence.
   i. Evidence to support or deny a static assumption must be obtained from the particular state or point in time specified by the assumption.
   ii. Evidence to support or deny a dynamic assumption must be gathered sequentially in time or process. The frequency of observation and the time span over which observations are taken must be consistent with the nature of the assumption.
   iii. Evidence to support or deny a realistic assumption must be gathered, processed, and presented in a manner as free from human evaluation and prejudice as possible. This is important, since what people do or do not believe is irrelevant to matters of fact. The inclusion of human evaluations only serves to reduce the reliability of the purported facts.
   iv. Evidence to support or deny a humanistic assumption (e.g., concerning the values or perceptions of some person or group) must be obtained from the physical, written, or verbal behavior of the person or group in question. Different groups of people may interpret the same reality in different ways and have different ideas about what is desirable or undesirable. The only valid means for determining what people believe or perceive is through the examination of their behavior. In particular, it is necessary to avoid the error of attributing one's own values to others or assuming that they perceive reality in the same way.
   v. Evidence to support or deny a law or principle must be gath-

ered under the circumstances in which the law or principle is assumed to apply. If the evidence is gathered under some other set of circumstances, it cannot be used either to prove or disprove the validity of the law in the case considered in the forecast.

vi. Evidence to support or deny the credibility of an expert must deal with his or her degree of knowledge regarding the case under consideration in the forecast. The expert's knowledge or lack of knowledge about some other case is not relevant.

If the evidence offered terms out not to be relevant to the assumption, the reliability of the assumption cannot be determined. If the evidence is relevant, it must then be checked for validity.

3. *The data.* Data may be used to support an assumption or as raw material from which a pattern is to be extracted (e.g., by some mathematical fitting procedure). By this point in the interrogation process, only data that are relevant either to an assumption or to a forecasting method should have survived the screening. For both kinds of evidence accuracy is an important issue.
   a. *How accurate is the data?* There are two possible sources of error in data: uncertainties of observation or measurement, and conscious or unconscious distortion.
      i. *Are there uncertainties of observation?* These are inherent in any observation technique. In observations of physical phenomena the uncertainties are usually known; in the social sciences they may be harder to evaluate. However, in either case it is necessary to estimate the uncertainties to determine whether the data are sufficiently accurate.
      ii. *Are there uncertainties of measurement?* These uncertainties arise from aggregation, rounding, and so on, and from sampling from a large population. Again it is necessary to estimate these uncertainties to determine whether the data are sufficiently accurate.
      iii. *Is there conscious distortion?* Have the observations been distorted or selected to fit preexisting human values? The agency that gathered the data must be identified, and its past actions examined to identify possible biases. If these are found, the data would be suspect.
      iv. *Is there unconscious distortion?* This may arise from the philosophy, culture, or ideology of the observer; it may originate in factors such as the "self-evident truths" of a culture, the unchallenged axioms of a scientific discipline, or the implicit assumptions of a common philosophy. Unconscious distortion

can show up in both the selection and classification of data, and it is often difficult to identify, especially when the forecast evaluator shares the unconscious assumptions of the data gatherers. However, it is important that this distortion be identified, especially when the data concern people who might not share the unconscious assumptions.

At this point evaluation is complete for data to be used for fitting models, extracting patterns, and so on. However, one additional question must be asked about data intended to support assumptions.

b. *Does the evidence presented tend to confirm or deny the assumption?* Assumptions about future conditions can never be denied or confirmed absolutely on the basis of available evidence. The evaluator must use judgment to determine whether to accept or reject an assumption on the evidence available. Even assumptions about the past cannot always be confirmed absolutely, since some of the evidence may have been lost or destroyed or it may cost more than it is worth to gather "all" the data. Thus, given the data available, the evaluator must determine whether it is sufficient to support or deny an assumption or whether the assumption must be evaluated as having indeterminable reliability.

This completes the description of the interrogation model. In summary, the model leads the evaluator through four steps:

1. Interrogation for need. The decision to be made is analyzed to determine what kind of forecast is required, including what should be forecast, how accurately it must be forecast, and over what time range.
2. Interrogation for underlying cause. Elements that have been identified in the previous step as required parts of the forecast are analyzed to identify causes of change.
3. Interrogation for relevance. The forecast itself is analyzed to determine how much information it contains about the factors previously identified as affecting the things to be forecast.
4. Interrogation for reliability. Those items found to be included and relevant are evaluated for reliability.

By asking this structured sequence of questions on the subject and the given forecast, the evaluator is led through an evaluation of the relevance and reliability of the latter. A forecast that is highly relevant to the subject in question and highly reliable (i.e., draws sound conclusions from the best data available) is very useful as an input to a decision. This is not an assertion that the forecast will come true; instead, it is an assertion that the decision maker is likely to do better taking the forecast into account. If the forecast turns out to be irrelevant or unreliable or both, then the decision maker is well advised to reject it. If the matter is sufficiently important and the available forecast is not sufficiently relevant or reliable, the decision maker is well advised to initiate the preparation of another, more useful, forecast.

3. Summary

The interrogation model has been presented in terms of its use by a decision maker to evaluate a forecast; however, it can also be used by forecasters. Forecasters should anticipate that their work will be evaluated by some means similar to the interrogation model. They should therefore plan their work from the beginning so that it will successfully pass the evaluation. A forecast should be focused on the needs of the user, incorporating all the relevant sources of change. The methods, assumptions, and data should be clearly spelled out and such that their reliability is verifiable. By applying the interrogation model to their own work while preparing a forecast, forecasters can increase the utility of their work and the likelihood that it will be accepted and used.

Problems

1. Why is it desirable to be able to evaluate a forecast before it is used, rather than waiting until the actual outcome is known?

2. Carry out an interrogation for underlying causes on each of the following situations and develop a relevance tree of causation for each:
   a. Number of automobiles in use in the United States.
   b. Number of first-class letters handled by the Post Office.
   c. Electrical power consumption in Central Africa.

3. The year is 1895. You are the U.S. Secretary of Agriculture. You have asked your staff to prepare a 30-year forecast of American farm production and acreage. The portion of this hypothetical forecast dealing with the acreage planted to oats is given below. Using the methods of this chapter, evaluate the forecast, carrying out all steps of the interrogation model. To what extent does your analysis exhibit 1895-style foresight, and to what extent does it exhibit hindsight?

   Our forecast of the demand for oats is based on the following considerations. The primary market for oats is as horse feed. Most farm workhorses receive a diet containing a mixture of feeds, primarily hay. However, the average farm horse consumes 31 bushels of oats annually. Since 1867, the average yield of farmland in oats has been about 26 bushels per acre. The number of horses has grown as

Problems *(continued)*

shown in the following table, which represents an excellent fit to an exponential growth law. Projecting this growth until 1925, there will be 53 million horses in the U.S. This will require 63 million acres planted to oats.

Horses on Farms in the United States

Year	Number of horses (thousands)
1867	6,280
1868	7,051
1870	7,633
1875	9,333
1880	10,357
1885	12,700
1890	15,732
1895	17,849

Source: Historical Statistics of the United States, Department of Commerce, Washington, D.C., 1976.

Chapter 19

Presenting the Forecast

1. Introduction

After the preparation of a forecast making use of one or more of the methods described earlier in this book, after having taken into account the application of the forecast as described in one of the chapters on applications, and after having made sure to avoid the pitfalls and mistakes described in the preceding chapters, there is still one more hurdle the forecaster must cross: The forecast must be accepted by a decision maker and used in the decision-making process.

This is the final step and the reason for all the work that came before. If the forecast is not accepted and used by a decision maker, all the work and resources that went into preparing it were wasted. No matter how interesting the work was to the forecaster, how elegant the procedures used, or how extensive the compilation of data that went into the forecast, it was all wasted effort if the forecast does not contribute to the making of an appropriate and timely decision.

The forecaster must remember that this criterion reflects the viewpoint of the decision maker, who is not a professional forecaster. All the professional niceties that mean so much to forecasters and their colleagues are of no consequence to the decision maker. It does the forecaster no good to complain that the decision maker "just didn't understand what I was trying to get across." This may well be true, but the fault is more likely that of the forecaster than that of the decision maker.

It may be that in some cases the forecaster will encounter a decision maker who rejects the concept of using rational and explicit forecasts as inputs to decisions. There are some decision makers of this type still in responsible positions; their number will grow fewer, however, as death and retirement take their toll, and it is unlikely that such a decision maker would have called on a forecaster for help in the first place. The progressive decision maker accepts all the help available in clarifying a tough decision problem. If a technological forecaster is asked to contribute to the decision-making process, it is because the decision maker appreciates the benefits that can be derived from a forecast of technological change. If one of these decision makers rejects a forecast after having asked for it, the fault is usually that of the forecaster. No matter how well the forecast was prepared in some way it was not presented properly.

There are two basic reasons why a forecast will not be used: The decision maker does not believe it to be useful or does not believe it to be credible. In this chapter we will look at some of the problems of presenting a forecast to a decision maker so that its utility and credibility are apparent.

2. Making the Forecast Useful

A forecast prepared by a technological forecaster will be stated in terms of some functional capability. It is a statement that a specific functional capability will (or could) have achieved a specified level at a definite time in the future. However, it must be remembered that forecasting does not actually place anything in the future. A graph, such as that shown in Figure 19.1, with historical data points and a best-fit curve projected to some future date must never be interpreted as though the projected line somehow lies in the future that has not yet arrived. The curve shown was implicit in the historical data and was extracted by a mathematical technique; it is connected with the future growth of technology only by an assumption about the relationship between past and future behaviors, an assumption that forms the basis for the forecast. The point is that the forecaster could simply have presented the decision maker with a tabulation of the historical data without a trend extraction or projection. The sole justification for the activity of extracting and projecting the trend implicit in the data is that it is more useful to the decision maker than a tabulation of historical data. The forecast is justified only to the extent that it is useful to the recipient, providing information that could not have been obtained as readily (or at all) from the historical data alone, and which is recognized as being of help in making a decision.

The decision maker will see the forecast as being useful to the extent that it is meaningful and understandable. If the forecast is incomprehensible, it is, of course, of no help. Even if it is understandable, it must still be meaningful in terms of a specific decision to be made. If the forecast does not illuminate the alternatives in a specific decision, leading to a choice of actions, it is not meaningful. These two properties of a good forecast will be discussed separately below.

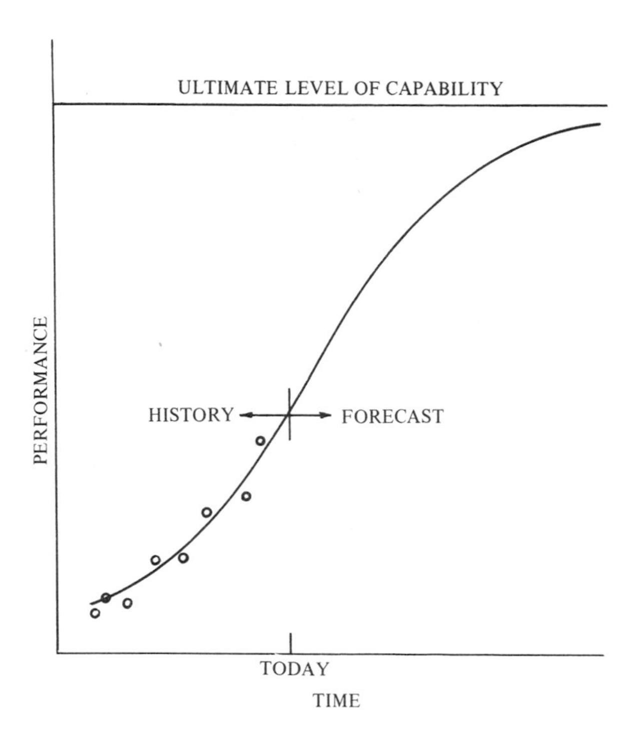

Figure 19.1. Forecast showing trend extracted from historical data.

Meaningful Forecasts. Former President Lyndon Johnson is reputed to have responded with the query, "Therefore, what?" following a briefing on the Middle Eastern situation (Hare, 1970). To the extent that a decision maker responds to a forecast with such a query, the forecast has not been meaningful; it has failed to get at whatever are considered to be the important issues. It may indeed have addressed the important issues, at least as seen by the forecaster, but this fact is not conveyed to the decision maker.

This points out the importance of tailoring the forecast to the needs of the decision maker in a specific situation. In other words, the forecast must be pertinent to the type of decision being made. How much control does the decision maker have over the technology being forecast? Will the decision affect the progress of that technology to a greater or lesser degree? Or is the decision maker trying to choose the appropriate response to technological change resulting from the decisions of others?

Depending on the type of decision situation, the forecaster will prepare either a conditional or an unconditional forecast; that is, the forecaster will state either that a certain event is going to come to pass regardless of actions taken by the decision maker or that the forecast is conditional upon the choice made by the decision maker. This can be especially important in forecasting the future growth of some technology. If the decision maker's organization has played a major role in establishing the trends that led to the achievement of the current level of functional capability in the area concerned, then the organization cannot drop out and assume that the trends will continue. On the other hand, if the organization has not contributed to the progress of that technology in the past, then the forecast will be based solely on assumptions about the behavior of others who have been contributing to its progress.

Consider, for instance, the following basic (and hypothetical) forecast: By 1990 half of all business travel will be replaced by long-distance videophone conversations. If this forecast were prepared for a decision maker in a telephone company, it would have to be a conditional forecast; that is, it would have to indicate that it was conditional on the decision maker taking the proper actions. In that case, the forecast might be phrased, "If we continue installation of new equipment at the present rate, by 1990 half of all business travel will be replaced by long-distance videophone conversations."

If, however, the forecast were being prepared for a decision maker in a large corporation concerned with the question of whether to expand its fleet of executive aircraft, it might be phrased, "At the present rate of improvement, long-distance videophone conversations will replace half of all business travel by 1990." From the standpoint of the recipient this forecast is unconditional; no action has to be taken to see it fulfilled. It is important that the forecaster state explicitly whether or not the forecast is conditional on certain actions undertaken by the decision maker receiving the forecast. In particular it must be made clear whether the decision maker's organization can simply go along for the ride or whether they must take some positive action to fulfill the forecast.

Another situation where even a well-prepared forecast can turn out to be meaningless from the decision maker's standpoint is that of the implied superiority of a technique or technical approach. Consider, for instance, the (hypothetical) forecast that low-cost composites (plastic and fiber) will be used in (1) secondary and primary commerical building structural elements by 1988 and (2) secondary and primary structures for highway bridges, airfield construction, and so on, by 1990.

It is not clear from the statement of the forecast whether the forecaster is saying that this capability will be available or that it will be superior to all other available techniques. Unless the decision maker knows which is the case, he or she does not know what action to take, if any at all. For instance, suppose this forecast were given to decision makers in the cement industry. They would naturally be concerned with the possibility that plastic and fiber composites might replace concrete in some structural applications; but they need to know whether it will be superior to concrete, and for what applications, before they can determine whether any action at all is needed, and if so, what action would be appropriate. Hence it is important that the forecaster make this point clear in the presentation of the forecast.

From time to time the forecaster will find it appropriate to forecast diminishing returns for a given technology; that is, the rate of progress will decline significantly during the period covered by the forecast. It is not enough, however, simply to state the forecast of diminishing returns; the cause must be identified and the implications of the fact made clear. Some common causes of diminishing returns in the progress of a technology are the following:

1. Imminence of a physical limit.
2. Progress slowed down by standardization, which puts significant constraints on possible new approaches.
3. Decrease in the demand for the technology, hence less effort expended on advancing it.
4. Breakdown in communications because the field has become very large, with many people working in it.
5. Continued growth at the previous rate requiring excessive resources.

There are other possible reasons in specific cases, but the foregoing cover a large proportion of the instances of diminishing returns. Not only is the cause of the diminishing returns important, but so are the implications. What must be done to restore the previous rate of progress? Some possible decision implications are the following:

1. Organizational changes to permit more rapid progress.
2. More resources to allow continued rapid progress.
3. Shift to another (specific) technology capable of more rapid progress.
4. Search for a new technical approach, as of yet unidentified, that is not faced with an imminent limit.
5. Abandonment of the existing technology, since the decline in the rate of progress is fundamental and cannot be reversed.

Note that the statement of these implications does not force a choice on the decision maker; it only indicates what must be done if the previous rate of progress is to be restored. The decision maker is still free to decide whether or not to restore the previous rate and to choose the most appropriate action under the circumstances. But an intelligent choice cannot be made unless one knows why diminishing returns have set in and what can be done to correct the situation. Hence it is important that the forecaster not stop with a forecast of diminishing returns but explain why the situation has arisen and what its implications are.

The points made above regarding meaningful forecasts have all tacitly assumed that the functional capability being forecast would be meaningful to the decision maker and that the forecaster's only concern is to present the forecast in such a way that the possible decisions are clearly illuminated; however, this is not always the case. Consider, for instance, a decision maker in a bank who is concerned with planning the bank's future purchases of computers. The decision maker wants to maintain adequate computer capability over the planning period but not be stuck with too many obsolescent computers. Therefore early purchases, which provide immediate capability, have to be balanced against waiting for better computers. Suppose a forecast of computer capability is presented in terms of technical parameters such as the ratio of memory size to access time. Unless the decision maker is a specialist in computers, this forecast is not going to be helpful at all; it must be translated into terms that are meaningful to the bank's operations, such as the number of deposits and withdrawals to be handled per day. This situation is in fact fairly general. There are many technologies where the functional capability to be forecast can best be expressed in some technical measure that is not likely to be meaningful to a decision maker who is not a specialist in the technology. In such cases it is necessary to carry out the forecast in the appropriate functional capability and then translate it into terms meaningful to the decision maker.

One of the best ways to render a forecast meaningless is to make it qualitative instead of quantitative. A forecast that states, "the performance of this class of devices will increase significantly," "This technique will be used more extensively," or, "This device will be used in a wide variety of applications," is not very helpful. Even if the forecast comes true, it is useless because it is so vague. Besides, the decision maker is probably already aware that performance will be improved and new applications found. The decision maker expects the professional forecaster to provide information that is not already known. In particular, this should include a quantitative estimate of the improvement in performance or the extent of use or application, as well as one of the degree of uncertainty in the forecast. Quantitative estimates of the nature and degree of change help a decision maker make a more appropriate and timely decision.

One situation that sharply outlines the problems of making a forecast meaningful is that of making several successive forecasts for the same date. For instance, suppose an organization has established the practice of maintaining a ten-year plan based on a ten-year forecast. Each year the forecast is redone completely, covering a ten-year period. Year 19XX, which was the tenth year the first time the forecast was prepared, becomes the ninth year after the first annual revision, the eighth year after the second annual revision, and so on. No matter what forecasting method is used, each version of the forecast will have a somewhat different description of conditions in the year 19XX. Suppose, for instance, that a growth curve is fitted to the performance of a technical approach to show how it will move toward its ultimate upper limit. The calculated upper limit remains the same year after year. As additional data points are collected, however, and included in the regression fit computations, a slightly different curve will be produced each year. (The only exception to this will be the unlikely circumstance where the new data point falls precisely on the curve computed the previous year.) What meaning should be attached to these changes in the forecast? Is new action or a reversal of previous action required of the decision maker as a result of the "revised" forecast? Clearly, if the underlying process has not changed and the original decision was the appropriate one for that underlying process, then no changes are required. Hence when this situation occurs, the presentation should stress that the forecast is basically unchanged and that the differences result simply from incorporating more data into the computations. The use of confidence limits can be helpful here in showing that the new data have about the scatter expected and that the new curve fits within the $Y\%$ confidence limits on the old curve [i.e., confidence limits calculated on the basis of $S(R, T_i)$, not on the basis of $S(F, T_i)$]. The percentage of confidence Y should be chosen appropriately to illustrate that the deviation from the old curve is a small one. Of course, if the new curve deviates significantly from the old one or if there is some indication that the underlying process has in fact changed, the forecaster is obliged to stress this fact. In this case there has been a significant change in the forecast, and the decision maker may be called upon to alter a previous decision. The same is true if there is a significant change in the calculated upper limit. The important point is that when a forecaster is required to produce a sequence of forecasts for a same point in time, a forecasting method or a means of presentation or both should be chosen that plays down unimportant variations or statistical fluctuations in the forecast and emphasizes significant differences which call for a change in the course of action being pursued. If the forecast is to be meaningful to the decision maker, the forecaster must show clearly whether the differences between the succeeding forecasts are significant or not.

Understandable Forecasts. The decision maker should not be in any doubt about precisely what future the forecaster is forecasting. The events in that future and their uncertainties and the qualifications (the if's, maybe's, possibly's, etc.) must be laid out clearly. The decision maker, after hearing or reading the forecast, should not be left wondering, "Now, what did he say?"

One of the biggest blocks to understanding is ambiguity. The forecaster inevitably deals with uncertainty and is therefore frequently tempted to make ambiguous forecasts. If this is done with the intent of hedging against being caught with an incorrect forecast, it is, of course, reprehensible; however, it is often done quite innocently. The forecaster is not certain what is going to happen, does not want to leave out any reasonable possibilities, finds it difficult to assign relative likelihoods to various possibilities, and therefore ends up with a forecast that is ambiguous. This forecast will then not be understood by the decision maker, who may then fail to take appropriate action.

In her study of the Pearl Harbor attack Roberta Wohlstetter (1962) describes a series of messages sent to General Short and Admiral Kimmel, the Army and Navy commanders, respectively, at Pearl Harbor during the year prior to the Japanese attack. These messages were based on information obtained by decoding Japanese messages to their own military forces and to their diplomatic officials in various parts of the world. Wohlstetter summarizes these messages as follows:

> Except for the alert order of June 17, 1940, none of the messages used the word 'alert.' They said, in effect, 'Something is brewing,' or 'A month may see literally anything,' or 'Get ready for a surprise move in any direction—maybe.' Phrases characterizing contingencies as 'possible' or 'barely possible' or 'strongly possible' or saying that they 'cannot be entirely ruled out' or are 'unpredictable' do not guide decision-making with a very sure hand. To say, for example, in the Army warning of November 27, 1941 that Japanese future action was unpredictable added nothing at all to the knowledge or judgment of General Short or Admiral Kimmel. Local pressures being what they were, General Short read a danger of sabotage into his final messages from Washington, while Admiral Kimmel read danger of attack by Japan on an ally or even on an American possession somewhere near the Hawaiian Islands. (Wohlstetter, 1962)

This example illustrates clearly the problems that can arise if a forecast is worded ambiguously. Even though the United States was in possession of decoded Japanese messages that provided reasonably clear signals that the Japanese were planning some offensive action, the value of this information was lost because the forecasts based on it, and supplied to the decision makers on the spot, were not worded unequivocally.

The forecaster is engaged in decoding signals not *from* the future but *about* the future. This process is not done for its own sake, but to provide information for decisions. The results of this decoding process should be presented as unambiguously as possible. When uncertainty is inevitable, the forecaster must estimate the degree of uncertainty as well as possible and present the forecast in such a way that the decision maker is aware of this uncertainty. Ambiguity should not be used to hide uncertainty.

One of the best means of presenting a forecast is the use of graphs. Quantitative forecasts lend themselves well to presentation in this fashion. Growth curves and trend curves, when presented graphically, can be grasped and understood readily by the decision maker whose time is limited; they are also easy to explain. Even intuitively derived forecasts,

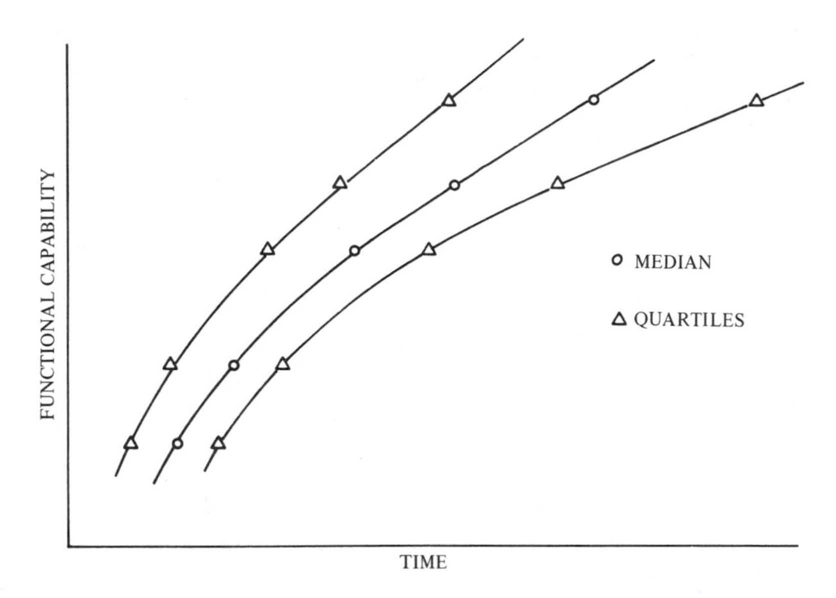

Figure 19.2 Graphical display of Delphi estimates.

Figure 19.3. Graphical display of Delphi estimates of discrete events.

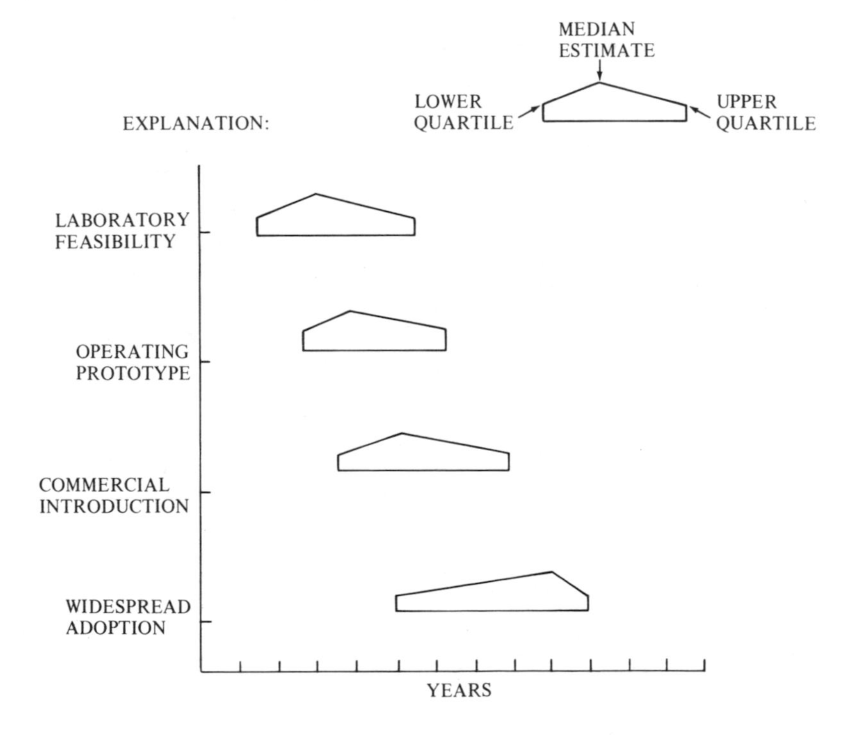

Table 19.1. Hypothetical Forecast of Aircraft Structural Materials for 19XX

Property	Level of capability	Implications
Yield strength	75% increase	Lighter structure for the same load-carrying capability
Corrosion resistance	20% increase	Longer life in the same corrosive environment
Low-temperature performance	Same as current	No change
High-temperature performance	40% increase in allowable temperature for short exposure; 10% increase for long exposure	Higher speeds; lighter structure at same speed
Fatigue life	50% increase	Longer service life for airframes: longer intervals between major overhauls

such as those obtained from a Delphi panel, can be presented in graphical form if one has obtained estimates of several levels of capability or estimates of the dates by which a technology will achieve different stages of innovation. Figure 19.2 illustrates a graphical display of Delphi estimates of the achievement of successive levels of functional capability. Figure 19.3 illustrates a graphical display of Delphi estimates of achievement of various stages of innovation. Graphical display also lends itself well to the display of degrees of uncertainty, whether these be in the form of calculated confidence limits or simply intuitive estimates.

Some items lend themselves well to tabular display, especially forecasts obtained for a single point in time. Table 19.1 illustrates a hypothetical forecast of structural materials for aircraft for the year 19XX. Each of the properties of the material is forecast separately in terms of its degree of improvement over the present level of functional capability. This format can be grasped readily by the decision maker who is pressed for time; it focuses on the essentials, presents these in quantitative terms, and describes the implications of the forecast. As shown, Table 19.1 is an unconditional forecast for a decision maker whose actions will not influence the outcome. In a conditional forecast it might be desirable to replace the final column with a description of the actions required by the decision maker if the forecast is to be fulfilled.

Another very useful method for making a forecast understandable is the scenario. This method is particularly applicable when a set of separate forecasts are to be combined into a composite one. In such cases the decision maker is concerned not with discrete events or individual technologies, but with a broad picture of the future. The scenario can provide a context in which the various options open to the decision maker can be placed so as to determine which one is the most satisfactory on an overall basis.

3. Making the Forecast Credible

A prerequisite that a forecast must meet if it is to have any impact on the decision-making process is that it be credible. The decision maker to whom it is presented must believe the forecast, but not, of course, in the sense that it will come true. (Progressive decision makers are well aware of the problems of uncertainty and recognize that their own actions may have some impact on the outcome of the situation being forecast.) The forecast must be believed in the sense that the forecaster appears to have extracted all or most of the relevant information from the past and has connected this with the future through an appropriate logical structure. In short, the decision maker must believe that the forecast presents the best picture of the future that can be obtained from existing knowledge.

Unfortunately, many forecasts that represent the good use of existing knowledge and the sound application of forecasting methods are rejected because the presentation robs them of credibility. The decision maker is always short of time. He or she knows that only part of all the decision information desired is available, as is the time required to thoroughly study the decision information given. In addition, the decision maker knows that arguments for and against each of the available courses of action will be presented. These arguments are often presented by people who have an interest in a particular course of action, and therefore each argument is calculated to make a specific course of action look good. Finally, these arguments will usually be couched in technical terms or will deal with technical matters. The decision maker does not always have the technical competence to judge these arguments in detail, and is aware of this. He or she therefore looks for quick means of assessing the validity of arguments or the credibility of people presenting the arguments. For instance, the decision maker may take the attitude, "If I know something about the field in which this person claims to be an expert, something that he ought to know but doesn't, I'll disregard everything that is said"—and such an attitude cannot be faulted. It is simply self-defense, a "quick and dirty" means of sorting out which experts should be believed and which should be ignored. Nevertheless, the existence of such an attitude on the part of the decision maker means that the work that went into a forecast can go for naught if the forecaster destroys his own credibility or that of the forecast through a mistake in presentation. Many of these mistakes do not really reflect on the quality or utility of a forecast and are therefore avoidable if the forecaster has done a good job up to that point. The most common of these credibility-destroying mistakes will be discussed in this section.

One such mistake is to present the forecast in a narrative instead of a quantitative form; a page of prose that boils down to the statement that things are good and are getting better is of no value. The decision maker needs to know in quantitative terms how good things are now and how good they are expected to be at some definite time in the future. Furthermore, there should be no need to hunt through paragraphs of high-sounding irrelevancies to find the few nuggets of quantitative information hidden there. A decision maker who is confronted with large quantities of prose narrative, even if it is based on a well-done forecast and contains some useful statistics, will most likely reject the whole thing on the basis of lack of credibility. If a good job has been done, the forecaster should have no difficulty presenting the forecast in a compact and quantitative form. To do otherwise would look like a cover-up for a shoddy job.

Even when the forecaster has presented the forecast in graphical form, it still may not be credible. There are three common failures in the graphic method of presentation that can cause a decision maker to doubt the credibility of the forecast, and perhaps even that of the forecaster. In fact, if the forecast is well done in the first place, these credibility-destroying mistakes will not emerge at all. Hence the appearance of one or more of them is a clue that the forecast was poorly done.

The first of these mistakes with graphics is shown in Figure 19.4. Here the forecaster has apparently extrapolated from a single point. If so, the forecast is, of course, worthless. If an adequate time series of historical data has been used, this data should appear on the graph. Omitting it in the mistaken notion that doing so unclutters the graphs and allows more room for the forecast (i.e., the future portion of the curve) is to invite the suspicion that there is no historical data backing up the forecast. A decision maker spotting a graphical display such as Figure 19.4 cannot be blamed for suspecting the worst and rejecting the forecast. Therefore a graphical presentation should include the complete time series of historical data, as well as the best-fit curve extrapolated into the future.

The second of the mistakes is the "northeast curve," illustrated in Figure 19.5. Here the forecaster has included some data points in the lower left corner of the graph but has covered them with a fuzzy band going generally northeast across the page; sometimes a wavy line is used instead of the fuzzy band, but the effect is the same. There is no indication of how the band was calculated, nor what level of confidence is to be associated with its upper and lower limits. The decision maker is fully justified in assuming that the fuzzy band represents a freehand drawing that is absolutely unreliable beyond the region covered by the historical

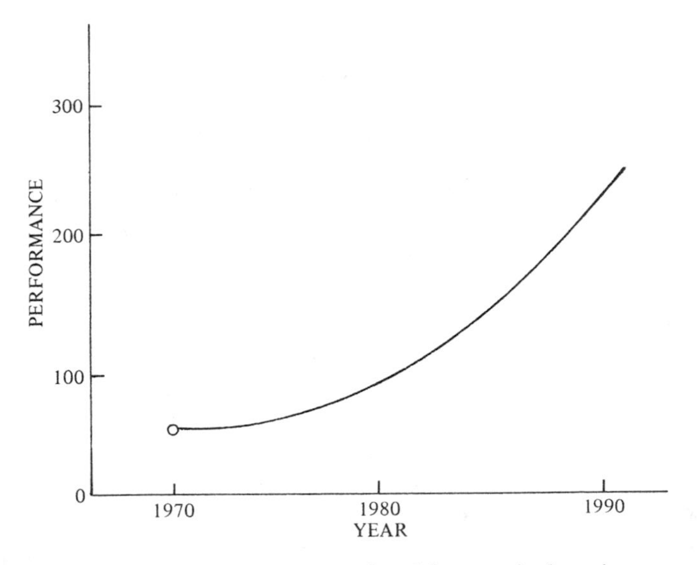

Figure 19.4. Forecast apparently extrapolated from a single point.

data. If a curve is fitted by some appropriate means, or confidence limits calculated, they should be shown and labeled clearly, and the method of curve fitting indicated.

The third of these mistakes is loss of nerve, illustrated in Figure 19.6. Here the forecaster has fitted a trend to a set of data, extended this trend

Figure 19.5. Forecast showing a "northeast curve."

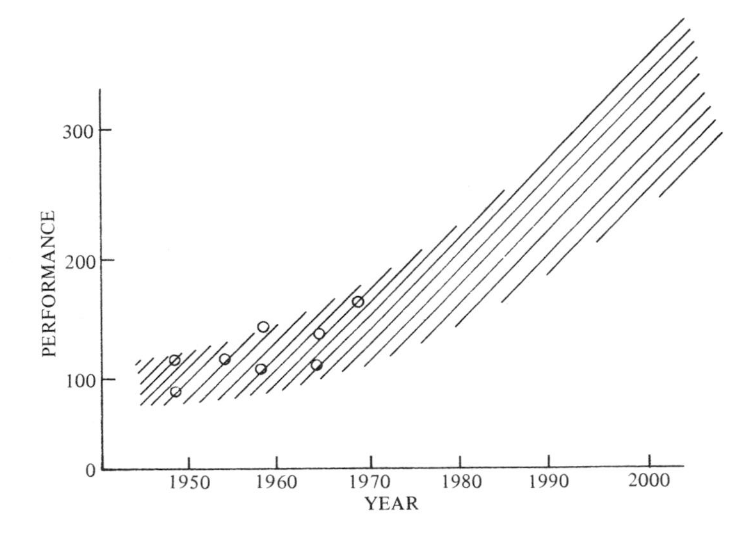

into the future, and then indicated a halt to progress. If, in fact, there is some fundamental limit to the possible level of functional capability that will bring progress to a halt, it should be indicated and explained; but if the forecast is simply presented as shown in Figure 19.6, the decision maker is very likely to assume that the forecaster simply lost faith and would not project a continuation of an obvious trend. This may well lead to the rejection of the entire forecast.

Any of these three mistakes, if actually representative of the way the forecast was prepared, is the graphical equivalent of a narrative forecast. Despite their quantitative appearance, the graphs provide no information at all and will be of no value to the decision maker. If they are used in an attempt to cover up a shoddy job of forecasting, this fact will undoubtedly be detected; if they are used by mistake, they can easily destroy the credibility of a well-prepared forecast.

Even if the forecaster avoids the mistakes described above, there is one more way of running into trouble. A situation that frequently occurs is that of anomalous data. In the set of historical data used to prepare the forecast there may be one or more points that do not seem to belong. This situation is illustrated in Figure 19.7, where the arrow indicates one point deviating markedly from the rest. The mistake, of course, is not in finding the anomalous data point, or even in including it in the historical time series; it is in not explaining it to the decision maker. A forecaster should never grind out a mathematical fit to the historical data without examining the anomalous data points. There is no magic in mathematical fitting techniques that can make up for bad or anomalous data, hence the

Figure 19.6. Forecast apparently showing loss of nerve on the part of the forecaster.

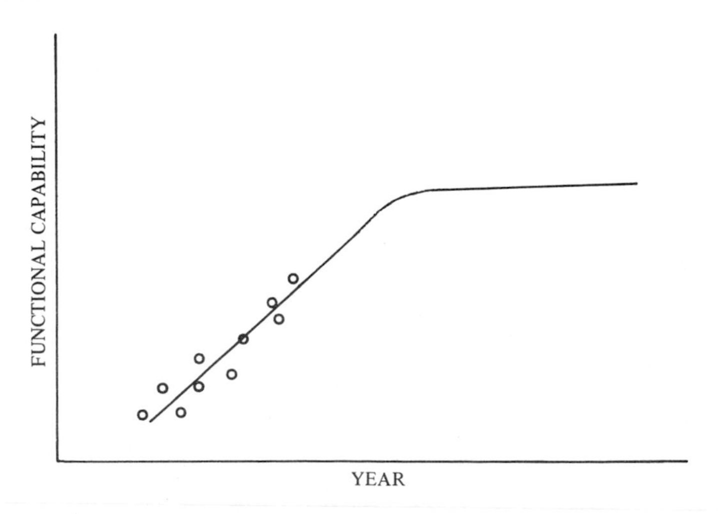

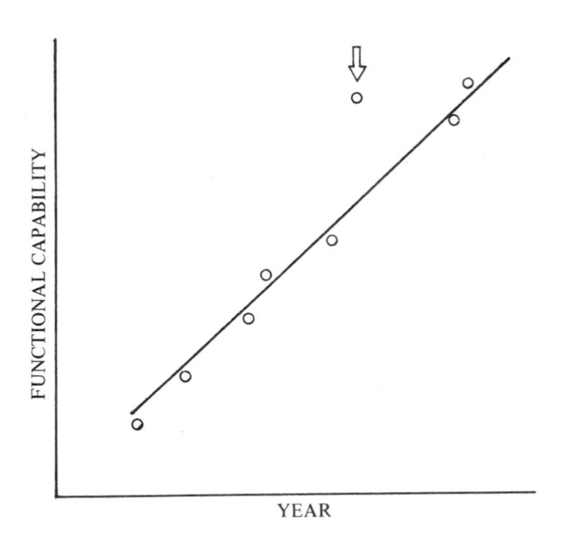

Figure 19.7. Forecast containing an apparently anomalous data point.

forecaster should determine whether these data points do belong with the rest. If so, they should be used, since they represent real deviations that might be repeated; if they do not belong, they should be eliminated from the calculations of the trend fit. If the forecaster fails to explain the anomalous data points, the decision maker may assume that a mathematical fitting procedure was applied blindly, without examining the data that went into the mathematical machinery. The following are some possible reasons for points appearing "too early": They represent unsuccessful freaks; they are suspect because of inadequate documentation of the conditions of measurement; they represent unexploited breakthroughs; they were "ahead of their time" for some reason not likely to be repeated. Similarly, points may appear to be "too late." Some of the reasons for this are the following: They represent development projects that were delayed for nontechnological reasons; they represent developments which did not have to meet stringent requirements and therefore did not fully utilize the available state of the art; they represent the use of an older and well-tested technology in order to satisfy extreme reliability requirements (e.g., amplifiers for undersea telephone cables use technology that is well behind the state of the art but which has demonstrated that it will operate without failure for many years). If one of these reasons applies or if there is some other reason for the existence of anomalous data points, the explanation should be given to the decision maker.

Another category of mistakes that destroys the credibility of a forecast is that of the apparent avoidance of responsibility. Whereas the decision maker rightly does not want the forecaster to make his decisions for him, he does want him to take the problem seriously; he wants to feel that the forecaster has a personal interest in the correctness of the decision being made and therefore would attempt to make the forecast as specific and useful as possible. If the forecaster attempts to avoid any responsibility for the forecast or for the correctness of the decision, he is not living up to his responsibilities. If he makes the mistake of appearing to try to avoid responsibility, he destroys the credibility of his forecast.

One of the ways of appearing to avoid responsibility is to take the attitude, "That's what the analysis showed." If the forecast appears unrealistic in some sense or seems to contradict common-sense notions of appropriateness, the decision maker may ask the forecaster to explain the peculiarities of the forecast. For the forecaster to respond to the effect that these are the results the analysis produced appears irresponsible. The forecaster cannot simply give the data, describe the forecasting technique used, and present the results. The historical data should have been selected on the basis that it is a true representation of what actually happened in the past; similarly, the forecasting method should have been selected on the assumption that it is the most appropriate one for the situation; and the results should have been examined to see if they appear reasonable. If the forecaster conveys the impression that the forecast was simply a mechanical routine, the application of a method to a set of data, without judgment, the decision maker cannot be blamed for refusing to believe the forecast.

Another way of appearing to avoid responsibility is to use multiple sets of assumptions. For instance, let us consider some process that is affected by both a growth rate and a decay rate. Population is one such process; the total number of people alive at some time in the future will depend on the birth and death rates in the intervening period. Similarly, the number of items of a technological device in use at some time in the future will depend on the production rate and the rate of failure or scrappage in the intervening period. When a process is governed by two such rates, the forecaster can assume different values for each rate. If two different growth rates and two different decay rates are assumed, a computer can be made to produce four separate and distinct forecasts. This in itself is not a mistake, however, the forecaster is obliged to choose the rates on a reasonable basis; that is, they appear to be the most likely rates, or they represent upper and lower bounds on likely rates, and so on. For the forecaster to say to the decision maker, "Here are several combinations of rates; you choose the one you like," appears irresponsible. The forecaster is likely to be better informed than the decision maker on which rates or ranges are most likely. Therefore the decision maker should not have to be responsible for selecting the appropriate rates on which to base a forecast. The same problem appears in the more general form of multiple assumptions. If the forecaster produces a range of forecasts, each based on a different set of assumptions, but makes no attempt to indicate their respective likelihoods, the decision maker may suspect that the forecaster is trying to avoid responsibility for the forecast. Thus the mistake of presenting several forecasts based on multiple assumptions without indicating their relative likelihoods can destroy the credibility of the forecast. It should be noted there are cases where it is appropriate to present the decision maker with forecasts based on multiple assumptions; one case is when the decision maker is better informed about the various rates of growth or decay, for instance, than the forecaster. This situation is highly unlikely to arise in practice, however, since in such a case the decision maker is unlikely to need a forecaster. A more common situation is that in which the decision maker has some control over the value that will be taken on by one of the factors; in such a case the real interest is in the effect of exercising his or her control. Then it is entirely appropriate for the forecaster to say, in effect, "If you do such and such, the results will be so and so." If all the relevant factors are outside the control of the decision maker, however, the forecaster must judge the likelihood of the various assumptions or risk seeing the forecast rejected as not credible.

A final way of appearing to avoid responsibility is to present a forecast with an escape hatch. The most common way of doing this to use progressive tenses in making statements, for example, "The rate of R&D expenditures in this field has been rising," "The number of engineers in this field is increasing," "Progress in this field has been continuing at a steady pace." These may all sound like forecasts, but they are not; they are statements about the past or present. The implication is that things will continue as they have been or are; but the forecaster does not actually say they will; and if they do not, the escape hatch is that the statement was true when it was made. It is not enough for one to observe that some trend is rising, one is also obliged to determine whether it is likely to continue to rise, and at what rate. If the forecaster makes a forecast on the basis of previous trends, he or she must state whether these trends will continue. If one gives the appearance of leaving an escape hatch, one cannot complain if the decision maker rejects the forecast as lacking credibility. Hence if a forecaster has done a job properly, the conclusions should be stated boldly so as to not make the mistake of appearing to leave a way out should the forecast turn out to be seriously wrong.

The final type of mistake that can diminish the credibility of a forecast is the omission of the data on which it is based. If a forecast is prepared in written form, the data should be included in an appendix or final section so that the continuity of the presentation is not broken. If the forecast is presented "live," the data should be readily available in case they are requested. In any case, it is good practice to include the data on which the forecast is based: If nothing else, it relieves the forecaster of the necessity to look it up again the next time he or she makes a forecast in the same subject area; more importantly, however, from the standpoint of our concern here, it increases the credibility of the forecast. It makes clear the degree of effort to which the forecaster has gone in order to make the forecast valid and useful. If, in fact, a good job has been done, it would be a mistake to detract from the credibility of the forecast by omitting the data. There is another consideration that must be taken into account: If the forecaster has rejected a historical data point for legitimate reasons, it should nevertheless be included in the historical data tables, with an explanation of why it was not used. If the data point is omitted, the forecaster runs the risk that the decision maker will happen to be aware of that very item and take its omission as an indication of sloppy historical research. Hence it is a mistake to omit even that historical data which was rejected for good reason. Including it, along with the reasons why it was not utilized in preparing the forecast, will increase the credibility of the forecast, since this serves as an indication of the thoroughness with which the historical research was carried out.

4. A Checklist for Forecasts

The two preceding sections have concentrated on mistakes that the forecaster is subjcct to and which may render good work useless. Concentration on avoiding these mistakes, which are both common and serious, can be very helpful in making the presentation of a forecast effective; however, it is also worthwhile to take a more positive approach to ensure that the forecast as a whole meets certain minimum criteria. Such a set of criteria is presented in this section.

These criteria are presented in the form of questions. They are in fact based on the interrogation model presented in Chapter 18; they represent questions that the alert decision maker will (or at least should) ask about a forecast. These questions may not be voiced explicitly, but they will at least implicitly be present in the decision maker's mind. If the manner of presentation of the forecast answers these questions satisfactorily, and the mistakes discussed in the previous section are avoided, then the forecaster may reasonably expect that the forecast will be accepted and used in the decision process.

1. *Why is this forecast needed?* In answering this question the forecaster must address the decision to be made and the alternatives open to the decison maker; the uncertainties that exist regarding each of the alternatives must also be identified. Then the relevance of the forecast to these alternatives and uncertainties must be demonstrated, which will mean showing that the forecast covers the appropriate subject area as well as the time range of interest and has the correct scope.

2. *What portion of the decision information needed is contained in the forecast?* First of all, the technological forecast never provides all the information needed for a decision. There are always considerations other than that of technological change that also bear on the decision; these may include, for instance, demographic or economic change. Hence the technological forecaster is called upon to provide only certain portions of the needed information. Therefore the presentation of the technological forecast should identify the following: (a) portions of the total set of decision information it provides; (b) portions that must be obtained from some other source; (c) how the latter interact with the technological forecast, and (d) portions that should be provided by a technological forecast but which are not contained in the one being presented. It should be noted that there may well be legitimate reasons for failure to include all the needed information. For instance, the forecast may just be an interim report on work not yet completed, in which case, all the information may not yet be available, or it may be a forecast that was originally prepared for some other purpose and which has now been adapted to the current needs. In any case, answering this question places the forecast in proper context; it shows how the forecast fits in with the other information available to the decision maker and how it provides information not available from any other source.
3. *What method was used to prepare the forecast?* The decision maker is not usually concerned with the technical details of the forecasting method used other than its strengths and weakness. The decision maker will want to know whether the method is really appropriate to the situation, whether a better method was available but not used, the kind of failures the method is prone to, and the type of surprises that lie in store should the forecast based on the method fail. The decision maker is also interested in the replicability of the method. For instance, if the forecast was obtained from a panel of experts, the decision maker may be concerned that another panel of equally eminent experts might produce a totally different forecast. In the forecast presentation the forecaster should attempt to satisfy the decision maker on these points by showing that the most appropriate method or combination of methods has been used or that the forecast represents the consensus of a cross section of the best experts available. In the latter case it is also appropriate to indicate whether there was a strong minority view, and the consequences should the minority turn out to be right. If the method is not an intuitive one, the forecaster should show that an adequate sample of historical data was obtained and that the method has worked satisfactorily for similar situations in the past.
4. *What are the assumptions on which the forecast is based?* One of the assumptions behind the forecast is the logical structure con-

necting the past with the future; this will already have been covered in point (3) above. Other assumptions may include estimates of numerical values for which adequate data are not available; assumptions about the behavior of other significant actors involved in the situation or about the continuation (or change) of certain governmental or organizational policies; other forecasts used as inputs to the technological forecast, such as economic or demographic forecasts; and assumptions about the success or failure of ongoing actions when these may have an impact on the events being forecast. If the assumptions behind the forecast are specifically identified, the decision maker will be able to evaluate them better; in particular, if a specific assumption seems incorrect, it can be changed and the forecast modified appropriately. When the forecaster does not specifically identify the assumptions and the decision maker has doubts, there may be no choice but to accept or reject the forecast as a whole.

5. *Why should the assumptions be accepted?* Whenever an assumption is used in place of a fact or in place of certain knowledge, presumably the opposite assumption, or at least a different one, could also be used. The forecaster must show that the assumptions used in the forecast are more reasonable, more plausible, more likely, or in some other way better than any alternative assumptions that might have been used. For instance, the forecaster might be able to place bounds, through physical considerations, on the possible value of some number; it may be demonstrated that the assumed future behavior of certain persons or groups is consistent with their past behavior; or it may be shown that the forecasts used as inputs were obtained from sources that have had a good batting average in the past. In short, the forecaster must show that the full range of possible assumptions has been examined and that the choice made from among them is based on the preponderance of the best evidence available.
6. *What are the sources of information and data?* The decision maker may be legitimately concerned that the forecast may be biased, deliberately or inadvertently, and that it may be invalid because of inappropriate data. By indicating the sources of the forecast's data, the forecaster can allay some or all of such apprehensions. If the data sources are authoritative, have no known bias, or have a known bias that can be discounted; if the data are relevant in the sense that they have been gathered from the appropriate time period, individuals, or groups; and if the data are complete, then the decision maker is more likely to accept the forecast. Thus if the forecaster has done a good job of assembling data for use in the forecasting method or for use in validating the assumptions of the forecast, the decision

maker's confidence in the forecast can be increased by identifying the sources of the data.

The questions in this checklist represent questions that any decision maker should want answered about a forecast before basing a decision on it. Likewise, they are questions that any forecaster should be prepared to answer with regard to a forecast. If the forecaster has done a good job, a presentation that answers these questions clearly should result in the acceptance and use of the forecast.

5. Summary

The ultimate test of a forecast is whether it is accepted and used in the decision-making process; if it is not used by a decision maker, the work that went into it was wasted.

Even when a forecast is well done, it may fail to be used because of its presentation. Hence it is important that the forecaster present the forecast in such a way that the quality of the work and the utility of the results will be readily apparent. In particular, the forecast must be useful and credible to the decision maker. If the latter does not see its utility or does not believe in it, it will not be accepted.

From the standpoint of utility, the forecast must be meaningful and understandable. The decision maker must see it as having meaning with respect to the decision to be made and must be able to understand it. The forecast must be more meaningful and understandable than the raw data that went into it. Graphical presentations can be very helpful in making the forecast understandable; however, there are a number of mistakes the forecaster can make that will render the forecast either meaningless or incomprehensible. If the forecaster is aware of these mistakes, they can be avoided more easily.

From the standpoint of credibility, the decision maker must believe that the forecast presents the best picture of the future which can be obtained on the basis of available data about the past, and knowledge of how the past is related to the future. Again, there are a number of common mistakes that can be made in presenting a forecast which destroy its credibility. Again, if the forecaster is aware of these mistakes, they can be avoided more readily.

From an overall standpoint, there are several questions that the decision maker will have about the forecast; these may be explicit or implicit. In either case the forecaster should make sure ahead of time that the presentation of the forecast will answer these questions. A checklist can be used to determine whether the presentation is satisfactory. If the presentation adequately answers all the questions in the checklist, it should be convincing to the decision maker.

Since the presentation of the forecast represents the last hurdle in the process of getting it accepted and used, the forecaster should give it adequate attention. Considerable time and effort have been expended in preparing the forecast. If the presentation is not convincing, all this work will go down the drain. The forecaster should therefore be willing to spend adequate time and effort on preparing a convincing presentation. By following the suggestions in this chapter, the chances of making a forecast convincing can be increased.

References

Hare, F. (1970). "How Should We Treat Environment." *Science* 167, 352–355.

Wohlstetter, R. (1962). *Pearl Harbor: Warning and Decision* (Stanford, CA: Stanford U. Press).

For Further Reference

Botez, Mihai C. (1981). "Anticipating Decision Makers' Attitudes: A Methodological Suggestion," *Technological Forecasting and Social Change* 19 (3), 257–264.

Problems

1. Consider the following forecasts. Under what circumstances would their presentation as given here be a mistake, and what would be the nature of the mistake?

   a. By 1990 most of the skin panels of aircraft will be made of high-strength, high-modulus composite materials consisting of strong fibers embedded in a binder.

   b. By 1988 most takeoffs and landings of commercial aircraft will be computer controlled, with the pilot serving as a monitor only.

   c. By 1995 space satellites will be used for manufacturing plants in order to take advantage of zero gravity, high vacuum, and abundant solar power.

2. You have obtained the following data for the growth in the functional capability of a specific technology. What is your forecast of the growth of this technology out to year 45, if there is no imminent limit to this growth and no radical changes seem likely in the industry that produces the technology? Prepare your forecast for presentation to a decision maker who has no influence on its progress.

Year	Capability	Year	Capability
2	105	15	223
3	122	20	259
6	142	21	300
11	165	24	349
14	192	29	406

3. A prospective decision depends on the level of functional capability to be achieved by five related technologies. The historical data on each of them is shown in the table below. You have been asked to present forecasts for each for years 60 and 70. You know of no reason for any of them to deviate from their past patterns of change. Prepare a forecast for each of them. In what form can your forecasts be presented most effectively?

Year	Technology				
	A	B	C	D	E
2	15	62	210	340	3000
5	24	64	190	420	3500
10	30	86	250	430	2600
14	54	130	230	540	2900
20	70	180	355	550	2200
23	110	185	360	600	2100
29	140	300	360	860	2400
34	280	360	480	800	1700
39	345	475	520	1100	2000
42	540	640	580	1000	1500
48	730	910	630	1400	1600

Appendix 1

Regression Analysis of Time Series

1. Introduction

Much of the work of the technological forecaster is carried out through the extrapolation of time series. In other forecasting fields the smoothing and extrapolation of time series has been developed to a high degree of sophistication. Economic forecasters in particular have developed techniques for extracting cyclical fluctuations such as seasonal variations in sales, and for taking into account longer-term swings of the business cycle. In addition, they have developed techniques such as moving averages and exponential smoothing, which allow them to detect changes in an underlying process by giving more weight to recent data and lesser weight to more remote data. For the technological forecaster, however, these refinements are luxuries that cannot be afforded. In contrast to the wealth of yearly, monthly, weekly, or even daily data available to the economic forecaster, the technological forecaster is lucky if even 10 or 15 data points spread over two or three decades can be found. And there is virtually no chance that the data points will be regularly spaced in time.

Because of these inadequacies in the data, the technological forecaster is restricted to fairly straightforward techniques for making extrapolations from time series data. The most common method is simple regression. In some cases, however, multiple regression on two or three variables may be appropriate, or polynomial regression might be of use. Only rarely will the technological forecaster have sufficient data to justify more sophisticated treatment.

The treatment of regression analysis given below is not intended to be exhaustive or thorough, instead, it is tailored to the practical needs of the technological forecaster. It will begin with an introduction to statistical analysis, which will be followed by an explanation of simple regression methods, including methods for estimating errors. These methods will next be extended to the fitting of a parabola to the data, and finally to multiple regression on two independent variables.

For each of these methods simple computational schemes will be given that are suitable for use with a hand-held calculator. These will allow efficient computation of the various statistical parameters when manual means must be used. Most computer centers also have programs for carrying out these methods. These programs should be used whenever possible, since they reduce the likelihood of error; they also demand less time, allowing the forecaster to make more detailed analyses than would have been possible if only manual computations were used. The computer programs for some of the methods are included in Appendix 4.

2. Statistical Analysis

The two basic concepts of statistical analysis are *population* and *sample*. We will discuss each of these in turn and then discuss their relationship. We will begin with the concept of population.

A population consists of a set of elements that have at least one characteristic in common. In some cases a population may be defined in terms of several characteristics; then, of course, every element of the population shares all the defining characteristics. Furthermore, a population is generally considered to be the set of all those elements that share the one or more common characteristics. Thus a population is usually defined in such terms as, "all the Xs that have characteristics Y_1, Y_2, ..."

To make the discussion more concrete we will consider some examples of populations. One example is all human beings. Here we have not stated any additional characteristics but have included all elements of a certain kind. As another example, consider all persons residing in the United States. All members of this population are members of the first one, but we have eliminated certain elements of the first population by requiring an additional characteristic that must be held by all elements of the second population. As a third example, consider the population consisting of all registered voters residing in the United States. Here again, by requiring a third characteristic, we have eliminated some additional elements of the previous population. Populations of interest, of course, need not consist of persons. We might be interested in studying any of the following populations: all the farms in Iowa, all the automobiles registered in the state of New Mexico, all the voting precincts in New York City. Furthermore, even when the population does consist of persons, the characteristics that define the population may involve only some common association rather than an intrinsic property of the elements or their location. Such a population, for instance, might be all the students in a particular high school. The point is that populations to be studied may be defined in any way that meets the interests of the person making the definition. In all cases, however, the definition can be cast in the form given above.

Even though a population must consist of elements with certain characteristics in common to meet the definition of a population, other characteristics on which the elements of the population differ may be defined. In fact, most of statistical analysis deals with the variability on one characteristic of populations that are identical by definition on one or more other characteristics. The characteristics on which populations may vary may be either qualitative or quantitative in nature. With regard to a qualitative characteristic, the population may be divided into two or more discrete categories, without there being any numerical relationship between the elements in one category and those in another. With regard to a quantitative characteristic, there is some scale on which the elements of the population can be measured. Thus it is possible to say that one element of the population is larger, with respect to the particular characteristic, than another.

The use of qualitative characteristics serves to subdivide a population into categories, or subpopulations, whose elements are identical with respect to one characteristic in addition to those that identify the larger population. For instance, a population consisting of persons can be subdivided into groups of males and females; the automobiles registered in New Mexico can be subdivided by manufacturer; and the farms in Iowa can be classified as being owner operated or tenant operated. The only quantitative factor that may be involved in the use of qualitative characteristics is the proportion of the total population falling into each category. In what follows we will concern ourselves with quantitative, not qualitative, characteristics of populations.

In the case of a quantitative characteristic, each element of the population can be measured on a one-dimensional scale with regard to the characteristic. This allows the elements of the population to be ordered with regard to their "size" on the quantitative characteristic. Furthermore, both differences and ratios of sizes can be established. Some examples of quantitative characteristics are as follows: The automobiles registered in New Mexico can be scaled according to their weight; the farms in Iowa can be scaled by acreage; and the students in a high school can be scaled by height.

Consider a population of high school students. Suppose we measure the height of each and round off the measurements to the nearest inch. For each student, then, we have a number, a height in inches. On this single characteristic of height the population can be described by a *distribution*, that is, a statement of the number of students having each height. This is known as a *discrete distribution,* since only certain discrete values of the quantitative characteristic are permitted.

Let the number of students having height i in. be n_i, and the total number of students be N. Then the proportion of the total student population having height i in. is given by n_i/N. This is also the probability that a student drawn at random from the population will be found to have height i. To indicate this alternative use as a probability, we will define the quantity $p(i) = n_i/N$. Note that $p(i)$ is a well-defined quantity for all possible values of i. It would be reasonable to assume that in an actual case $p(0)$ would be found to equal zero, and likewise for $p(100)$. However, there will be some range of values for i for which all or most of the $p(i)$ are positive numbers.

Now what is the mean height of all the students in the high school? We will designate this quantity $\mu(i)$. The most common way of computing $\mu(i)$ would be to add the heights of all the students and divide by N. Another possibility is a weighted average, computed by multiplying each possible height by the number of students with that height, adding the products, and dividing by N. In equation form this becomes

$$\mu(i) = \frac{1}{N} \sum_i in(i). \tag{A1.2.1}$$

As a further possibility, the mean could be computed by multiplying each possible height by the proportion of students with that height and summing the products. In equation form this becomes

$$\mu(i) = \sum_i ip(i). \tag{A1.2.2}$$

Each of these methods of computation will produce the same mean value. The first method given is the most common, since it is computationally the simplest. However, the last method can be extended to a more general concept that has considerable utility in statistical analysis; this is the *expected value* or *expectation*.

Let $x(i)$ be any reasonably well-behaved function of i. Then we define the expected value of x, denoted $E(x)$, as

$$E[x(i)] = \sum_i x(i)p(i). \tag{A1.2.3}$$

In the simple case where $x(i)$ is just i, this expression reduces to the preceding one; that is, the expected value is a more general concept than the mean, which reduces to the mean in this simple case. Put another way, the mean is the expected value of the heights i.

The population mean $\mu(i)$ in a sense measures the center of gravity of the population distribution. In addition to a measure of the center of gravity of a distribution, it is useful to have a measure of the dispersion about the mean. To obtain such a measure we take the deviation of each i from the mean $i - \mu$, square these, and then take the expected value of the squared deviations. In equation form this becomes

$$\sigma^2 = \sum_i (i - \mu)^2 p(i). \tag{A1.2.4}$$

This quantity σ^2 is known as the *variance*. Its square root σ is known as the *standard deviation*. As with the mean, there are computationally simpler ways to obtain the variance. The usual form for computation of the variance is

$$\sigma^2 = \frac{1}{N} \sum_i (i - \mu)^2 n(i); \tag{A1.2.5}$$

that is, each deviation is included in the sum as many times as the value i appears in the population.

We will now take up some additional properties of the expected value, properties that we will need later. Assume that instead of measuring just one characteristic of a population, we measure two characteristics. We then have two separate discrete distributions describing the same population. For each individual in the population we have two numbers i and j as measurements of the two characteristics. Let $n(i)$ be the number of individuals having a measurement of i on the first characteristic; let $m(j)$ be the number of individuals having a measurement of j on the second characteristic, let $l(i, j)$ be the number of individuals having measurements of both i on the first characteristic and j on the second. We can then define the following probabilities:

$$p(i) = n(i)/N$$

is the probability of an individual, drawn at random, having a measurement i on the first characteristic;

$$q(j) = m(j)/N$$

is the probability of an individual, drawn at random, having a measurement j on the second characteristic;

$$r(i, j) = l(i, j)/N$$

is the *joint probability* that an individual drawn at random will have both i on the first characteristic and j on the second. By definition, of course,

$$\sum_j r(i, j) = p(i)$$

and

$$\sum_i r(i, j) = q(j).$$

Now let us consider the sum of the measurements of the two characteristics on each individual; that is, for each individual we make measurements i and j and take the sum $i + j$. What is the mean $\mu(i + j)$ and the variance $\sigma^2(i + j)$ of the sum? What is the relationship between these quantities and the same quantities for the two characteristics taken separately? In answering these questions we will make use of the joint probability distribution of the two characteristics.

The sum of the measurements of the two characteristics for each individual will be designated as k. There will, of course, be a distribution of values of k, in that for each value of k there will be a certain number of individuals for whom the sum of $i + j$ is equal to that value. We can then define a probability $s(k)$. This is the probability that if an individual is drawn at random from the population, the measurements on the two characteristics will sum to k. Formally, we can write the expected value of the sum as

$$E(k) = \sum_k ks(k).$$

But the probability $s(k)$ can be written as

$$s(k) = \sum_{i+j=k} r(i, j),$$

where the sum is over all pairs (i, j) that add up to k. Thus we can rewrite the expected value of k as

$$E(k) = \sum_k \sum_{i+j=k} (i + j)r(i, j).$$

We can expand this double sum as shown in the following array:

$$\begin{array}{ll}
(\text{terms with } k = 0) & (0 + 0)r(0, 0) + \\
(\text{terms with } k = 1) & (0 + 1)r(0, 1) + (1 + 0)r(1, 0) + \\
(\text{terms with } k = 2) & (0 + 2)r(0, 2) + (1 + 1)r(1, 1) + (2 + 0)r(2, 0) + \\
\vdots & \vdots \\
(\text{terms with } k = K) & (0 + K)r(0, K) + \cdots + (K + 0)r(K, 0),
\end{array}$$

where K is the maximum value of k.

This array can be regrouped by combining terms involving the same value of i. The result is

$$\begin{array}{ll}
(\text{terms with } i = 0) & \sum_j (0 + j)r(0, j) + \\
(\text{terms with } i = 1) & \sum_j (1 + j)r(1, j) + \\
\vdots & \vdots \\
(\text{terms with } i = K) & \sum_j (K + 0)r(K, 0).
\end{array}$$

This can be written more compactly as

$$\sum_i \sum_j (i + j) r(i, j)$$

or

$$\sum_i \sum_j i r(i, j) + \sum_i \sum_j j r(i, j).$$

This, in turn, can be written

$$\sum_i i \sum_j r(i, j) + \sum_j j \sum_i r(i, j)$$

or

$$\mu(i) + \mu(j).$$

Thus

$$E(i + j) = E(i) + E(j). \tag{A1.2.6}$$

If a is a constant, then

$$\begin{aligned} E(ai) &= \sum_i aip(i) \\ &= a \sum_i ip(i) \qquad \text{(A1.2.7)} \\ &= a\mu(i) \end{aligned}$$

Thus taking the expected value is a linear operation.

Finally, by methods similar to those used for the case of the mean, it can be shown that

$$\sigma^2(i + j) = \sigma^2(i - j) = \sigma^2(i) + \sigma^2(j) \tag{A1.2.8}$$

and

$$\sigma^2(ai) = a^2\sigma^2(i). \tag{A1.2.9}$$

In the preceding paragraphs we dealt with discrete distributions; however, we can also consider a *continuous distribution*. Suppose that in the example of the high school students the heights had not been rounded off to the nearest inch, but had been measured with any degree of precision desired. Under this assumption any measured height can always be extended to more decimal places if desired. Hence it is not meaningful to talk about the probability that a student, drawn at random, will have a height exactly equal to any given number. Instead we consider a *probability density*. This is the probability that a student, drawn at random, will be found to have a height within a specified interval. For instance, in the example with heights rounded off to the nearest inch, we could say that the probability density associated with the interval $i - \frac{1}{2} \leq x < i + \frac{1}{2}$ is $p(i)$. However, we are interested in other intervals as well, especially infinitesimal intervals. So we define a *probability density function* $f(x)$, which is the probability that a student drawn at random will be found to have a height between $x + dx$; that is, it is the probability associated with the infinitesimal interval whose lower end is x.

In populations that have a continuous distribution, instead of the summations used with the discrete distribution, we use integration. In particular,

$$\int_{-\infty}^{\infty} f(x)\,dx = 1, \qquad \text{(A1.2.10)}$$

$$\int_{-\infty}^{\infty} xf(x)\,dx = \mu(x), \qquad \text{(A1.2.11)}$$

$$\int_{-\infty}^{\infty} (x - \mu)^2 f(x)\,dx = \sigma^2(x). \qquad \text{(A1.2.12)}$$

If $g(x)$ is any reasonably well-behaved function of x, we define the expected value of $g(x)$ as

$$\int_{-\infty}^{\infty} g(x)f(x)\,dx = E[g(x)]. \qquad \text{(A1.2.13)}$$

And taking the expected value is still a linear operation, so that

$$E[g(x) + h(x)] = E[g(x)] + E[h(x)] \qquad \text{(A1.2.14)}$$

and

$$E[ag(x)] = aE[g(x)], \qquad \text{(A1.2.15)}$$

where a is a constant.

One of the most common and widely used continuous probability distributions is the *normal* or *Gaussian* distribution. The probability density for this distribution is given by

$$f(x) = \frac{1}{\sqrt{2\pi}\sigma} e^{-(x-\mu)^2/2\sigma^2}. \qquad \text{(A1.2.16)}$$

This distribution is extremely useful because of the *central limit theorem*, which states that if several independent random variables are added, almost without regard to the probability distributions of the individual variables, the sum tends to have the normal distribution. This fact will be of use to us later.

If t is a normally distributed variable, with mean zero and standard deviation equal to unity, its probability density is given by

$$f(t) = \frac{1}{\sqrt{2\pi}} e^{-t^2/2}. \qquad \text{(A1.2.17)}$$

The probability that a value of t, drawn at random, will be between T and $-T$ is given by

$$P(-T \leq t \leq T) = \sqrt{\frac{2}{\pi}} \int_0^T e^{-t^2/2}\, dt. \qquad \text{(A1.2.18)}$$

Tables of the normal probability distribution are widely available; in addition, calculators can be used to compute these probabilities. The values, whether obtained from tables or from a calculator, give the probability that a value drawn at random will have a deviation from zero smaller than the t value. However, in regression analysis we are interested in the reverse question, that is, How wide an interval about the mean must we take to assure that we have a given probability of including an element drawn at random? This information is provided for selected probability values in the confidence interval table (Table A2.2) of Appendix 2. The last row of this table, for cases with more than 30 degrees of freedom, is equivalent to the normal distribution. For example, suppose we have a population that is normal, with mean 5 and standard deviation 2. How wide an interval about the mean has a 50% probability of including an element drawn at random? Reading across the bottom row to the column headed 50%, we obtain a t value of 0.674. Thus the interval 5 $\pm$ 2 $\times$ 0.674 or 5 $\pm$ 1.348 or the interval (3.652, 6.348), has a 50% probability of including an element drawn at random from a normally distributed population with mean 5 and standard deviation 2.

Next we turn to a consideration of samples. If we have a population and are interested in some characteristic of it (other than the ones that define it as a population), we might measure that characteristic for each element of the population. However, in many cases this is either impossible or impractical because of expense or difficulty. In these cases we can take a sample from the population. The sample is a set of elements of the population, usually much fewer in number than the whole population. The sample is drawn randomly, which means that some selection procedure is used assuring that each element of the population is as likely to be selected for the sample as is any other element. Most of statistical analysis deals with questions of what we can say about the parent population on the basis of measurements made on the sample, and the degree of assurance with which we can make these statements.

A sample will consist of a set of elements, x_i, with a total of n elements. As with the population, we can compute a mean value. This is

$$M(x) = \frac{1}{n} \sum_i x_i. \qquad \text{(A1.2.19)}$$

What is the relationship between the sample mean $M(x)$ and the population mean $\mu(x)$? We can determine this by taking the expected value of the sample mean, making use of the properties of the expected value we obtained earlier:

$$
\begin{aligned}
E[M(x)] &= E\left(\frac{1}{n}\sum_i x_i\right) \\
&= \frac{1}{n}E(\sum_i x_i) \\
&= \frac{1}{n}\sum_i E(x_i) \\
&= \frac{1}{n}[n\mu(x)] = \mu(x);
\end{aligned} \tag{A1.2.20}
$$

that is, the expected value of the sample mean is the population mean. We can then use the sample mean as an estimate of the population mean.

Since we know that the expected value of the sample mean is equal to the population mean, we can ask what the dispersion of the sample mean about the population mean is. That is, if we took a large number of samples from the same parent population and computed the mean of each sample, how would the sample means vary about the population mean? To answer this we will compute the variance of the sample mean. Noting that the sample mean is a sum of the form

$$M(x) = (1/n)(x_1 + x_2 + \cdots + x_n),$$

and making use of the rule about the variance of a sum, we see that the variance of the sample mean is

$$
\begin{aligned}
\sigma^2[M(x)] &= 1/n^2[\sigma^2(x) + \sigma^2(x) + \cdots + \sigma^2(x)] \\
&= \sigma^2/n.
\end{aligned} \tag{A1.2.21}
$$

Taking the square root, we get the standard deviation

$$\sigma[M(x)] = \sigma/\sqrt{n}. \tag{A1.2.22}$$

For reasons that will become apparent later, this is also known as the *standard error* of the mean.

We can also compute a variance for the sample:

$$S^2(x) = \frac{1}{n}\sum_i [x_i - M(x)]^2. \tag{A1.2.23}$$

In computing this, we take the deviations from the sample mean, since the population mean is assumed to be unknown. We use the sample mean as an estimate of the population mean. Note that by expanding the term in the brackets above, it can be shown that the following relation holds:

$$S^2(x) = \frac{1}{n}\sum_i (x_i^2) - M^2(x). \tag{A1.2.24}$$

This second form is much simpler from a computational standpoint, since it avoids the necessity of subtracting the sample mean from each element before squaring it. It can be used very conveniently on a calculator.

What is the relationship between the sample variance and the population variance? To determine this we need to develop two relationships based on earlier results. First, we have

$$\sigma^2(x) = E(x - \mu)^2.$$

After expansion this becomes

$$\sigma^2(x) = E(x^2) - \mu^2.$$

Rearranging terms, we have

$$E(x^2) = \sigma^2(x) + \mu^2.$$

Second, we note that

$$\sigma^2/n = E[M^2(x) - \mu^2].$$

After expansion this becomes

$$\sigma^2/n = E[M^2(x)] - \mu^2.$$

Rearranging terms, this becomes

$$E[M^2(x)] = \sigma^2/n + \mu^2.$$

Now we evaluate the expected value of the sample variance:

$$E[S^2(x)] = E\left(\frac{1}{n}\sum_i (x_i^2) - M^2(x)\right). \tag{A1.2.25}$$

Making use of the two relationships obtained above, we have

$$\begin{aligned} E[S^2(x)] &= \sigma^2 - \sigma^2/n \\ &= [(n - 1)/n]\sigma^2; \end{aligned} \tag{A1.2.26}$$

that is, the expected value of the sample variance is slightly less than the population variance. It is said to be a *biased estimate* of the population variance, since on the average it is in error by a known amount in a known direction. We define an unbiased estimate as follows:

$$\overline{S}^2(x) = [n/(n - 1)]S^2(x). \tag{A1.2.27}$$

In what follows we will use an upper bar to designate an estimate that has been modified to remove bias. Paralleling the formulation given above for simple computation of the standard deviation of the sample, the following formula will permit simple computation of the unbiased estimate of the population standard deviation:

$$\overline{S}^2(x) = \sum_i x_i^2 - nM^2(x)/(n - 1). \tag{A1.2.28}$$

To obtain the standard deviation we take the square root of this quantity.

Let us consider this matter of a biased estimate from another viewpoint. We have stated that our sample of n elements was chosen randomly, which means that each of the n elements was chosen independently of the others. Now suppose we deliberately set out to pick a sample whose sample mean is equal to some value we specify, such as a constant a. We may choose the first $n - 1$ elements of our sample independently. However, the nth element cannot be chosen independently. It is now completely specified. Given the $n - 1$ elements we have already selected, and the specified mean value of a, the nth element is fixed in value. Thus if we choose to obtain a sample with a specified mean value, there are only $n - 1$ independent elements in that sample. By analogy with physical and mechanical systems in which the parts are restricted to slide, rotate, or vibrate only in certain specified ways, we say that our sample has $n - 1$ *degrees of freedom*. A sample whose mean is not specified and that has n independent elements, of course, has n degrees of freedom. The point is that by estimating the population mean from the sample mean, we have used up one degree of freedom. Therefore, when we used the estimate of the mean in computing the standard deviation, we introduced a bias in the estimate. To correct for this bias we must divide the sum of squared deviations from the mean by the number of degrees of freedom, rather than by the number of elements in the sample. This point will come up again in cases where we reduce the number of degrees of freedom by more than one.

To illustrate the use of the above material, let us consider the following example. We have taken a sample of elements from some population: x_1, x_2, . . . , x_{50}. The sample mean $M(x)$ is equal to 50. We also compute $S^2(x) = 400$. Correcting for bias, we obtain

$$\overline{S}^2(x) = (\tfrac{50}{49}) \times 400 = 408.1633.$$

We can then compute the estimated standard deviation of the population as

$$\overline{S}(x) = \sqrt{408.1633} = 20.2031.$$

Now what can we say about the mean of the original population from which we drew this sample? Our estimate is that it is 50, but we know that it is highly unlikely to be exactly 50, since another sample might well have given us a different sample mean. In fact, we may have obtained an unfortunately chosen sample whose mean differs considerably from the population mean. However, we can make a probabilistic statement about the population mean. First we compute the standard error of the sample mean:

$$\overline{S}(M) = 20.2031/\sqrt{50} = 2.8571.$$

Since the mean of a sample tends to have the normal probability distribution, regardless of the distribution describing the original population, we can make use of the last row of Table A2.2 in Appendix 2 to determine the probability that our sample mean lies within a certain range of the true mean of the population. The probability that the range $M \pm t\bar{S}(M)$ includes the true mean is the probability associated with the t value. We can thus construct the following set of intervals for the true mean; these are known as *confidence intervals*, with *confidence* meaning the probability that the true value is found in the interval.

Range	Interval	Probability
$M \pm 0.256\bar{S}(M)$	(49.2685824, 50.7314176)	20%
$M \pm 0.685\bar{S}(M)$	(48.0743145, 51.9256854)	50%
$M \pm 1.645\bar{S}(M)$	(45.3000705, 54.6999295)	90%
$M \pm 2.576\bar{S}(M)$	(42.6401104, 57.3598896)	99%

Thus we see that if we assert that the true mean is between about 48.07 and 51.93, we stand one chance in two of being wrong. We can reduce our risk of being wrong by increasing the width of the confidence interval. However, the wider the confidence interval, the less useful it is for whatever our original purpose was in desiring to estimate the population mean. Thus there is a tradeoff between the precision with which we need to be able to estimate the population mean, and the degree of risk we are willing to take of being wrong. In a typical case of making an estimate for a population on the basis of a sample, the tradeoff between the required precision of the estimate (e.g., shortness of the confidence interval) and the tolerable risk of being wrong determines the sample size required. In the usual case in technological forecasting, however, the sample size is fixed by considerations beyond the control of the forecaster. In this case, however, the methods just described can still be used to obtain a confidence interval. This will be described in more detail later.

It was stated above that, because of the central limit theorem, sums of independent random variables tend to be normally distributed. We made use of this fact in the case of the mean value of a sample. However, this situation is strictly true only for large samples. For moderately large samples of 30 or more elements the approximation is quite good and can be used with negligible error. For samples under 30, however, the error is too large to be neglected. In these cases, another probability distribution is used—Student's t distribution. This distribution is tabulated for selected probability values in the first 30 rows of Table A2.2. These table entries are used just as the table entries in the last row were to construct the confidence intervals in the preceding example. The t values are selected from the row corresponding to the appropriate number of degrees of freedom.

Before leaving the simple theory of random sampling and going on to regression methods, we will take up one more property of the standard deviation of the sample. Let us take a sample of n elements from some population $x_1, x_2, \ldots, x_n$. Let c be any constant. Form the deviations $x_i - c$. Now consider the sum

$$D^2 = \sum_i (x_i - c)^2,$$

that is, the sum of squared deviations from the constant c. What value of c will minimize this sum? We will use the procedure of taking the partial derivative of D^2 with respect to c and equating to zero

$$\frac{\partial D^2}{\partial c} = \sum_i \frac{\partial}{\partial c}(x_i - c)^2$$

$$= \sum_i 2(x_i - c).$$

Equating to zero,

$$\sum_i (x_i - c) = 0,$$

or

$$\sum_i = nc,$$

$$c = \frac{1}{n}\sum_i x_i = M(x);$$

that is, for any sample the sum of squared deviations from the sample mean is smaller than the sum of squared deviations from any other constant. We will make use of an analogous relationship in regression methods, where we will also wish to minimize the sum of squared deviations.

Before leaving the subject of statistical analysis, one other topic deserves mention. It has nothing to do with regression methods but will be of value in other aspects of technological forecasting and in the use of statistical methods of analysis of data.

It involves the use of statistics other than the mean and the standard deviation to describe a sample. The mean of a sample is often referred to as a *measure of central tendency*. The variance is often referred to as a *measure of dispersion*. These, however, are not the only such measures. Another measure of central tendency is the *median*. If the elements of a sample are arrayed in order, from smallest to largest, the median is the value of the middle element in the array. If the sample has an odd number of elements, the median is defined unambiguously by this definition. If the sample has an even number of elements, then there is no middle element; instead, there are two elements that share a "middle" position. If these two elements have the same value, the median is taken to be their common value. If these two elements have different values, then the median is not defined. Any value between the two can be chosen as the median. By convention, a value halfway between them is usually chosen. The following examples illustrate these concepts. Suppose a sample is obtained that has the values 1,3,3,4,7,8, and 9. The median for this sample is 4. If a sample has the values 1,3,4,6,7, and 9, the median can have any value between 4 and 6. By convention, it would normally be taken as 5.

The median can also be defined as a value such that half the sample falls above it, and half below it. (Note that this definition holds whether the sample size is even or odd.) A generalization of this definition of the median is that of quartiles, deciles, or, in general, *m*-tiles. Just as the median divides the sample into two equal halves, quartiles divide it into four equal parts. For instance, suppose a sample has the values 1,1,3,4,6,7,8,8,9,11, and 13. The median is 7. The lower or first quartile is 3 and the upper or third quartile is 9. Note that the median is also the second quartile, but this designation for the median is rarely used. If the exact value of a quartile is not precisely defined by this definition, the ambiguity is resolved in the same manner as for medians. For instance, if the sample 1,1,3,5,6,7,7,8,9,9,11,13 is obtained, the median is 7, the lower quartile is 4, and the upper quartile is 9.

Deciles are defined in a manner analogous to the quartiles; they divide the sample into ten equal parts. Similarly, *m*-tiles divide the sample into *m* equal parts.

Once the median and the *m*-tiles are defined, it becomes possible to use them to define a measure of dispersion. For instance, if a sample has been divided into quartiles, a measure of dispersion is the *interquartile range* (IQR), which is defined as the value of the upper quartile minus the value of the lower quartile. For instance, in the last example the IQR is $9 - 4 = 5$. This can be considered as the length of the middle half of the sample and provides some estimate of the degree of dispersion of the sample about the median. An interdecile range can also be defined, which is the value of the ninth decile minus the value of the first decile.

The medians and *m*-tiles are frequently easier to compute than the mean and variance, and for small samples are frequently just as useful for describing the "center" and "spread" of a sample. They are found to be useful in connection with the Delphi procedure described in Chapter 2.

3. Simple Regression

Regression analysis is concerned with the way in which one variable changes when one or more other variables change, and the extent to which changes in the one can be explained by changes in the others. In simple regression only two variables are involved. The values of one of these, known as the independent variable, may be chosen in one of several ways: at random from a population of possible values, systematically in some sort of experimental design intended to explore a specified range of variation, or they may simply be "given," being the only data available because of limitations of historical or organizational data sources. With each of the values of the independent variable, however obtained, there is associated a value of the dependent variable. This is not selected in any way whatsoever but is determined simply by the independent variable and whatever relation connects the two.

In many cases the values of the independent variable are chosen by some random process; that is, there is a population of values of the independent variable and we select a sample from this population by some random method. Most of the techniques of regression analysis deal with this situation. They are designed to determine what can be said about the relationship of the dependent and independent variable and to allow probabilistic statements of the nature described in the preceding section. In the case of technological forecasting, however, this situation simply does not apply. If we have collected data about a certain technological capability and the years in which various levels of that capability were achieved, by no stretch of the imagination can the set of years be viewed as a random sample from a population of possible years. Hence most of the mathematical machinery developed for correlation and regression analysis is of no use to the technological forecaster. We will therefore deal only with the methods that are useful in cases where the independent variable must be considered a "given," and which will allow us to develop the manner in which the dependent variable is affected by changes in the independent variable.

To illustrate the situation more specifically, consider Table A1.1. It shows values of the independent variable T and corresponding values of the dependent variable Y. We may think of Y as the value of some technological parameter as embodied in a specific machine and T as the year in which the machine was introduced. Simply observing Table A1.1 gives the impression that with the passage of time there is a general upward

Table A1.1. *Y* Versus *T*

T	Y	T	Y
1	10	16	40
2	9	17	34
4	6	19	39
5	10	20	50
7	17	22	51
8	23	23	49
10	30	25	50
11	23	26	50
13	26	28	61
14	36	29	66

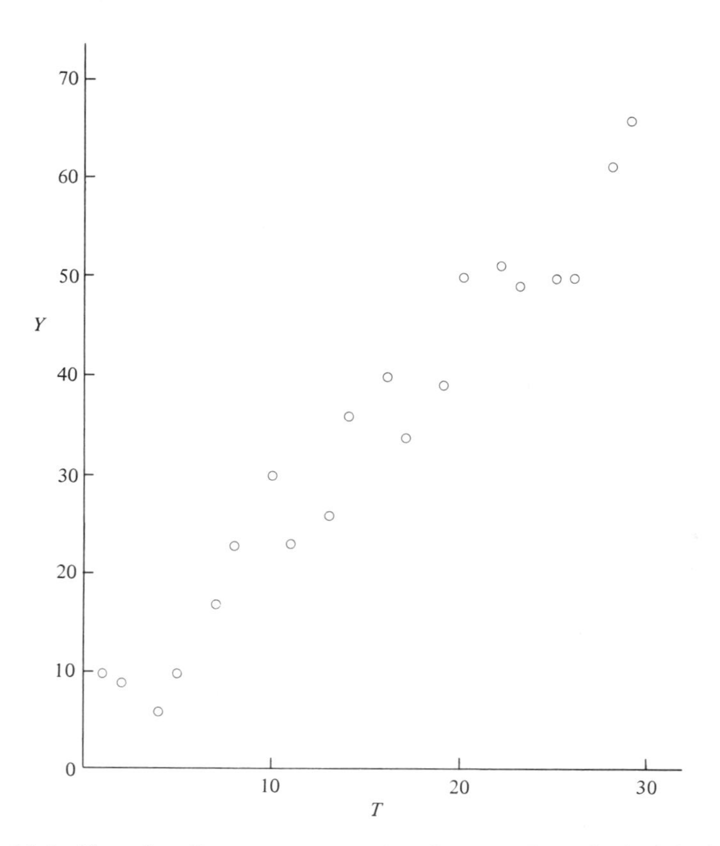

Figure A1.1. Plot of performance versus time for some hypothetical device (data from Table A1.1).

trend in technological achievement. This can be displayed more clearly by plotting Y against T as shown in Figure A1.1.

We will assume, then, that there is a tendency for Y to increase with increasing T. Stated more precisely, we will assume that the relationship between Y and T is expressed by

$$Y = A + BT,$$

which is the equation of a straight line. The measured values of Y do not fall on such a straight line. We will assume that the deviations are the result of random influences that we have not taken into account. Stating this more precisely, we will assume that the ith measured value of Y, at T_i, is described by the equation

$$Y_{im} = A + BT_i + e_i,$$

where the e_i are errors drawn at random from some population of possible errors. We will assume that the parent population of errors is normally distributed and that $E(e_i) = 0$; that is, the errors have a mean value of zero. Our problem, then, is to estimate the values of A and B on the basis of the measured Y_{im}, and to do this despite the corrupting effects of the errors e_i.

There are a number of ways by which we could fit a straight line to our set of data points. The simplest, perhaps, is to plot the points on a graph and then take a straightedge and draw in an "eyeball" fit. One major disadvantage of this method is that it is not objective. Two analysts given the same set of data will probably produce two different lines. To obtain an objective fit to the data independent of the idiosyncrasies of individual analysts, we want some mathematical procedure that is clearly specified and gives the same result for all users. However, there is a large number of mathematical curve-fitting techniques, each one of which naturally gives the same results no matter who uses it. So we need some further criterion for choosing the curve-fitting method to use. Our choice is based on the desire to make use of the material given in the preceding section, so that once the "best-fit" curve has been obtained, we can make probabilistic statements about data points not contained in our original sample.

In the case of the sample for which a characteristic was measured on a single dimension, we chose the sample mean as a measure of the "expected" behavior of the sample, a value that minimized the sum of the squared deviations of the sample elements. By analogy, in the two-dimensional case we choose as our measure of the "expected" behavior of the sample a line that minimizes the sum of the squared deviations of the values of the dependent variable. That is, we wish to minimize the quantity

$$\text{SSD} = \sum_i (Y_{im} - Y_{ic})^2,$$

where Y_{im} is the measured value of Y at T_i, Y_{ic} is the value calculated from the equation of the line that describes the relationship between the dependent and independent variable, and SSD stands for the sum of the squared deviations.

We will use the same approach as we did in showing that the sample mean minimized the sum of the squared deviations. Since we must evaluate two parameters A and B, we will take the partial derivatives of the SSD with respect to both, equate to zero, and solve simultaneously:

$$\text{SSD} = \sum_i (Y_{im} - A - BT_i)^2,$$

$$\frac{\partial \text{SSD}}{\partial A} = -2 \sum_i (Y_{im} - A - BT_i),$$

$$\frac{\partial \text{SSD}}{\partial B} = -2 \sum_i T_i(Y_{im} - A - BT_i).$$

Equating to zero, we have the two so-called *normal equations*

$$\sum_i Y_{im} = nA + B\sum_i T_i, \qquad \text{(A1.3.1.)}$$

$$\sum_i Y_{im}T_i = A\sum_i T_i + B\sum_i T_i^2, \qquad \text{(A1.3.2)}$$

where n is the number of elements in the sample. Solving these two equations, we find that

$$B = (n\sum_i Y_{im}T_i - \sum_i T_i \sum_i Y_{im})/[n\sum_i T_i^2 - (\sum_i T_i)^2], \qquad \text{(A1.3.3)}$$

$$A = \sum_i T_{im}/n - B\sum_i T_i/n. \qquad \text{(A1.3.4)}$$

Other forms of this equation are also used, but these forms are particularly convenient for use on a desk calculator, since the only computations required involving the original variables are

$$\sum_i Y_{im}, \quad \sum_i T_i, \quad \sum_i T_i^2, \quad \text{and} \quad \sum_i Y_{im}T_i.$$

Other forms involving the computation of $M(Y)$ and $M(T)$ and then summation of the squared deviations from these values may be used on a digital computer, but they are inconvenient for manual use.

The equation relating Y_{ic} to T_i is known as the *regression equation*. When plotted on a graph, it is referred to as the *regression line*. Finally, the slope of the regression line B is known as the *regression coefficient*; it is one measure of the relationship between Y and T. In particular, it measures the degree of change in Y associated with changes in T.

We carry out the necessary calculations on the data given above and obtain

$$\sum_i T_i = 300,$$

$$\sum_i T_i^2 = 5990,$$

$$\sum_i Y_{im} = 680,$$

$$\sum_i Y_{im}T_i = 13180.$$

Substituting these values in the formulas, we find that $A = 4$ and $B = 2$, so that the least-squares best-fit equation is

$$Y = 4 + 2T.$$

This line, along with the original data, is plotted in Figure A1.2. The data points are scattered about the regression line, but because of the way we

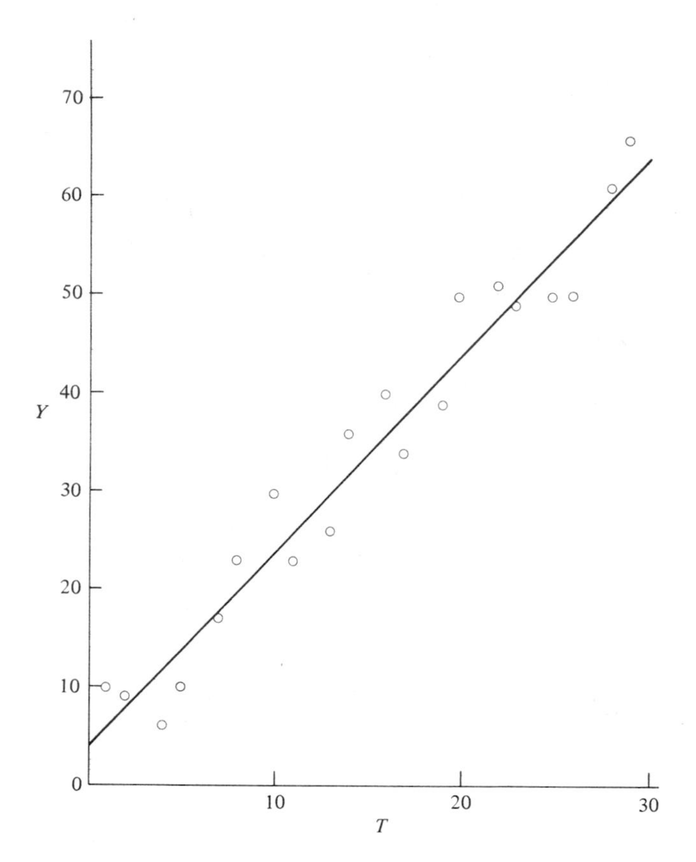

Figure A1.2. Regression line fitted to the data of Table A1.1.

have determined the line, the mean deviation of the data points from the line is zero, and the sum of squared deviations is smaller than for any other possible line. Now we would like to say something regarding the scatter of additional points, if any, that might lie in the region covered by our data, or even more important, regarding the scatter of points about an extension of the line into a region not covered by our data.

To begin with, we consider the mean-square deviation of the measured values of Y from their mean:

$$S^2(Y) = \frac{1}{n} \sum_i [Y_{im} - M(Y)]^2. \tag{A1.3.5}$$

We will add and subtract the same quantity Y_{ic} inside the brackets:

$$S^2(Y) = \frac{1}{n} \sum_i [Y_{im} - Y_{ic} + Y_{ic} - M(Y)]^2.$$

We group the first two and last two terms inside the brackets and carry out the squaring operation:

$$S^2(Y) = \frac{1}{n} \sum_i (Y_{im} - Y_{ic})^2 + \frac{2}{n} \sum_i (Y_{im} - Y_{ic})[Y_{ic} - M(Y)] + \frac{1}{n} \sum_i [Y_{ic} - M(Y)]^2.$$

The second term, involving a cross product, can be shown to be identically zero. We thus have

$$S^2(Y) = \frac{1}{n} \sum_i (Y_{im} - Y_{ic})^2 + \frac{1}{n} \sum_i [Y_{ic} - M(Y)]^2. \qquad \text{(A1.3.6)}$$

We have partitioned the mean-square deviations of Y_{im} from their mean into two parts. The first is recognizable as the mean-square deviation of the Y_{im} from the regression line. This will be designated as $S^2(Y \cdot T)$, and its square root $S(Y \cdot T)$ is known as the *standard error of estimate* of the regression of Y on T. It plays a role analogous to the standard deviation of a sample in statistical analysis. The second term is the mean-square deviation of the *calculated* values of Y from $M(Y)$. It will be designated $S^2(Y_c)$. (Note that $Y_{im} - Y_{ic}$ is known as the *residual* at X.)

The term $S^2(Y_c)$ is the portion of the mean-square deviation of Y, $S^2(Y)$, which is said to be "explained" by the regression equation. The ratio $S^2(Y_c)/S^2(Y)$ is the proportion of the original mean-square deviation from $M(Y)$ explained by the regression equation and is known as the *coefficient of determination*; it is designated $r^2(YT)$. Clearly, if the regression equation is a "good fit" to the original data, it will explain a large proportion of the original mean-square deviation from the mean. If $r^2(YT)$ is equal to unify, then all the points Y_{im} would lie on the regression line.

The square root of the coefficient of determination $r(YT)$ is known as the *correlation coefficient* between Y and T. The sign of the correlation coefficient is important, since a positive sign means that an increase in T is accompanied by an increase in Y; a negative sign means that an increase in T is accompanied by a decrease in Y. In these cases Y and T are said to be *positively* or *negatively* correlated, respectively. Values of $r(YT)$ near 1 or -1 indicate a strong degree of correlation between Y and T. However, this does not necessarily imply that a large change in T will be accompanied by a large change in Y. On the contrary, it means that the measured points are clustered closely about the regression line. A large value of r, indicating small deviations from the regression line, might still be accompanied by a small value of B, indicating that Y does not change much, even for large changes in T. Conversely, a small value of r, indicating a large amount of scatter about the regression line, might also be accompanied by a large value of B, indicating that Y changes significantly, even for small changes in T. Thus r and B, although related, are not substitutes for each other nor measures of the same thing. Frequently, it is useful to compute both to get a complete picture of the relation between Y and T.

In practice, $r(YT)$ is not computed in the way implied above, as the square root of the ratio $S^2(Y_c)/S^2(Y)$. In fact, the procedure is the reverse. The quantity $r(YT)$ may be computed directly from the formula

$$r(YT) = \frac{(n \sum_i T_i Y_{im} - \sum_i T_i \sum_i Y_{im})}{\{[n \sum_i (T_i^2) - (\sum_i T_i)^2][n \sum_i (Y_{im}^2) - (\sum_i Y_{im})^2]\}^{1/2}} . \tag{A1.3.7}$$

Many of the terms appearing in this formula also have to be computed for the coefficients of the regression equation, thus cutting down on the amount of computational effort required. In fact, if A and B have already been computed, $r(YT)$ can be obtained by the formula

$$r(YT) = B \left(\frac{n \sum_i (T_i^2) - (\sum_i T_i)^2}{n \sum_i (Y_{im}^2) - (\sum_i Y_{im})^2} \right)^{1/2} \tag{A1.3.8}$$

One then computes $S^2(Y \cdot T)$ as $S^2(Y)[1 - r^2(YT)]$.

As with the standard deviation of a sample, $S^2(Y \cdot T)$ is a biased estimate of the true scatter of the data points about the regression line. However, since we have estimated two parameters A and B from the sample, we have used up two degrees of freedom. The unbiased estimate of the standard error of the sample scatter about the regression line is given by

$$\overline{S}^2(Y \cdot T) = nS^2(Y \cdot T)/(N - 2). \tag{A1.3.9}$$

Having obtained an unbiased estimate of the scatter, we are now in a position to make probabilistic statements about the behavior of points not contained in the original sample, especially points about extensions of the regression line.

Our primary concern is with making a forecast. We are thus interested in extending the regression line into a region in which we have no data. In the example above, for instance, we might well be interested in forecasting the level of the technological parameter Y that will be achieved at some specific time T greater than $T = 29$, for example, $T = 35$ or $T = 40$. We make the forecast by assuming the regression equation holds for values of T greater than $T = 29$, that is, we extend or extrapolate the regression line into the future. Our forecast of Y at T_{40} is obtained by substituting $T = 40$ into the equation, and in this case the calculated value of $Y_{40} = 84$. But the points did not all fall on the regression line in the past; we should therefore not expect them all to fall on the regression line in the future. In short, there will be some error in our forecast. We next consider how to make probabilistic estimates of the size of the error.

Errors in the forecast arise from two sources: the scatter of points about the regression line and errors in the line itself. The errors in the regression line in turn arise from two sources: errors in the height of the line and errors in the slope of the line. We will start with the errors in the regression line.

Consider first the question of the height of the regression line. We have $Y_{ic} = A + BT_i$. If we sum the values of Y_{ic} for all the points T_i, and divide by the number of points, we have

$$M(Y_{ic}) = A + BM(T_i).$$

The mean value of T_i, $M(T_i)$, is simply a constant determined by the selection of points available for the analysis. B, the regression coefficient, is the slope of the line. The height of the line is fixed by the constant A, and errors in A are determined by errors in the mean value of Y_{ic}. In the case of simple statistical analysis we found that the standard error of the mean was equal to the debiased standard error of the sample divided by the square root of the sample size. The situation is analogous here, with the standard error of the sample replaced by the standard error of estimate of the regression of Y on T; that is,

$$\overline{S}[M(Y_c)] = \overline{S}(Y \cdot T)/\sqrt{n}. \tag{A1.3.10}$$

The error in the slope of the line, or the error in the regression coefficient, also involves the standard error of estimate. In addition, it is inversely proportional to the spread of points T_i. This is intuitively understandable: the greater the span of time over which data points are gathered, the smaller the error in the slope of the regression line. If the same number of data points are clustered in a short interval of time, the possible error in the slope would be expected to increase. The expression for the standard error of the regression coefficient is

$$\overline{S}(B) = \overline{S}(Y \cdot T)/(\sum_i [T_i - M(T)]^2)^{1/2}. \tag{A1.3.11}$$

For ease of computation, this can be rewritten as

$$\overline{S}(B) = \sum\sqrt{n}\,\overline{S}(Y \cdot T)/[n \sum_i (T_i^2) - (\sum_i T_i)^2]^{1/2}. \tag{A1.3.12}$$

This makes use of an expression that has already been calculated in the calculation of $\overline{S}(Y \cdot T)$.

Since the squares of standard errors are akin to variances, they may be added directly. Thus we may obtain the standard error for the regression line $\overline{S}(R, T_i)$ as

$$\overline{S}(R, T_i) = \{\overline{S}^2[M(Y_c)] + \overline{S}^2(B)[T_i - M(T_i)]^2\}^{1/2}. \tag{A1.3.13}$$

Note that this standard error is not a single number but a function of T_i that increases with deviations in either direction from $M(T_i)$. This is to be expected, since it involves an error in the slope of the regression line, which passes through the point $[M(T_i), M(Y_c)]$. For the example above we can compute

$$\bar{S}(Y \cdot T) = 4.1633.$$

$$\bar{S}(B) = 0.1077,$$

$$\bar{S}[M(Y_c)] = 0.9310.$$

We can thus set up a table of computations, Table A1.2, for the standard error of the regression line as a function of T_i. We would like to plot a pair of 50% confidence limits about the regression line; that is, we will plot a pair of curves such that it is equally likely that the "true" regression line be inside or outside the band they form. In asserting that the line is inside, we are taking a risk of one chance in two of being wrong. Entering Tabie A2.1, at the row for 18 degrees of freedom, we find that a 50% confidence interval is ±0.688 standard errors from the mean. So we will draw our confidence limits at ±0.688 standard errors about the regression line. The upper and lower confidence limits are given in Table A1.2 and plotted in Figure A1.3, which shows the original data, the regression line,

Table A1.2. Confidence Limits on the Regression Line

		Confidence limits	
T_i	$\bar{S}(R, T_i)$	Upper 50%	Lower 50%
1	1.7720	7.2192	4.7808
2	1.6814	9.1568	6.8432
4	1.5067	13.0366	10.9634
5	1.4236	14.9794	13.0206
7	1.2865	18.8727	17.1273
8	1.1979	20.8242	19.1758
10	1.0755	24.7399	23.2601
11	1.0258	26.7057	25.2943
13	0.9555	30.6574	29.3426
14	0.9372	32.6448	31.3552
16	0.9372	36.6448	35.3552
17	0.9555	38.6574	37.3426
19	1.0258	42.7057	41.2943
20	1.0755	44.7399	43.2601
22	1.1979	48.8242	47.1758
23	1.2685	50.8727	49.1273
25	1.4263	54.9794	53.0206
26	1.5067	57.0366	54.9634
28	1.6814	61.1568	58.8432
29	1.7720	63.2192	60.7808

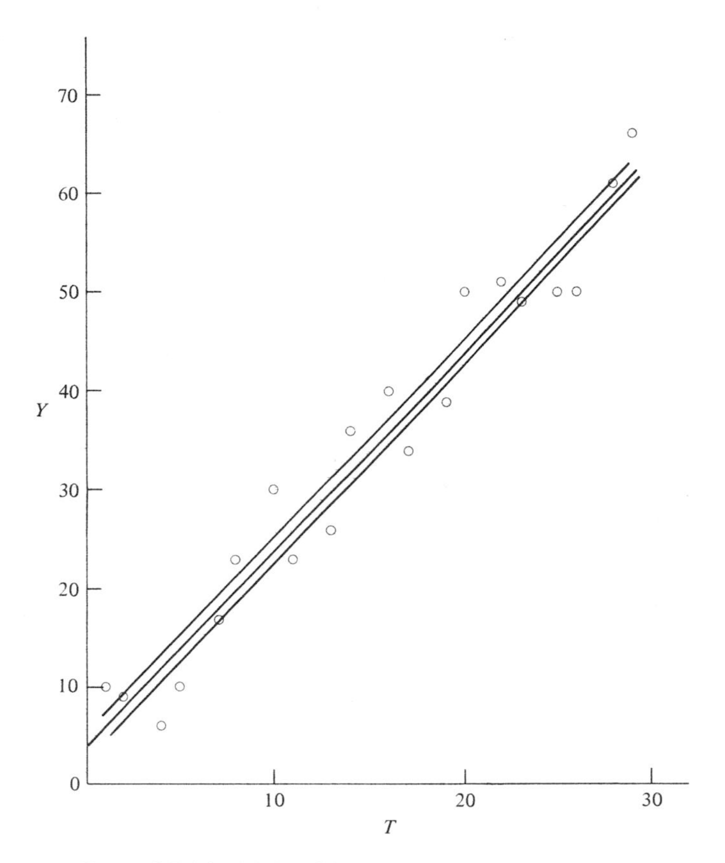

Figure A1.3. Data of Table A1.2, with the regression line and the 50% confidence limits.

and the 50% confidence limits on the regression line. Both from Figure A1.3 and Table A1.2, we see that the limits are symmetrical about $M(T)$. This is to be expected, since the variable term in the expression for $\overline{S}(R, T_i)$ is symmetrical about $M(T)$.

Now that we have examined the errors due to the regression line itself, we consider the final source of error in our forecast, the scatter of points about the regression line. We already have the standard error of this scatter, since it is simply the standard error of estimate. Thus we can obtain the standard error of the forecast as

$$\overline{S}(F, T_i) = [\overline{S}^2(Y \cdot T) + \overline{S}^2(R, T_i)]^{1/2}. \qquad \text{(A1.3.14)}$$

This can be expressed more simply as

$$\overline{S}(F, T_i) = \overline{S}(Y \cdot T)\left(1 + \frac{1}{n} + \frac{n[T_i - M(T_i)]^2}{n \sum_i (T_i^2) - (\sum_i T_i)^2}\right)^{1/2}. \qquad \text{(A1.3.15)}$$

Table A1.3. Confidence Limits on the Forecast

		Confidence limits	
T_i	$\bar{S}(F, T_i)$	Upper 50%	Lower 50%
1	4.5247	9.1130	2.8870
2	4.4900	11.0891	4.9108
4	4.4275	15.0462	8.9538
5	4.4000	17.0272	10.9728
7	4.3522	20.9943	15.0057
8	4.3522	22.9806	17.0194
10	4.3000	26.9584	21.0416
11	4.2878	28.9500	23.0500
13	4.2715	32.9388	27.0612
14	4.2675	34.9360	29.0640
16	4.2675	38.9360	33.0640
17	4.2715	40.9338	35.0612
19	4.2878	44.9500	39.0500
20	4.3000	46.9584	41.0416
22	4.3322	50.9806	45.0194
23	4.3522	52.9943	47.0057
25	4.4000	57.0272	50.9728
26	4.4275	59.0462	52.9538
28	4.4900	63.0891	56.9109
29	4.5247	65.1130	58.8870
30	4.5617	67.1385	60.8615
33	4.6859	73.2239	66.7761
35	4.7791	77.2880	70.7120
38	4.9331	83.3940	76.6060
40	5.0447	87.4708	80.5292
45	5.3515	97.6819	90.3181

For the example above, the 50% confidence limits on the forecast are given in Table A1.3 and plotted in Figure A1.4, which shows the original data, the regression line extrapolated to $T = 45$, and the confidence limits drawn ± 0.688 standard errors about the regression line. This means that if we use the extrapolated regression line as our forecast, 50% of the time we can expect that the actual value of the technological parameter Y, achieved at some time after $T = 29$, will be found to fall outside the confidence limits given. We can increase the likelihood of enclosing the actual values by going to a broader confidence interval, say 70 or 95%.

The standard error of the regression estimate has another use besides that of obtaining the standard error of the forecast. If one carries out the calculations for B on any set of data whatsoever, a value of B will be obtained, whether there is an actual trend or not. If there is no trend, the calculations should, of course, give a value of zero for B; however, because of sampling variations, sometimes a nonzero value of B will be obtained. Thus it is usually a good idea to subject the calculated value

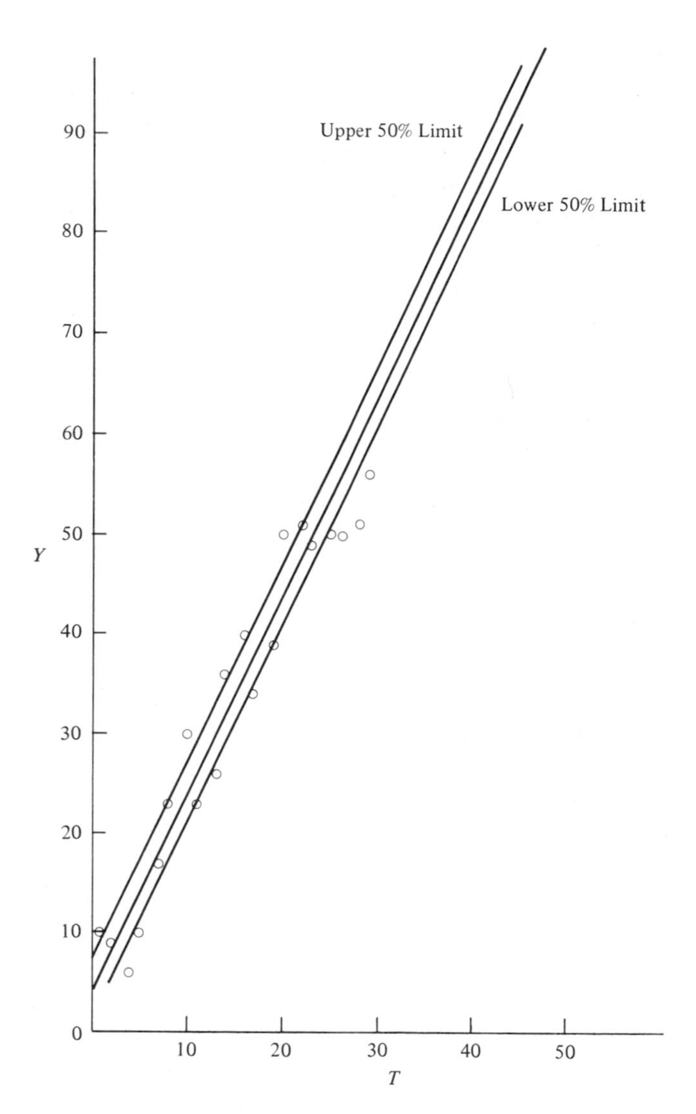

Figure A1.4. Data of Table A1.3, with the regression line and the 50% confidence limits extrapolated into the future.

of B to a statistical significance test. The ratio of $B/\bar{S}(B)$ is the number of standard errors B is away from zero. Using Table A2.1, we can determine the probability of obtaining a value of B that deviates from zero by as much as or more than the value we calculated from our sample, if in fact the true value were zero (i.e., no actual trend). In the sample above the ratio $B/\bar{S}(B)$ is 18.54. Entering Table A2.2 the row for 18 degrees of freedom, we find that if the true value of B were zero, we would have a probability of only 1% that the value of B would exceed 2.878 times its standard error. Since it is extremely unlikely that we would get a deviation as big as the one we did, we can feel confident that our calculated B is significantly different from zero and that the trend is real.

One final point about the linear regression example is worth mentioning. While the least-squares regression procedure has many advantages, as described above, it has one serious shortcoming: It is very sensitive to errors in the data. For instance, suppose that the value Y_{im} corresponding to $T = 28$ is inadvertently misrecorded as 16 instead of 61 (transpositions of this kind are a fairly common mistake in data copying); we find that the computed values of A and B are altered to 7.6405 and 1.6703, respectively. These are quite significant changes, arising from only one error in the data. This serves to emphasize the importance of care in carrying out the computation of the regression coefficients. The regression procedure is not very tolerant of errors and can easily give a seriously wrong answer.

4. Parabolic Regression

In the preceding section we covered methods for fitting a straight line to a set of data. However, in some cases the relationship between Y and T may be curved rather than linear. In such cases a parabola rather than a straight line may provide a better fit to the data. In this section we will consider fitting equations of the form

$$Y_{ic} = A + BT_i + CT_i^2.$$

Again, we will use a least-squares fit in order to be able to make probabilistic statements about the results. The procedure is the same as the one we followed previously. We form the sum

$$\sum_i (Y_{im} - Y_{ic})^2 = \sum_i (Y_{im} - A - BT_i - CT_i^2)^2.$$

Taking partial derivatives with respect to A, B, and C and equating these to zero, we obtain the following normal equations:

$$\sum_i Y_{im} = nA + B\sum_i T_i + C\sum_i T_i^2, \quad \text{(A1.4.1)}$$

$$\sum_i T_i Y_{im} = A\sum_i T_i + B\sum_i T_i^2 + C\sum_i T_i^3, \quad \text{(A1.4.2)}$$

$$\sum_i T_i^2 Y_{im} = A\sum_i T_i^2 + B\sum_i T_i^3 + C\sum_i T_i^4. \quad \text{(A1.4.3)}$$

These equations might be solved algebraically in symbolic form, thus giving formulas for the coefficients, as was done in the linear regression case. Alternatively, they might be solved numerically in each specific case, either by repeated substitution or by Cramer's rule. However, we will manipulate them into another form before solving them, since we can reduce later computational work by doing so.

First, we solve the first equation for A obtaining

$$A = \frac{1}{n}\sum_i Y_{im} - \frac{1}{n}B\sum_i T_i - \frac{1}{n}C\sum_i T_i^2. \qquad \text{(A1.4.4)}$$

The expression for A is substituted into the remaining two equations, and after some manipulation they can be put in the forms

$$n\sum_i T_iY_{im} - \sum_i Y_{im}\sum_i T_i$$
$$= B[n\sum_i T_i^2 - (\sum_i T_i)^2] + C(n\sum_i T_i^3 - \sum_i T_i\sum_i T_i^2), \qquad \text{(A1.4.5)}$$

and

$$n\sum_i T_i^2Y_{im} - \sum_i T_i^2\sum_i Y_{im}$$
$$= B(n\sum_i T_i^3 - \sum_i T_i\sum_i T_i^2) + C[n\sum_i T_i^4 - (\sum_i T_i^2)^2]. \qquad \text{(A1.4.6)}$$

To solve these equations we must evaluate the following quantities from the data:

$$\sum_i Y_{im},\quad \sum_i T_iY_{im},\quad \sum_i T_i^2Y_{im},\quad \sum_i T_i,\quad \sum_i T_i^2,\quad \sum_i T_i^3,\quad \sum_i T_i^4.$$

These two equations can be solved readily once numerical values are substituted. With B and C evaluated, these values can then be substituted in the equation for A.

Next we evaluate the standard error of estimate of the regression of Y on T and T^2. In this case the most convenient way of obtaining it is to evaluate it directly. We have

$$\sum_i (Y_{im} - Y_{ic})^2 = \sum_i (Y_{im} - A - BT_i - CT_i^2)^2,$$
$$= \sum_i Y_{im}^2 + nA^2 + B^2\sum_i T_i^2 + C^2\sum_i T_i^4 - 2A\sum_i Y_{im}$$
$$- 2B\sum_i T_iY_{im} - 2C\sum_i T_i^2Y_{im} + 2AB\sum_i T_i$$
$$+ 2AC\sum_i T_i^2 + 2BC\sum_i T_i^3.$$

(A1.4.7)

Since we have evaluated three constants from the sample, we have used up three degrees of freedom. Thus

$$\overline{S}^2(Y \cdot T, T^2) = \sum_i (Y_{im} - Y_{ic})^2/(n - 3). \qquad \text{(A1.4.8)}$$

To evaluate the standard errors of the two coefficients B and C and that of the forecast, we solve the following two pairs of simultaneous equations for the constants C_{11}, C_{12}, C_{21}, and C_{22}. Actually, since $C_{12} = C_{21}$, only three of the four need be solved for:

$$\begin{aligned} n &= [n \sum_i T_i^2 - (\sum_i T_i)^2]C_{11} + [n \sum_i T_i^3 - \sum_i T_i^2 \sum_i T_i]C_{12}, \\ 0 &= (n \sum_i T_i^3 - \sum_i T_i^2 \sum_i T_i)C_{11} + [n \sum_i T_i^4 - (\sum_i T_i^2)^2]C_{12}, \end{aligned} \tag{A1.4.9}$$

and

$$\begin{aligned} 0 &= [n \sum_i T_i^2 - (\sum_i T_i)^2]C_{21} + (n \sum_i T_i^3 - \sum_i T_i^2 \sum_i T_i)C_{22}, \\ n &= (n \sum_i T_i^3 - \sum_i T_i^2 \sum_i T_i)C_{21} + [n \sum_i T_i^4 - (\sum_i T_i^2)^2]C_{22}. \end{aligned} \tag{A1.4.10}$$

Note that the right-hand sides of these equations, except for the constants to be solved for, are identical with the right-hand sides of the equations to be solved for B and C. Thus if the equations are solved by means of determinants, very little additional computation is required to obtain the four constants above. Once these constants are obtained, we immediately have

$$\bar{S}(B) = \bar{S}(Y \cdot T, T^2)\sqrt{C_{11}}, \tag{A1.4.11}$$

$$\bar{S}(C) = \bar{S}(Y \cdot T, T^2)\sqrt{C_{22}}, \tag{A1.4.12}$$

$$\begin{aligned} \bar{S}(F, T_i) = \bar{S}(Y \cdot T, T^2) \Bigg[1 &+ \frac{1}{n} + C_{11}\left(T_i - \frac{1}{n}\sum_i T_i\right)^2 \\ &+ C_{22}\left(T_i^2 - \frac{1}{n}\sum_i T_i^2\right)^2 + \cdots \\ &+ 2C_{21}\left(T_i - \frac{1}{n}\sum_i T_i\right)\left(T_i^2 - \frac{1}{n}\sum_i T_i^2\right)\Bigg]^{1/2}. \end{aligned} \tag{A1.4.13}$$

To illustrate the method we will work out an example. Assume that we have the values Y_{im} measured at T_i as given in Table A1.4. Carrying out the summations required for solution of the normal equations, we have the following:

$$\sum_i Y_{im} = 1228.1, \qquad \sum_i T_i = 240,$$

$$\sum_i T_i Y_{im} = 23460.0, \qquad \sum_i T_i^2 = 4440,$$

$$\sum_i T_i^2 Y_{im} = 498127.2, \qquad \sum_i T_i^3 = 91800,$$

$$\sum_i T_i^4 = 2{,}049{,}672.$$

Table A1.4. Parabolic Regression Example

T_i	Y_{im}	T_i	Y_{im}
1	3.1	16	61.6
5	25.5	17	87.9
8	50.4	18	106.4
9	35.1	20	88.0
10	28.0	21	107.1
12	70.4	22	134.4
13	63.9	25	145.5
14	49.6	29	171.1

Substituting these values into the two simultaneous equations for B and C, we obtain the following results:

$$80{,}616 = 13{,}440B + 403{,}200C,$$

$$2{,}517{,}271.2 = 403{,}200B + 13{,}081{,}152C.$$

We will solve these equations by Cramer's rule. We recall that to evaluate a 2×2 determinant we proceed as follows. The determinant is written

$$\begin{vmatrix} a_{11} & a_{12} \\ a_{21} & a_{22} \end{vmatrix} .$$

The value of this determinant is $a_{11}a_{22} - a_{12}a_{21}$. Thus the determinant of the coefficients of B and C is (13,440)(13,081,152) − (403,200)(403,200) = 13,240,442,880.

To evaluate B we replace the coefficients of B in the equation above by the numbers to the left of the equality sign. The value of the determinant formed in this way is (80,616)(13,081,152) − (403,200)(2,517,271.2) = 39,586,401,792.

The value of B is then found by dividing this result by the determinant of the original coefficients:

$$39{,}586{,}401{,}792/13{,}240{,}442{,}880 = 3.0.$$

Replacing the coefficients of C by the numbers to the left of the equality sign, and following the same procedure, we obtain $C = 0.1$.

We next evaluate the constants C_{ij}, required for determining standard errors, from the following equations:

$$16 = 13{,}440C_{11} + 403{,}200C_{12},$$

$$0 = 403{,}200C_{11} + 13{,}081{,}152C_{12},$$

and

$$0 = 13{,}440C_{21} + 403{,}200C_{22},$$

$$16 = 403{,}200C_{21} + 13{,}081{,}152C_{22}.$$

Since the coefficients of the C_{ij}s are the same as the coefficients of B and C in the equations above, we need to calculate only three more determinants to obtain the values of these constants. We obtain

$$C_{11} = 0.0158,$$

$$C_{12} = C_{21} = -0.00048723,$$

$$C_{22} = 0.00001624.$$

Using the values of B and C, we can solve for A, obtaining $A = 4$. With all three coefficients of the equation, we can obtain

$$\overline{S}(Y \cdot T, T^2) = 12.93772,$$

$$\overline{S}(B) = 1.6262,$$

$$\overline{S}(C) = 0.0521,$$

$$\overline{S}(A) = 3.2344.$$

The calculated 50% confidence limits are listed in Table A1.5 and plotted in Figure A1.5. We also determine the significance of the regression coefficients:

$$A/\overline{S}(A) = 1.23671, \quad 50\% \geq S \geq 10\% \text{ (13 d.f.)}$$

$$B/\overline{S}(B) = 1.8447, \quad 10\% \geq S \geq 1\% \text{ (13 d.f.)}$$

$$C/\overline{S}(C) = 1.91939, \quad 10\% \geq S \geq 1\% \text{ (13 d.f.)}$$

Table A1.5

			50% Confidence limits	
T_i	Y_{ic}	$\bar{S}(F \cdot T, T^2)$	Lower limit	Upper limit
1	7.1	16.4912	−4.4449	18.4449
5	21.5	14.2558	11.6065	31.3935
8	34.4	13.6818	24.9049	43.8951
9	39.1	13.6156	29.6508	48.5492
10	44.0	13.5862	34.5712	53.4288
12	54.4	13.5876	44.9702	63.8298
13	59.9	13.5991	50.4622	69.3378
14	65.6	13.6089	56.1554	75.0446
16	77.6	13.6108	68.1541	87.0459
17	83.9	13.6028	74.4597	93.3403
18	90.4	13.5925	80.9668	99.8332
20	104.0	13.5913	94.5676	113.4324
21	111.1	13.6193	101.5482	120.4518
22	118.4	13.6827	108.5042	127.4958
25	141.5	14.2388	131.1183	150.8817
29	175.1	16.4289	163.5983	186.4017
35	221.5	24.0340	214.3204	247.6796
45	341.5	47.7759	307.8435	374.1565

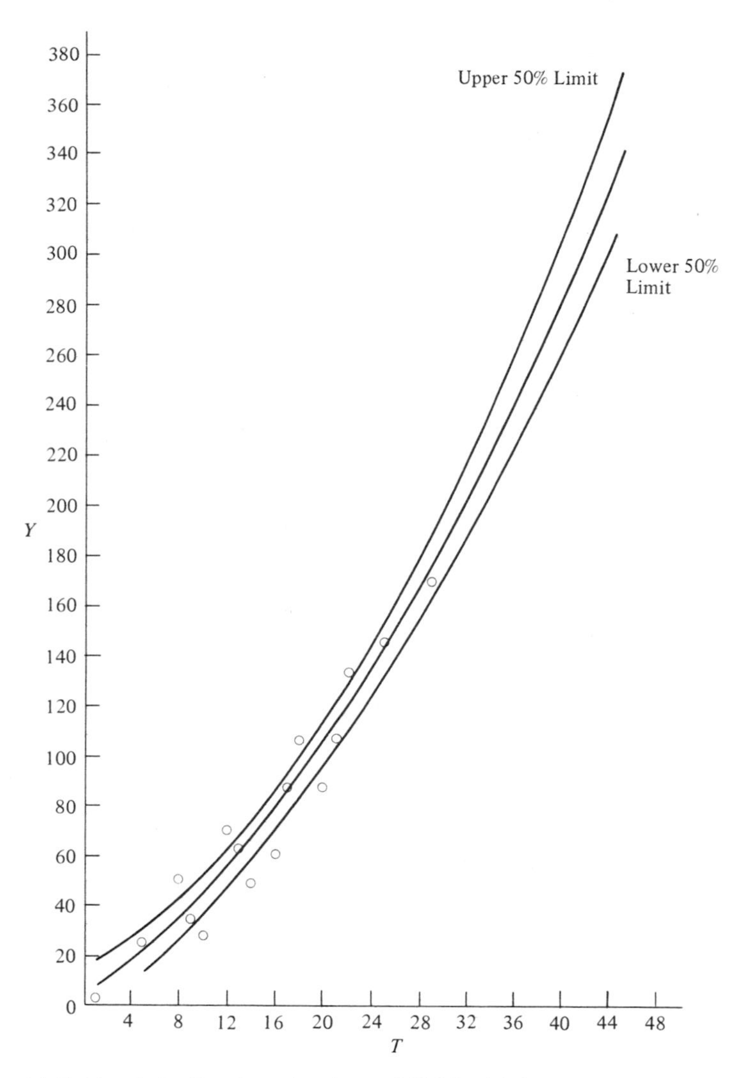

Figure A1.5. Parabola fitted to the data of Table A1.4.

We see that eight of the 16 data points in the original sample fall outside the 50% confidence limits. Also, from the third column of Table A1.5 we see that the standard error of the forecast does not have a minimum at $T = M(T)$, as was the case in the linear regression example. This means that the confidence band is not symmetrical about $T = M(T)$. Examination of equation (A1.4.13) shows that this will normally be the case, since the expression for the standard error involves three variable terms, which in general will not all be equal to zero simultaneously. Furthermore, the cross-product term involving both T and T^2 will change sign at $T = M(T)$

and $T = M(T^2)$. Hence the size of the standard error of the forecast is a rather complicated function of T.

The width of the confidence band at $T = 45$, 66.3130, is almost 20% of the forecast value of 341.5. While a sharper forecast might be desired, it must be remembered that $\pm 10\%$ is "not too bad," in the sense that the technological forecaster frequently has to be satisfied with even broader confidence limits. The additional data points that would increase the number of degrees of freedom in the standard error of estimate, and which then might decrease the width of the confidence interval, simply cannot be obtained. Of course, if additional data were badly corrupted by error, it would not help to improve the sharpness of the forecast. In this example the sum of squares due to regression (i.e., explained by the regression equation) is 33031.72, and the sum-of-squares deviation from the regression line is 2176.0. Thus the regression equation explains 93.41% of the sum-of-squares deviation from $M(Y)$. If additional data points did not alter this percentage significantly, the width of the confidence band would not be reduced.

One final point should be mentioned. If the coefficient of the squared term is not statistically significant, the regression analysis should be repeated without the squared term, that is, a linear regression should be obtained. The point is that if the squared term is included, the constant term and the coefficient of the first power term will be different from the equivalent coefficients in a linear regression. Hence if the forecaster carries out a parabolic regression and finds that the coefficient of the squared term is statistically insignificant, it is not sufficient merely to drop the squared term and retain the constant and the coefficient of the first power term. The analysis must be redone to obtain the correct coefficients for the linear regression.

5. Multiple Linear Regression

In the preceding two sections we have considered the situation where the variable to be forecast depended on only a single variable. In technological forecasting, of course, this single variable is usually time. We have considered the cases of simple linear dependence and of dependence involving both the first and second powers of time. In this section we will depart from dependence on a single variable. The *multiple* in the section heading indicates that the variable to be forecast may depend on two or more independent variables. The *linear,* however, means that we will consider only linear dependence between the dependent and independent variables. Specifically, this means that we will not consider terms in our regression equation involving higher powers, or cross products, of the independent variables. In addition, we will limit ourselves to the case of two independent variables. In most cases of interest to the technological forecaster, one of these variables will be time. The other variable, however, may be almost anything that appears to have some impact on the growth of technology.

We will be dealing with regression equations of the form

$$Y = A + BT + CX.$$

As before, we will estimate the coefficients A, B, and C, using the method of least squares. The sum of the squared deviations between the measured and calculated values of Y is

$$\sum_i (Y_{im} - Y_{ic})^2 = \sum_i (Y_{im} - A - BT_i - CY_i)^2.$$

Taking partial derivatives, we obtain

$$\frac{\partial}{\partial A} \sum_i (Y_{im} - Y_{ic})^2 = -2 \sum_i (Y_{im} - A - BT_i - CX_i),$$

$$\frac{\partial}{\partial B} \sum_i (Y_{im} - Y_{ic})^2 = -2 \sum_i T_i(Y_{im} - A - BT_i - CX_i),$$

$$\frac{\partial}{\partial C} \sum_i (Y_{im} - Y_{ic})^2 = -2 \sum_i X_i(Y_{im} - A - BT_i - CX_i).$$

Equating to zero, we have the following normal equations:

$$\sum_i Y_{im} = nA + B \sum_i T_i + C \sum_i X_i, \qquad \text{(A1.5.1)}$$

$$\sum_i T_i Y_{im} = A \sum_i T_i + B \sum_i T_i^2 + C \sum_i X_i T_i, \qquad \text{(A1.5.2)}$$

$$\sum_i X_i Y_{im} = A \sum_i X_i + B \sum_i T_i X_i + C \sum_i X_i^2. \qquad \text{(A1.5.3)}$$

To solve these equations we must obtain the following quantities from the data:

$$\sum_i Y_{im}, \qquad \sum_i X_i,$$

$$\sum_i T_i Y_{im}, \qquad \sum_i X_i T_i,$$

$$\sum_i Y_i Y_{im}, \qquad \sum_i X_i^2,$$

$$\sum_i T_i, \qquad \sum_i T_i^2.$$

Later we will also need $\sum_i Y_{im}^2$.

These quantities could be substituted in the normal equations and the coefficients solved for directly. However, as in the previous section, we will follow an alternate procedure that will reduce the amount of computation required in the long run. First we solve for A:

$$A = \frac{1}{n}\sum_i Y_i - \frac{1}{n}B\sum_i T_i - \frac{1}{n}C\sum_i Y_i. \qquad \text{(A1.5.4)}$$

Substituting this value for A in the other two equations, we obtain two simultaneous equations:

$$n\sum_i T_iY_{im} - \sum_i T_i \sum_i Y_{im} = B\,[n\sum_i T_i^2 - (\sum_i T_i)^2] + C(n\sum_i T_iX_i - \sum_i T_i\sum_i X_i), \qquad \text{(A.1.5.5)}$$

$$n\sum_i X_iY_{im} - \sum_i X_i \sum_i Y_{im} = B(n\sum_i T_iX_i - \sum_i T_i\sum_i X_i) + C[n\sum_i X_i^2 - (\sum_i X_i)^2]. \qquad \text{(A.1.5.6.)}$$

As in the preceding section, the right-hand sides of these equations are also used in equations involving constants that will be used to obtain the standard errors of the regression coefficients. These equations are

$$\begin{aligned} n &= [n\sum_i T_i^2 - (\sum_i T_i)^2]C_{11} + (n\sum_i T_iX_i - \sum_i T_i\sum_i X_i)C_{12}, \\ 0 &= (n\sum_i T_iX_i - \sum_i T_i\sum_i X_i)C_{11} + [n\sum_i X_i^2 - (\sum_i X_i)^2]C_{12}, \end{aligned} \qquad \text{(A1.5.7)}$$

and

$$\begin{aligned} 0 &= [n\sum_i T_i^2 - (\sum_i T_i)^2]C_{21} + (n\sum_i T_iX_i - \sum_i T_i\sum_i X_i)C_{22}, \\ n &= (n\sum_i T_iX_i - \sum_i T_i\sum_i X_i)C_{21} + [n\sum_i X_i^2 - (\sum_i X_i)^2]C_{22}. \end{aligned} \qquad \text{(A1.5.8)}$$

Once we have the determinant of the right-hand side of these three pairs of equations, the coefficients can be solved for quite rapidly. As before, $C_{12} = C_{21}$, and it need be solved for only once.

In the preceding sections we obtained the standard error of estimate by obtaining the values Y_{ic}, subtracting them from the corresponding Y_{im}, and summing the squares of the differences. In the case of multiple linear regression an easier method is available. The standard error of estimate may be obtained as follows:

$$\bar{S}^2(Y \cdot T, X) = \frac{[n\sum Y_{im}^2 - (\sum Y_{im})^2] - B(n\sum T_iY_{im} - \sum Y_{im}\sum T_i) - C(n\sum X_iY_{im} - \sum Y_{im}\sum X_i)}{n(n-3)}. \qquad \text{(A1.5.9)}$$

The standard errors of the coefficients can then be obtained from the following equations:

$$\bar{S}(A) = \frac{\bar{S}(Y \cdot T, X)}{\sqrt{n}}, \tag{A1.5.10}$$

$$\bar{S}(B) = S(Y \cdot T, X)\sqrt{C_{11}}, \tag{A1.5.11}$$

$$\bar{S}(C) = S(Y \cdot T, X)\sqrt{C_{22}}, \tag{A1.5.12}$$

The standard error of the forecast is given by the following expression:

$$\begin{aligned}\bar{S}(F \cdot T_i, X_i) = \bar{S}(Y \cdot T, X) \Bigg[&\frac{n+1}{n} + C_{11}\left(T_i - \frac{1}{n}\sum_i T_i\right)^2 \\ &+ C_{22}\left(X_i - \frac{1}{n}\sum_i X_i\right)^2 + \cdots \\ &+ 2C_{21}\left(T_i - \frac{1}{n}\sum_i T_i\right)\left(X_i - \frac{1}{n}\sum_i X_i\right)\Bigg]^{1/2}.\end{aligned} \tag{A1.5.13}$$

The *coefficient of multiple correlation* can be computed by the expression

$$R(Y \cdot TX) = \left(\frac{B(n\sum T_i Y_{im} - \sum T_i \sum Y_{im}) + C(n\sum X_i Y_{im} - \sum X_i \sum Y_{im})}{n\sum Y_{im}^2 - (\sum Y_{im})^2}\right)^{1/2}. \tag{A1.5.14}$$

The square of the quantity is the *coefficient of multiple determination*. These two coefficients are analogous to the correlation coefficient and coefficient of determination in the simple linear regression case. In particular, the coefficient of multiple determination is the proportion of the variance of Y_{im} explained by the regression equation.

To illustrate the use of this technique we will consider the following example. Let us assume that Y represents some level of technological capability, such as productivity per person-hour in a specific industry. One would assume that this should increase with time, since production technology in any industry can eventually take advantage of technological advances elsewhere in the economy. However, one would also expect it to advance more rapidly in those companies with higher R&D expenditures. We will then assume the following model for productivity per person-hour for a specific company in a specific industry:

$$Y = A + BT + CX,$$

where T is time and X is the average annual R&D expenditure of the company.

To attempt to evaluate the coefficients in the regression equation we assume that the following data have been gathered on four different companies in a particular industry. The four companies have annual R&D expenditures of 2, 4, 6, and 8 units, respectively. Productivity data have been obtained on all four companies in each of the years 1, 3, 5, and 7. The entries in the table are the Y_{im}, which presumably contain the information we want, plus normally distributed errors arising from other causes we have neglected, as well as measurement errors. For use in the formulas the pairs (T, X) can be numbered arbitrarily from 1 to 16.

	X			
T	2	4	6	8
1	13	16	20	25
3	16	19	23	28
5	22	25	29	34
7	31	34	38	43

Before continuing with the example, some observations should be made. The array of data given above represents a complete experimental design. We have two variables at four levels, for a total of 16 combinations. We show data for each of the combinations of levels. This simplifies the example and allows us to concentrate on the technique rather than the problems of data gathering. However, in practice it is very unusual to have a complete experimental design of this nature. There are two major reasons for this.

The first reason for not having a complete experimental design is the frequent problem of missing data. Only very rarely is the technological forecaster able to design an experiment in which the data collection can proceed according to some plan. Usually one just has to make use of whatever historical data happen to be available. Quite frequently the available data have not been collected in exactly the form needed, and often some of the pieces have not been collected at all. This may not have any serious impact on a forecaster's work, provided that most of the data points needed are available and there is no systematic bias involved in the missing data points. As an example of the problem that can arise from the selective deletion of data, assume an extreme case where all the data points except those on the main diagonal (upper left to lower right) of the array above are missing. In such a case it would be impossible to separate the effects of time and of R&D expenditure. In the jargon of the statistician, these two variables would be *confounded*. In a less extreme case the effect of confounding could be reduced, but it still might be present. If, for example, the data points above the main diagonal were missing (i.e., the companies with high R&D expenditures failed to collect data in the earlier years), there would still be some confounding. The effect of time would receive too much weight in the final result, and the effect of R&D expenditure would probably be underestimated. This would most likely show up as a large standard error for the coefficient C. The forecaster will usually be faced with the problem of missing data, and this problem may or may not have a serious impact on the results, depending on how the missing data points are distributed.

The other major reason for not having a complete experimental design may be the cost or other difficulties in gathering a complete set of data. In the example with two variables at four levels we already have 16 data points. If each variable had been taken at eight levels, there would have been 64 data points. This might have been too expensive to permit a complete job of data collection. In this case what is known as a "partially balanced incomplete block design" might be used. This is a scheme for the deliberate deletion of certain data points (i.e., not gathering data for those points at all), with the deletions made in accordance with a pattern designed to have minimal impact on the accuracy of the regression analysis. Most such incomplete designs are intended for use in obtaining the linear regression relationships without a reduction in accuracy, sacrificing information about higher-order interactions; that is, the effect of cross products and higher powers is deliberately confounded with the error term. If there is reason to believe that the effect of these higher-order terms is negligible, then this confounding can result in a significant decrease in the data requirements, with no reduction in the accuracy of the analysis. [The methods for obtaining partial experimental designs are beyond the scope of this presentation; the interested reader should consult a standard work on the subject, such as *Experimental Designs* by Cochran and Cox (1957).] Once the data have been obtained from such a partial design, the analysis is exactly the same as for a complete design.

Continuing with our example, we obtain the following quantities for substitution in the various formulas:

$$\sum T_i = 64, \qquad \sum X_i T_i = 320,$$

$$\sum T_i^2 = 336, \qquad \sum Y_{im} = 416,$$

$$\sum X_i = 80, \qquad \sum T_i Y_{im} = 1904,$$

$$\sum X_i^2 = 480, \qquad \sum X_i Y_{im} = 2240,$$

$$\sum Y_{im}^2 = 11896.$$

We then need to solve the following sets of simultaneous equations:

$$3840 = B(1280) + C(0),$$

$$2560 = B(0) + C(1280);$$

$$16 = C_{11}(1280) + C_{12}(0);$$

$$0 = C_{11}(0) + C_{12}(1280);$$

and

$$0 = C_{21}(1280) + C_{22}(0),$$

$$16 = C_{21}(0) + C_{22}(1280).$$

The determinant of the right side is 1,638,400. Carrying out the solutions, we obtain in the following values:

$$B = 3,$$

$$C = 2,$$

$$A = 4,$$

$$\bar{S}(B) = 0.196114,$$

$$\bar{S}(C) = 0.196114,$$

$$\bar{S}(A) = 0.4385,$$

$$R(Y \cdot TX) = 0.9813.$$

The resulting equation is

$$Y = 4 + 3T + 2X.$$

This equation explains $100 \times (0.9813)^2$ or 96.3% of the variance of the original data, hence is a very good fit. A plot of the regression equation would be a surface in three dimensions. A representation of this three-dimensional surface is shown in Figure A1.6. While representations such as this are sometimes useful from a conceptual standpoint, they are not as useful as a graph in the two-dimensional case, since it is difficult to show the regression surface, the confidence surfaces, and the original data points without the drawing becoming too cluttered to be readable.

A graphic representation is not needed, however, to make use of the regression equation for forecasting purposes. It is possible to forecast productivity in any of the four hypothetical companies in our example if they maintain their current annual R&D expenditures over the period up to the date forecast. The productivity of any other company in the same industry could likewise be forecast if the annual R&D expenditure were

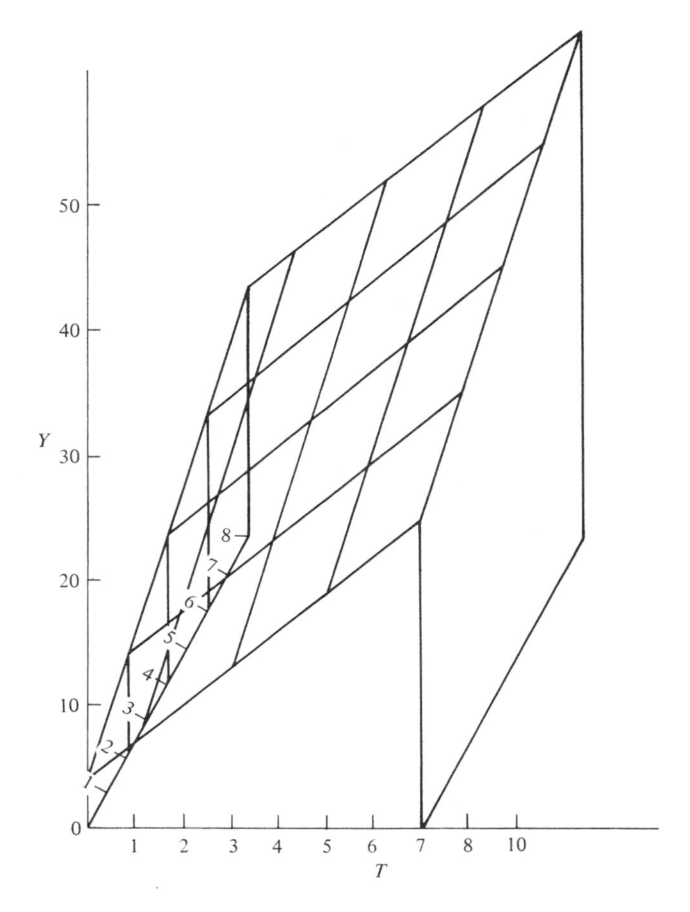

Figure A1.6. Surface for the multiple linear regression example.

known. However, this model does not attempt to take into account the lag between expenditure and payoff and is therefore not suitable for forecasting the effects of changed R&D expenditures. A more complex model could be derived, taking this lag into account if the data were available. The analysis would be much the same, except that productivity in a particular year would be related to the R&D expenditure in some previous year.

As with the case of parabolic regression, if one of the coefficients arising in a multiple linear regression turns out to be statistically insignificant, the regression should be repeated with the variable corresponding to the coefficient deleted. Inclusion of the insignificant variable will certainly distort the value of the constant term and may distort the coefficients of other variables correlated with it.

6. Summary

The preceding sections have provided an introduction to statistical analysis and regression methods. The methods given are adequate for most of the practical problems that will be encountered by the technological forecaster. However, we have barely scratched the surface. Those interested in purusing more advanced methods should consult Ezekiel and Fox (1959) and Spurr and Bonini (1967).

As pointed out earlier, computer programs for carrying out these analyses are widely available. Any good computation center should have in its library at least one program for each of the types of analysis described above. Whenever possible, these programs should be used, since the forecaster's time is more effectively used in the analysis and interpretation of results than in the computation of regression coefficients by manual methods. However, the use of computerized regression techniques should not become a ritual carried out without thought. The forecaster should understand the model of the true situation that underlies each of the regression methods and be certain that in any specific case the model is at least a plausible description of the reality and that there are no reasons to doubt the applicability of the model.

In the practical application of these methods another word of caution should be given. In the examples given above we have consistently used 50% confidence intervals about the forecasts. It should be emphasized that there is nothing sacrosanct about 50%. Other examples used this value mainly for consistency in making comparisons. There was no intent to imply that this is the right value, which forecasters should always use. The percentage of confidence used in a specific case will depend on the nature of the decision to be made. As always, there is a tradeoff between the narrowness of the confidence interval (i.e., the "sharpness" of the forecast) and the risk of being wrong. The forecaster can decrease the risk of being wrong only by broadening the confidence interval. If the interval is too broad, however, it will be useless for decision-making purposes. Thus there is no "right" or "customary" percentage of confidence that should be used habitually. The whole purpose of these techniques is to estimate, with some degree of precision, the risk involved in making a certain forecast. They are not intended to serve as an escape hatch or minimize the forecaster's risk of being found wrong through an expedient that renders the forecast useless for decision-making purposes.

References

Cochran, William G., and Gertrude M. Cox (1957). *Experimental Designs* (New York: John Wiley & Sons).

Ezekiel, Mordecai, and Karl A. Fox (1959). *Methods of Correlation and Regression Analysis* (New York: John Wiley & Sons).

Spurr, William A., and Charles P. Bonini (1967). *Statistical Analysis for Business Decisions* (Homewood, IL: Richard D. Irwin).

Problems

1. Determine the mean and standard deviation for each of the following samples:
a. 1, 2, 3, 4, 5, 6, 7, 8, 9, 10.
b. 10, 22, 24, 42, 36, 77, 99, 96, 89, 85.
c. 6, 72, 91, 14, 36, 69, 40, 93, 61, 97.
d. 8, 3, 76, 7, 6, 27, 98, 18, 1, 79.

2. Determine the median and the interquartile range for the following samples:
a. 99, 19, 19, 74, 6, 81, 21, 84, 44, 11.
b. 91, 36, 63, 97, 1, 44, 10, 42, 5, 18.
c. 40, 55, 64, 35, 58, 32, 44, 22, 18, 94.

3. Consider the following set of values for X and Y.

X	Y
1	1
3	3
7	5
9	7

Regress Y on X. Determine the constant term, the regression coefficient, the correlation coefficient $\overline{S}(Y \cdot X)$, the standard error of the forecast, and the standard errors of the constant term and the regression coefficient.

Appendix 2

Statistical Tables

Table	Subject
A2.1.	Confidence Interval Widths
A2.2.	Table of Random Numbers

Table A2.1. Confidence Interval Widths

Degrees of freedom	Percentage of confidence (%)			
	20	50	90	99
1	0.325	1.000	6.314	63.657
2	0.289	0.816	2.290	9.925
3	0.277	0.765	2.253	5.841
4	0.271	0.741	2.132	4.604
5	0.267	0.727	2.015	4.032
6	0.265	0.718	1.943	3.707
7	0.263	0.711	1.895	3.499
8	0.262	0.706	1.860	3.355
9	0.261	0.703	1.833	3.250
10	0.260	0.700	1.812	3.169
11	0.260	0.697	1.796	3.106
12	0.259	0.695	1.782	3.055
13	0.259	0.694	1.771	3.012
14	0.258	0.692	1.761	2.977
15	0.258	0.691	1.753	2.947
16	0.258	0.690	1.746	2.921
17	0.257	0.689	1.740	2.898
18	0.257	0.688	1.734	2.878
19	0.257	0.688	1.729	2.861
20	0.257	0.687	1.725	2.845
21	0.257	0.686	1.721	2.831
22	0.256	0.686	1.717	2.819
23	0.256	0.685	1.714	2.807
24	0.256	0.685	1.711	2.797
25	0.256	0.684	1.708	2.787
26	0.256	0.684	1.706	2.779
27	0.256	0.684	1.703	2.771
28	0.256	0.683	1.701	2.763
29	0.256	0.683	1.699	2.756
30	0.256	0.683	1.697	2.750
>30	0.256	0.674	1.645	2.576

Table A2.2. Table of Random Numbers

58326	04037	58118	33915	40940	28298	59353	26703	15362	32794	09595	38241	58395	67886	61204
00659	05849	12085	31532	55099	48335	04193	93955	32640	75557	07968	97567	86000	87907	45368
98010	97825	76974	55292	40356	07136	01376	31669	18673	95546	97968	86347	28149	94741	91287
13903	07263	24167	35403	67092	13869	99365	68452	56191	11496	03631	11304	74801	49805	33930
18629	52704	17537	95358	12867	42538	70727	30630	15677	15655	44093	45479	64522	27371	21583
13282	11665	72198	20521	37056	90491	46472	47921	25682	11752	60568	91984	01576	68970	22868
63973	19111	55632	84918	94362	81129	35444	28919	71181	10570	82978	31390	74724	40675	74167
56378	98683	92352	27339	38073	69535	11942	25357	73807	54506	57579	06474	29320	79162	03027
28223	05939	29898	98877	91445	31408	04755	77218	77481	09630	50311	96433	83279	98622	29345
23798	28438	39177	73224	32781	50312	36260	46509	45515	45636	00954	56637	18694	20295	57637
32010	74537	13569	42767	52378	04759	69088	55870	39327	36983	82530	20453	89196	60988	94580
36132	29496	92775	79958	69649	53646	41260	16309	97943	23729	81442	12208	70685	64029	58043
96090	87794	06300	54964	64519	73610	80005	60882	60298	54313	20366	59010	40604	65213	65321
66134	38515	61706	79588	18568	09390	41149	27585	95907	18006	48405	03894	89004	73870	65099
08837	96682	38351	05380	47902	68187	42964	37733	09342	74354	94000	58814	84589	61527	32569
33506	34291	86741	85310	75954	13829	76662	83976	74961	65850	78808	75585	44243	47504	00533
58880	21184	82789	59123	30776	45751	54695	97326	22973	02316	04723	84923	65046	70529	33048
34273	63910	90928	74292	24101	27229	00926	96724	77672	23461	67639	58517	06542	49446	88857
56318	77917	92135	71951	82265	31277	41543	88576	59433	13756	83117	59276	04018	61305	42388
34772	10242	79724	32509	71947	48925	93547	64068	96053	08307	90861	23611	41829	31433	44822
39293	60133	73007	72115	86140	13896	97572	75940	47826	54413	22962	63542	39862	30375	38651
74187	39245	76746	25194	66171	99198	40211	69965	11266	49241	13006	40462	99356	56683	86462
67104	13050	75516	88891	02298	28082	94537	17267	30973	80273	79919	61412	18445	05685	71495
08982	33256	76055	77337	53324	07968	39346	45235	04940	46446	63459	83084	39567	92933	48991
39890	80308	50987	88575	11096	85777	69307	49545	82890	79064	79020	22397	00904	01222	94560
49845	43961	18955	92178	31715	34383	26195	65683	17727	16875	69841	91393	45122	66340	44304
38064	20520	09063	33865	61707	46133	82441	40618	30185	20996	78051	43085	80749	88630	19893
25667	86998	54994	02931	05528	24217	78918	18737	68123	64606	39919	21496	75586	46886	37135
38531	46176	50141	84987	70866	58495	04347	89499	25427	69445	18356	94020	36095	93570	21434
41710	72771	99779	38581	99477	33301	97969	00926	32329	53932	57179	81979	42453	85862	95602

Appendix 3
Historical Data Tables

Table	Subject
A3.1.	U.S. Electric Power Production
A3.2.	Low Temperatures Achieved in the Laboratory
A3.3.	Speed Trend of U.S. Military Aircraft
A3.4.	Transport Aircraft Characteristics
A3.5.	U.S. Population Since 1790
A3.6.	Illumination Efficiency
A3.7.	Solid-Propellant Rocket Engine Performance
A3.8.	Digital Computer Performance
A3.9.	Gross Weight of U.S. Single-Place Fighter Aircraft
A3.10.	Characteristics of Integrated-Circuit Gates for Electronic Computers
A3.11.	Speeds Achieved by Experimental Rocket Aircraft
A3.12.	Papers Published on Masers and Lasers
A3.13.	Steam Engine Efficiencies
A3.14.	Power-Generation Efficiency—Public Utilities
A3.15.	Telephones per 1000 Population
A3.16.	Conversion of U.S. Merchant Marine from Sail to Mechanical Power
A3.17.	Conversion of U.S. Merchant Marine from Wood to Metal
A3.18.	Percent of Dwellings with Electric Power
A3.19.	Energy Consumption from Various Sources
A3.20.	Deep Versus Surface Coal Mining
A3.21.	Ratio of Wing Span to Length of U.S. Combat Aircraft
A3.22.	Maximum Thrust of Liquid-Propellant Rocket Engines
A3.23.	Automobiles in the United States
A3.24.	U.S. Space Payloads Successfully Injected into Orbit
A3.25.	Characteristics of STOL and VTOL Fixed-Wing Aircraft
A3.26.	Characteristics of Helicopters
A3.27.	Activities of U.S. Scheduled Air Carriers
A3.28.	U.S. Total Gross Consumption of Energy Resources by Major Sources

Table	Subject
A3.29.	Measurements of the Velocity of Light
A3.30.	Accuracy of Time Measurement
A3.31.	Historical Data on General Aviation
A3.32.	National Income and Telephone and Radio Usage in 42 Countries
A3.33.	Residential Consumption of Electricity in the United States
A3.34.	Installed Horsepower in Motor Vehicles and Total Number of Motor Vehicles in the United States
A3.35.	Highest Electrical Transmission Voltages Used in North America
A3.36.	Aircraft Speeds
A3.37.	Official Aircraft Speed Records
A3.38.	Accuracy of Astronomical Angular Measure
A3.39.	Accuracy of Measurements of Mass
A3.40.	Efficiency of Incandescent Lights
A3.41.	Lumens From 40-W Fluorescent Lights
A3.42.	Growth of the Cable Television (CATV) Industry: 1952–1978
A3.43.	Magnitude of Engineering Projects
A3.44.	U.S. Aircraft Production
A3.45.	Characteristics of Reciprocating Aircraft Engines
A3.46.	Installed Hydroelectric Generating Capacity
A3.47.	Inanimate Nonautomotive Horsepower
A3.48.	Mill Consumption of Man-Made Versus Natural Fibers
A3.49.	Locomotives in Service
A3.50.	Installed Steam-Generating Capacity
A3.51.	Maximum Steam Turbine Capacity
A3.52.	Maximum Hydroturbine Capacity

Table A3.1. U.S. Electric Power Production (Millions of Kilowatt-Hours)

Year	Production
1945	271,255
1946	269,609
1947	307,400
1948	336,808
1949	345,066
1950	388,674
1951	433,358
1952	463,055
1953	514,169
1954	544,645
1955	629,010
1956	684,804
1957	716,356
1958	724,752
1959	797,567
1960	844,188
1961	881,495
1962	946,526
1963	1,011,417
1964	1,083,741
1965	1,157,583
1966	1,249,444
1967	1,317,301
1968	1,436,029
1969	1,552,757
1970	1,639,771
1971	1,718,000
1972	1,853,000
1973	1,965,000
1974	1,968,000
1975	2,003,000
1976	2,125,000
1977	2,212,000
1978	2,286,000

Source: Statistical Abstract of the United States, Department of Commerce (various years).

Table A3.2. Low Temperatures Achieved in the Laboratory

Year	Experimenter	Temperature (°K)
1721	Fahrenheit	273
1799	de Morveau	240
1834	Thilorier	216
1837 (approx.)	Faraday	163
1883	Wroblewski and Olszewski	137
1885	Wroblewski and Olszewski	121
1885	Wroblewski and Olszewski	48
1898	Dewar	20.4
1899	Dewar	14
1908	Kamerlingh-Onnes	4
1909	Kamerlingh-Onnes	0.8
1933	Giauque and de Haas (independently)	0.25
1933	de Haas, Wiersma, and Kramers	0.085
1934	de Haas and Wiersma	0.031
1935	de Haas and Wiersma	0.0044
1949	De Klerk	0.0029
1950	De Klerk	0.001
1952	Kurti	0.00002

Sources: Encyclopedia Brittanica; Asimov's *Bigraphical Encyclopedia of Science and Technology; National Bureau of Standards Circular* 519 (1951); Garret, *Magnetic Cooling*, 1954; Burton, Smith, and Wilhelm, *Phenomena at the Temperature of Liquid Helium.*

Table A3.3. Speed Trend of U.S. Military Aircraft

Year of first delivery	Airplane	Maximum speed (mph)
1909	Wright Bros. B.	42
1916	Curtiss JN-4	80
1918	(Nieuport 27 C.1)	110
1918	(Spad XIII C.1)	135
1921	Boeing MB-3A	141
1924	Curtiss PW-8	161
1925	Curtiss P-1	163
1927	Boeing PW-9C	158
1929	Curtiss P-6	180
1929	Boeing P-12	171
1933	Boeing P-26A	234
1934	Martin B10-B	212
1937	Boeing YB-17	256
	Seversky P-35	281
1938	Curtiss P-36A	300
1939	Curtiss P-40	357
1940	North American B-25	322
	Bell P-39C	379
1941	Martin B-26	315
	Republic P-43	350
1942	Republic P-47D	420
	North American P-51A	390
1943	North American P-51B	436
1945	Lockheed P-80A	578
1946	Republic XP-84A	619[a]
1948	North American F-86A	671[a]
1950	Boeing B-47A	600
1953	Convair F-102A	860
1954	McDonnell F-101C	1200
1956	Convair B-58	1330
1958	Lockheed F-104A	1404[a]
1961	North American B-70	1800
1965	Lockheed SR-71	1800
1967	Convair F-111	1900

Sources: James C. Fahey, *U.S. Army Aircraft 1908–1946* (New York: Ships and Aircraft), 1946. Performance after 1953 from *Aviation Week* **90** (March 9, 1969), p. 186.

[a] World record performance.

Table A3.4. Transport Aircraft Characteristics

Year	Speed (mph)	Payload (tons)	Length (ft)	Span (ft)	Number of passengers	Ton-miles per hour	Passenger-miles per hour	Span/length	Aircraft
1925	95	—	—	—	—	—	—	—	Fokker F-IV
1926	111	—	49.8	74.0	11	—	1,221	1.486	Ford 4-AT-B
1927	116	—	—	—	—	—	—	—	Fokker Trimotor
1928	148	—	49.8	77.8	13	—	1,924	1.562	Ford 5-AT-B
1933	161	—	—	—	—	—	—	—	Curtis T-32
1933	200	—	51.2	74.0	10	—	2,000	1.445	Boeing 247D
1934	192	—	38.6	55.0	10	—	1,920	1.425	Lockheed 10 Electra
1934	213	—	62.0	85.0	14	—	2,982	1.371	Douglas DC-2
1934	225	—	37.0	50.0	8	—	1,800	1.351	Vultee V-1A
1935	220	2.8	64.5	95.0	21	605.0	4.620	1.473	Douglas DC-3
1938	228	—	74.6	107.3	33	—	7,524	1.438	Boeing 307B
1940	241	—	76.3	108.1	36	—	8,676	1.417	Curtis C-46
1940	272	—	49.8	65.5	14	—	3,808	1.315	Lockheed Lodestar
1940	275	8.6	93.9	117.5	42	2,378.8	11,550	1.251	Douglas DC-4
1941	264	—	—	—	—	—	—	—	Curtis CW-20
1946	329	—	95.1	123.0	64	—	21,056	1.293	Lockheed 649
1946	—	—	71.3	93.3	40	—	—	1.309	Martin 202
1947	315	15.5	100.6	117.5	68	4,882.5	21.420	1.168	Douglas DC-6
1947	347	4.7	74.7	91.8	40	1,630.9	13,880	1.229	Convair CV-240
1948	375	20.5	110.3	141.3	134	7,687.5	50,250	1.281	Boeing C-97A
1949	375	—	110.3	141.3	86	—	32,250	1.281	Boeing 377
1950	312	—	74.6	93.3	40	—	12,480	1.251	Martin 404

1950	370	20.0	113.6	123.0	92	7,400.0	34,040	1.083	Lockheed 1049
1951	314	—	79.2	105.3	44	—	13,816	1.330	Convair CV-340
1952	265	87.0	130.4	174.1	200	23,055.0	53,000	1.335	Douglas C-124C
1954	409	—	108.9	117.5	95	—	38,855	1.079	Douglas DC-7
1955	360	18.3	97.8	132.6	92	6,588.0	33,120	1.356	Lockheed C-130A
1955	400	12.3	112.3	127.5	95	4,920.0	38,000	1.135	Douglas DC-7C
1956	377	16.0	113.6	150.0	92	6,032.4	34,684	1.320	Lockheed 1649A
1956	300	25.0	—	—	—	7,500.0	—	—	Douglas C-133A
1958	579	47.6	150.5	142.4	189	27,560.4	109,431	0.946	Douglas DC-8
1958	450	10.8	104.5	99.0	98	4,860.0	44,100	0.947	Lockheed 188 Electra
1959	300	45.4	157.5	179.7	—	13,620.0	—	1.141	Douglas C-133B
1959	597	48.4	152.9	145.7	189	28,894.8	112,833	0.953	Boeing 707-320B
1959	615	11.3	129.3	120.0	110	6,949.5	67,650	0.928	Convair 880
1959	622	21.6	136.2	130.8	140	13,435.2	87,080	0.960	Boeing 720B
1961	600	44.5	134.5	130.8	126	26,700.0	75,600	0.972	C-135
1963	571	35.4	145.0	159.9	154	20,213.4	87,934	1.103	Lockheed C-141A
1963	632	14.7	133.2	108.0	114	9,290.4	72,048	0.811	Boeing 727
1968	550	110.0	247.8	222.8	—	60,500.0	—	0.899	C-5A
1968	1,550	16.5	215.5	94.5	140	25,575.0	217,000	0.438	Tupolev TU-144
1969	640	110.0	231.3	195.7	490	70,400.0	313,600	0.846	Boeing 747
1969	1,354	14.0	203.7	83.8	100	18,956.0	135,400	0.411	Concorde
1970	575	53.2	182.1	165.3	270	30,590.0	155,250	0.908	Douglas C-10
1970	610	48.5	177.7	155.3	256	29,585.0	156,160	0.874	Lockheed 1011
1972	570	38.1	175.9	147.1	336	21,717.0	191,520	0.836	Airbus 300

Source: Aviation Week, annual Forecast and Inventory issues, various years.

Table A3.5. U.S. Population Since 1790

Year	Population
1790	3,929,214
1800	5,308,483
1810	7,239,881
1820	9,638,453
1830	12,866,020
1840	17,069,453
1850	23,191,876
1860	31,443,321
1870	39,818,449
1880	50,155,783
1890	62,947,714
1900	75,994,575
1910	91,972,266
1920	105,710,620
1930	122,775,046
1940	131,669,275
1950	150,697,361
1960	178,464,236
1970	203,849,000
1971	206,076,000
1972	208,088,000
1973	209,711,000
1974	211,207,000
1975	212,748,000
1976	214,446,000
1977	216,058,000
1978	217,874,000
1979	219,699,000

Source: Statistical Abstract of the United States, Department of Commerce, various years.

Table A3.6. Illumination Efficiency

Year	Source	Lumens per Watt
1850	Paraffin candle	0.1
1879	Edison's first lamp	1.6
1892	Acetylene lamp	0.7
1894	Cellulose filament	2.6
1901	Mercury arc	12.7
1907	Tungsten filament	10.0
1913	Inert gas fill	19.8
1928	Sodium lamp	20.0
1935	Mercury lamp	40.0
1942	Fluorescent lamp	55.0
1962	Gallium arsenide diode	180.0

Source: Encyclopedia Brittanica, 1964 edition.

Table A3.7. Solid-Propellant Rocket Engine Performance

Year	Engine or vehicle	Thrust (lb)	Impulse (10^6 lb-sec)
1956	Sergeant	50,000	1.19
1958	Algol	105,000	3.625
1959	Polaris	62,000	10.0
1961	Minuteman 1st stage	155,000	10.0
1961	AGC FW-1	—	35.0
1962	AGC FW-4	300,000	40.0
1964	TV-412	1,500,000	170.0
1965	Titan III	1,100,000	90.0
1965	AGC-260 (120 in.)	3,500,000	380.0
1967	AGC-260	5,800,000	840.0
1969	7-Segment (120 in.)	1,000,000	140.0

Table A3.8. Digital Computer Performance

Year	Computer	Memory capacity (bits)/access time (μsec)	Matrix inversion time (min)
1946	Eniac	—	5.0
1951	Edvac	0.0011	7.5
1951	Univac I	0.0011	5.2
1951	MIT Whirlwind	0.0011	0.028
1952	IAS Johnniac	—	1.1
1953	IBM 701	0.013	0.22
1953	Univac 1103A	0.040	0.71
1955	IBM NORC	0.014	0.10
1956	IBM AN/FSQ-7 (SAGE)	0.0390	0.047
1956	IBM 704	0.11	0.20
1958	Philco 2000/210	0.17	0.074
1959	IBM 7090	0.50	0.059
1960	Philco 2000/211	1.2	0.047
1960	CDC 1604	0.50	0.038
1960	Univac LARC	0.93	0.015
1961	IBM Stretch	3.4	—
1963	Univac 1107	0.93	0.045
1963	Philco 2000/212	2.39	0.021
1964	CDC 6600	8.3	0.0013
1965	Univac 1108	12.5	0.0079
1966	IBM 360/67	17.7	0.012
1967	B-8500	24.5	—
1967	CDC 6800	31.6	0.00030

Source: Dr. Robert U. Ayres, personal communication.

Table A3.9. Gross Weight of U.S. Single-Place Fighter Aircraft

Year	Aircraft	Gross weight (1000s of pounds)
1918	Nieuport 27 C.1	1.3
1918	Spad XIII C.1	2.0
1921	Boeing MB-3A	2.5
1924	Curtiss PW-8	3.2
1925	Curtiss P-1	2.8
1927	Boeing PW-9C	3.2
1929	Curtiss P-6	3.2
1930	Boeing P-12B	2.6
1933	Boeing P-26A	3.0
1937	Seversky P-35	5.6
1938	Curtiss P-36A	6.0
1939	Curtiss P-40	7.2
1940	Bell P-39C	7.2
1941	Lockheed P-38	15.3
1941	Republic P-43	7.8
1942	Republic P-47D	14.5
1942	North American P-51	9.0
1943	North American P-51B	11.8
1943	Bell P-63A	10.0
1945	Lockheed P-80A	11.7
1946	Republic YP-84	16.5
1948	North American F-86A	13.8
1952	Republic F-84F	25.0
1953	North American F-86F	17.0
1953	North American F-100	28.0
1954	McDonnell F-101	49.0
1956	Convair F-102	27.0
1958	Lockheed F-104	22.0
1958	Republic F-105	48.0
1959	Convair F-106	35.0
1963	McDonnell F-4[a]	54.0
1967	Convair F-111A[a]	91.0

Source: James C. Fahey, *U.S. Army Aircraft 1908–1946* (New York: Ships and Aircraft), 1946; Jane's *All the World's Aircraft* (various editions); *Aviation Week*, inventory issues (various years).

[a] Two-place aircraft.

Table A3.10. Characteristics of Integrated-Circuit Gates for Electronic Computers

Year	Technology	Delay (nsec)	Power (mW)	Speed–power product (pJ)
1968	ECL-III	1.1	60	66
1970	S/TTL	3	20	60
1971	ECL-1000	2	25	50
1972	LA/TTL	10	2	20
1973	NMOS	100	0.1	10
1975	IIL	10	0.1	1
1977	EEIC	0.25	2	0.5

Source: Computer Design, July 1978, p. 69.

Table A3.11. Speeds Achieved by Experimental Rocket Aircraft

Aircraft	Date	Speed (mph)
X-1	14 October 1947	703.134
D-558 II	7 August 1951	1238.0
	20 November 1953	1328.0
X-2	July 1956	1900.0
X-15	17 September 1959	1350.0
	4 August 1960	2196.0
	21 April 1961	3074.0
	15 May 1961	3307.0
	23 June 1961	3603.0
	9 November 1961	4093.0
	27 June 1962	4159.0

Table A3.12. Papers Published on Masers and Lasers

Year	Masers	Lasers	Optical pumping	Light/ coherence	Total
1968	52	1330	124	97	1603
1966	75	1197	58	61	1391
1965	66	730	56	47	899
1964	64	655[a]	59		778
1963	69	305	37		411
1962	53	115	37		205
1961	78		30		108
1960	66		21		87
1959	34				34
1958	37[b]				37
1957	3				3

Source: Number of papers as listed in *Physical Abstracts* for years shown.

[a] 1964 and prior, listed as Optical Masers.
[b] 1958 and prior, carried under heading Electromagnetic Oscillations.

Table A3.13. Steam Engine Efficiencies

Year	Steam engine	Efficiency
1698	Savery	0.5
1712	Newcomen	1.3
1770	Watt	2.8
1796	Watt	4.1
1830	Cornish	10.0
1846	Cornish	15.0
1890	Triple expansion	18.0
1910	Parsons turbine	20.0
1950	Steam turbine	30.0
1955	Steam turbine	38.0

Source: Hans Thirring, *Energy for Man* (Bloomington, Indiana: Indiana University Press), 1958.

Table A3.14. Power-Generation Efficiency—Public Utilities

Year	Pounds of coal (per kilowatt-hour)	Kilowatt-hours (per pound of coal)
1979	0.979	1.021
1978	0.987	1.013
1977	0.969	1.032
1976	0.948	1.055
1975	0.952	1.050
1970	0.909	1.100
1965	0.858	1.166
1960	0.88	1.136
1955	0.95	1.053
1950	1.19	0.840
1945	1.30	0.769
1940	1.34	0.743
1935	1.44	0.694
1930	1.60	0.625
1925	2.03	0.493
1920	3.05	0.328

Sources: Historical Statistics of the United States, U.S. Bureau of the Census, Washington, D.C.; *Statistical Abstract of the United States,* U.S. Bureau of the Census, Washington, D.C.

Table A3.15. Telephones per 1000 Population

Year	Number of telephones
1979	793
1978	769
1977	744
1976	718
1975	695
1970	584
1967	518.3
1966	498.7
1965	478.2
1960	407.8
1955	337.2
1950	280.8
1945	198.1
1940	165.1
1935	136.4
1930	162.6
1925	144.6
1920	123.4
1915	103.9
1910	82.0
1905	48.8
1900	17.6
1895	4.8
1890	3.6
1885	2.7
1880	1.1
1876	0.1

Source: Historical Statistics of the United States, U.S. Bureau of the Census, Washington, DC.

Table A3.16. Conversion of U.S. Merchant Marine from Sail to Mechanical Power

Year	Sail (thousands of tons)	Power (thousands of tons)	Percentage of power
1965	8	19,730	99.95
1960	23	23,553	99.90
1955	40	26,792	99.85
1950	82	28,327	99.71
1945	115	30,247	99.62
1940	200	11,353	98.26
1935	441	12,535	96.60
1930	757	13,757	94.78
1925	1,125	14,976	93.01
1920	1,272	13,823	91.57
1915	1,384	5,944	81.11
1910	1,655	4,900	74.36
1905	1,962	3,741	65.59
1900	1,855	2,658	58.50
1895	1,965	2,213	52.96
1890	2,109	1,859	46.84
1885	2,374	1,495	38.64
1880	2,366	1,212	33.87
1875	2,585	1,169	31.14
1870	2,363	1,075	31.26
1860	4,486	868	16.0
1850	3,010	526	15.0
1840	1,978	202	10.0
1830	1,127	64	5.0
1820	1,258	22	1.7

Source: Historical Statistics of the United States, U.S. Bureau of the Census, Washington, DC.

Table A3.17. Conversion of U.S. Merchant Marine from Wood to Metal

Year	Wood (thousands of tons)	Metal (thousands of tons)	Percentage of metal
1965	1,198	25,318	95.48
1960	1,397	27,184	95.11
1955	1,622	28,336	94.58
1950	1,952	29,263	93.74
1945	1,915	20,898	94.16
1939	2,473	12,159	83.09
1935	2,185	12,469	85.08
1930	2,554	13,514	84.10
1925	2,907	14,499	83.29
1920	3,876	12,448	76.25
1915	3,085	5,305	63.23
1910	3,391	4,117	54.83
1905	3,607	2,850	44.13
1900	3,572	1,593	30.84
1895	3,666	970	20.92
1890	3,798	627	14.16
1885	3,836	430	10.07

Source: Historical Statistics of the United States, U.S. Bureau of the Census, Washington, DC.

Table A3.18. Percentage of Dwellings with Electric Power

Year	Percentage
1955	98.4
1950	94.0
1945	85.0
1940	78.7
1935	68.0
1930	68.2
1925	53.2
1920	34.7
1917	24.3
1912	15.9
1907	8.0

Source: Historical Statistics of the United States, U.S.Bureau of the Census, Washington, DC.

Table A3.19. Energy Consumption from Various Sources (Trillions of Btu)

Year	Coal	Petroleum	Natural gas	Hydropower central station equivalent	Wood
1979	15,100	37,000	19,900	2,876	
1978	13,900	38,000	20,000	2,902	
1977	14,000	37,200	19,900	2,300	
1976	13,700	35,200	20,400	2,954	
1975	12,800	32,700	20,000	3,125	
1970	12,922	29,614	22,029	2,650	425
1965	11,908	23,241	16,098	2,058	577
1960	10,140	20,067	12,699	1,657	832
1955	11,540	17,524	9,232	1,407	1,037
1950	12,913	13,489	6,150	1,440	1,171
1945	15,972	10,110	3,973	1,486	1,261
1940	12,535	7,730	2,726	917	1,358
1935	10,634	5,668	1,974	831	1,397
1930	13,639	5,895	1,969	785	1,455
1925	14,706	4,280	1,212	701	1,533
1920	15,504	2,676	827	775	1,610
1915	13,294	1,418	673	691	1,688
1910	12,714	1,007	540	539	1,765
1900	6,841	229	252	250	2,015
1890	4,062	156	257	22	2,515
1880	2,054	96			2,851
1870	1,048	11			2,893
1860	518	3			2,641
1850	219				2,138

Sources: Historical Statistics of the United States, U.S. Bureau of the Census, Washington, DC; *Statistical Abstract of the United States,* U.S. Bureau of the Census, Washington, DC.

Table A3.20. Deep Versus Surface Coal Mining (Millions of Short Tons)

Year	Deep coal mining	Surface coal mining
1978	242	423
1977	266	425
1976	294.9	383.8
1975	292.8	355.6
1970	338.8	264.1
1965	332.7	179.4
1960	284.9	130.6
1955	343.5	121.2
1950	392.8	123.5
1945	467.6	110.0
1940	417.6	43.2
1935	348.7	23.6
1930	447.7	19.8
1925	503.2	16.9
1920	559.8	8.86
1915	439.8	2.83
1914	421.4	1.28

Source: Mineral Yearbook, U.S. Bureau of Mines, 1980 edition.

Table A3.21. Ratio of Wing Span to Length of U.S. Combat Aircraft

Year of first delivery	Aircraft	Span (ft)	Length (ft)	Span/length
1921	Boeing MG-3A	26.8[a]	20.0	1.43
1922	Curtiss NBS-1	81.5[a]	42.7	1.90
1924	Curtiss PW-8	35.8[a]	22.5	1.58
1925	Curtiss P-1	34.6[a]	22.9	1.51
1927	Boeing PW-9C	35.2[a]	23.0	1.53
1929	Curtiss P-8	34.6[a]	23.5	1.47
	Boeing P-12	33.0[a]	20.0	1.65
1932	Keyston B-4A	82.2[a]	48.9	1.68
1933	Boeing P-26A	28.0	23.9	1.17
1934	Martin B-10B	70.5	44.8	1.57
1937	Boeing YB-17	103.9	68.3	1.51
	Douglas B-13	89.5	56.7	1.58
	Seversky P-34	36.0	25.1	1.43
1938	Curtiss P-36A	37.4	28.5	1.31
1939	Lockheed YP-38	52.0	37.9	1.37
	Curtiss P-40	37.4	31.8	1.18
1940	North American B-25	67.5	51.1	1.32
	Bell P-39C	34.0	30.1	1.13
1941	Martin B-26	65.0	56.0	1.16
	Convair B-24D	110.0	66.3	1.65
	Republic P-43	36.0	28.5	1.26
1942	Martin B-26B	71.0	58.3	1.22
	Republic P-47D	40.8	36.0	1.13
	North American P-51A	37.0	32.3	1.14
1943	Bell P-63A	38.3	32.7	1.17
1944	Boeing B-29	141.3	99.0	1.43
1945	Lockheed P-80A	39.0	34.5	1.13
1946	Republic YP-84	36.9	36.5	1.01
1947	Convair B-36	230.0	163.0	1.41
1948	North American F-86A	37.0	37.0	1.00
1949	Boeing B-47A	116.0	107.0	1.08
1952	Republic F-84F	33.5	43.4	0.77
1953	Convair F-102A	38.0	68.3	0.56
	NAA F-100	38.0	47.0	0.81
1954	Boeing B-52A	185.0	152.8	1.21
	McDonnell F-101C	40.0	67.5	0.59
1956	Convair B-58	57.0	97.0	0.59
1958	Republic F-105B	35.0	64.0	0.55
	Lockheed F-104A	22.0	58.8	0.40
1959	Convair F-106	38.0	70.0	0.54
1963	McDonnell F-4	38.0	58.0	0.66
1964	North American B-70	105.0	185.0	0.57
1967	Convair F-111	63/32	72.0	0.88/0.44

[a] Equivalent monoplan span.

Table A3.22. Maximum Thrust of Liquid-Propellant Rocket Engines

Year	Engine	Thrust (lb)
1942	AL-1000 (JP)	1,000
1943	X35AL-6000 (AJ)	6,000
1945	CORPORAL-E-HW	19,000
1948	XLR-10-RM-2	20,750
1949	XLR-59-AJ-1	90,000
1952	XLR-43-NA-3	120,000
1953	XLR-71-NA-1	240,000
1956	XLR-83-NA-1	415,000
1960	XLR-109-NA-3	500,000
1961	F-1	1,500,000
1963	F-1A	1,522,000

Source: Roderick W. Clarke, "Innovation in Liquid Propellant Rocket Technology," unpublished doctoral dissertation, Stanford University, Stanford, CA, 1968.

Table A3.23. Automobiles in the United States

Year	Number of automobiles (thousands)	Number of automobiles per capita
1900	8	0.000105
1905	77	0.000918
1910	458	0.004956
1915	2,332	0.023192
1920	8,131	0.076371
1925	17,481	0.150916
1930	23,035	0.187159
1935	22,568	0.177351
1940	27,465	0.208140
1945	25,793	0.194692
1950	40,334	0.266699
1955	52,136	0.317316
1960	61,559	0.342009
1965	75,258	0.388298
1970	89,200	0.437579
1975	106,700	0.501532
1976	110,400	0.514815
1977	112,300	0.519768
1978	116,600	0.535172
1979	120,500	0.548478

Sources: Historical Statistics of the United States, U.S. Bureau of the Census, Washington, DC. *Statistical Abstract of the United States,* U.S. Bureau of the Census, Washington, DC.

Table A3.24. U.S. Space Payloads Successfully Injected into Orbit

Year	Successes	Failures	Total
1957	0	1	1
1958	7	10	17
1959	11	8	19
1960	16	13	29
1961	29	12	41
1962	52	7	59
1963	37	8	45
1964	54	5	59
1965	62	6	68
1966	70	4	74
1967	59	3	62
1968	45	3	48
1969	40	1	41
1970	28	1	21
1971	29	4	33
1972	30	0	30
1973	23	2	25
1974	25	2	27
1975	28	2	30
1976	26	1	27
1977	24	2	26
1978	31	0	31
1979	16	0	16

Source: TRW Space Log, 1978/1979, TRW, Inc.

Table A3.25. Characteristics of STOL and VTOL Fixed-Wing Aircraft[a]

Year	Aircraft	Payload (lb)	Empty weight (lb)	Gross weight (lb)	Cruise speed (mph)	Range (miles)
1947	U-1A	1,051	3,341	5,100	103	148
1951	U-1A	1,912	4,680	8,000	106	250
1957	U-3A	671	3,154	4,830	157	425
1958	C-7A	6,219	18,355	28,500	123	250
1958	U-10A	1,000	2,010	3,400	144	330
1961	O-2A	696	3,212	4,850	118	340
1961	B-941S	22,045	29,675	58,422	215	310
1964	C-8A	13,843	23,157	41,000	180	250
1964	XC-142A	8,000	25,610	41,500	207	275
1965	OV-10A	3,600	6,969	9,908	230	228
1965	CL-84	4,600	8,685	14,500	260	115
1967	X-22A	1,500	11,000	14,600	213	445
1976	YC-15	62,000	105,000	216,680	449	920
1977	XV-15	3,400	9,750	14,460	230	512

[a] Data are design payload and design gross weight, most economical cruise speed, and range with given payload and gross weight at given cruise speed.

Table A3.26. Characteristics of Helicopters[a]

Year	Aircraft	Payload (lb)	Empty weight (lb)	Gross weight (lb)	Cruise speed (mph)	Range (miles)
1947	UH-13	465	1,435	2,200	69	125
1948	HH-43B	2,939	4,604	9,150	87	91
1949	UH-19A	1,605	5,013	8,100	83	157
1952	CH-21A	1,200	8,266	10,955	87	174
1953	CH-21B	2,249	8,786	13,500	80	135
1953	CH-37A	7,510	20,690	31,000	100	69
1954	CH-34	3,980	7,630	13,000	85	145
1956	UH-1A	800	3,950	5,864	101	72
1958	CH-46	2,400	11,585	21,400	130	320
1962	CH-47A	10,367	17,878	33,000	106	108
1962	CH-54A	12,590	19,110	38,000	95	130
1963	CH-3C	2,400	12,248	22,050	125	313
1964	CH-53A	8,000	22,444	39,713	150	130
1965	AH-1G	1,927	6,073	9,500	172	357
1967	CH-47C	6,400	21,464	46,000	158	230
1968	UH-1N	3,161	5,549	10,000	115	273
1969	OH-58	866	1,464	2,768	117	299
1969	AH-1J	1,790	6,610	10,000	207	359
1974	YUH-61A	5,924	9,750	17,962	167	370
1974	YCH-53E	30,000	32,048	65,828	173	306
1974	YUH-60A	4,077	10,624	16,825	167	373
1975	YAH-64	2,284	9,500	13,950	180	359

[a] Data are design payload and design gross weight, most economical cruise speed, and range with given payload and gross weight at given cruise speed.

Table A3.27. Activities of U.S. Scheduled Air Carriers (Foreign and Domestic Flights)[a]

Year	Revenue passenger-miles	Available seat-miles	Revenue aircraft-miles
1950	10,243	16,842	477
1955	24,351	38,574	780
1956	27,625	43,674	869
1957	31,261	51,059	976
1958	31,499	53,115	973
1959	36,372	59,247	1,030
1960	38,863	65,567	998
1961	39,831	71,857	970
1962	43,760	82,612	1,010
1963	50,362	94,845	1,095
1964	58,494	106,316	1,189
1965	68,676	124,320	1,354
1966	79,889	137,844	1,482
1967	98,720	174,733	1,742
1968	113,997	216,524	2,147
1969	125,420	250,846	2,385
1970	131,719	264,904	2,418
1971	135,652	279,823	2,378
1972	152,406	287,411	2,376
1973	161,957	310,597	2,448
1974	162,919	297,006	2,258
1975	162,810	303,006	2,241
1976	178,988	322,822	2,320
1977	193,219	345,566	2,419
1978	236,900	369,000	2,520
1979	261,700	416,000	2,788

Source: Aviation Week, various issues.

[a] All figures in millions.

Table A3.28. U.S. Total Gross Consumption of Energy Resources by Major Sources (Trillions of Btu)

Year	Anthracite	Bituminous and lignite	Natural gas, dry[a]	Petroleum[b]	Total fossil fuels	Hydropower[c]	Nuclear power	Total gross energy inputs[d]
Historical year								
1947	1,224.2	14,599.7	4,518.4	11,367.0	31,709.3	1,459.0	—	33,168.3
1948	1,275.1	13,621.6	5,032.6	12,558.0	32,487.3	1,507.0	—	33,994.3
1949	957.6	11,673.1	5,288.5	12,120.0	30,039.2	1,565.0	—	31,604.2
1950	1,013.5	11,900.1	6,150.0	13,489.0	32,552.6	1,601.0	—	34,153.6
1951	939.8	12,285.3	7,247.6	14,848.0	35,320.7	1,592.0	—	36,912.7
1952	896.6	10,971.4	7,760.4	15,334.0	34,962.4	1,614.0	—	36,576.4
1953	711.2	11,182.1	8,156.0	16,098.0	36,147.3	1,550.0	—	37,697.3
1954	683.2	9,512.2	8,547.6	16,138.0	34,881.0	1,479.0	—	36,360.0
1955	599.4	11,104.0	9,232.0	17,524.0	38,459.4	1,497.0	—	39,956.4
1956	609.6	11,340.8	9,834.4	18,624.0	40,408.8	1,598.0	—	42,006.8
1957	528.3	10,838.1	10,416.2	18,570.0	40,352.6	1,568.0	1.2	41,921.8
1958	482.6	9,607.6	10,995.2	19,214.0	40,299.4	1,740.0	1.5	42,040.9
1959	477.5	9,595.9	11,990.3	19,747.0	41,810.7	1,695.0	2.2	43,507.9
1960	447.0	9,967.2	12,698.7	20,067.0	43,179.9	1,775.0	5.5	44,960.4
1961	403.8	9,809.4	13,228.0	20,487.0	43,928.2	1,628.0	17.0	45,573.2
1962	381.0	10,159.7	14,120.8	21,267.0	45,928.5	1,780.0	23.0	47,731.5
1963	361.0	10,722.0	14,843.0	21,950.0	47,876.0	1,740.0	33.0	49,649.0
1964	365.8	11,295.0	15,647.5	22,385.8	49,694.1	1,873.0	34.0	51,601.1
1965 preliminary	327.7	12,030.0	16,136.1	23,209.3	51,703.1	2,050.0	38.0	53,791.1
Projected years								
1970	309.0	14,251.0	19,374.0	27,275.0	61,209.0	2,193.0	874.0	64,276.0
1975	280.0	16,865.0	22,360.0	31,875.0	71,380.0	2,422.0	1,803.0	75,605.0
1980	250.0	19,290.0	25,455.0	35,978.0	80,973.0	3,026.0	4,076.0	88,075.0

Source: Bureau of Mines Information Circular 8384, "An Energy Model for the United States, Featuring Energy Balances for the Years 1947 to 1965 and Projections and Forecasts."

[a] Excludes natural gas liquids.

[b] Petroleum products including still gas, liquefied refinery gas, and natural gas liquids.

[c] Represents projections of outputs of hydropower and nuclear power converted to theoretical energy inputs at projected rates of pounds of coal per kilowatt-hour at central electric stations. Excludes inputs for power generated by nonutility plants, which are included within the other consuming sectors.

[d] Gross energy is that contained in all types of commercial energy at the time it is incorporated in the economy, whether the energy is produced domestically or imported. Gross energy comprises inputs of primary fuels (or their derivatives) and outputs of hydropower and nuclear power converted to theoretical energy inputs. Gross energy includes the energy used for the production, processing, and transportation of energy proper.

Table A3.29. Measurements of the Velocity of Light

Author	Year	Velocity (km/sec)	Standard deviation (km/sec)
Corni-Helmert	1875	299,990	300
Michelson	1879	299,910	75
Newcomb	1883	299,860	45
Michelson	1883	299,853	90
Perrotin	1902	299,901	104
Rosa-Dorsey	1906	299,784	15
Michelson	1927	299,798	22
Mittelstaedt	1928	299,786	15
Michelson, Pease, and Pearson	1933	299,774	6
Anderson	1937	299,771	15
Huttel	1937	299,771	15
Anderson	1941	299,776	9
Bergstrand	1951	299,793.1	0.32
Mackenzie	1953	299,792.4	0.5

Source: Encyclopedia Britannica, 1964 edition.

Table A3.30. Accuracy of Time Measurement

Year	Inventor	Error (sec/day)
1656	Huygens—first pendulum clock	10
1721	Graham—cylinder escapement	5.5
1726	Harrison—temperature compensation	1
1761	Harrison—bimetallic strip, reduced friction	0.22
1835	Robinson—barometric compensation	0.1
1893	Reifler	0.01
1923	Shortt—free pendulum in vacuum	0.0035

Sources: H. T. Pledge, *Science Since 1500* (New York: Philosophical Library), 1947; and *Encyclopedia Brittanica*.

Table A3.31. Historical Data on General Aviation

Year	Number of aircraft	Aircraft shipments	Fuel consumed (millions of gallons)	Flying time (millions of hours)
1978	198,700	16,456	NA	39.4
1977	184,300	16,920	1176	35.8
1976	178,300	15,447	1021	33.9
1975	168,475	14,057	907	32.0
1974	161,502	14,165	886	31.4
1973	153,540	13,645	786	30.0
1972	145,010	9,774	977	27.0
1971	131,148	7,466	734	25.5
1970	131,743	7,283	759	26.0
1969	130,806	12,457	690	29.8
1968	124,237	13,698	610	22.9
1967	114,186	13,577	541	21.6
1966	104,706	15,747	486	18.9
1965	95,442	11,852	378	16.2
1964	88,742	9,336	307	15.4
1963	85,088	7,569	285	14.8
1962	84,121	6,697	264	14.0
1961	80,632	6,778	257	13.4
1960	76,549	7,588	246	13.0
1959	68,727	7,689	221	11.9
1958	67,839	6,414	209	11.3
1957	65,289	6,118	213	10.6

NA = not available.

Sources: Aerospace Facts and Figures, Aviation Week and Space Technology, New York; *Statistical Abstract of the United States,* U.S. Bureau of the Census, Washington, DC.

Table A3.32. National Income and Telephone and Radio Usage in 42 Countries

Country	GNP per capita	Telephones per capita	Radios per capita
Australia	7,479	0.41	0.739
Austria	8,522	0.324	0.291
Belgium	10,750	0.316	0.414
Bolivia	745	0.023	0.096
Brazil	1,015	0.041	0.147
Canada	8,459	0.617	0.997
Chile	740.7	0.0206	0.0766
China	360	0.0052	0.0469
Colombia	646	0.0540	0.112
Costa Rica	1,598	0.0687	0.189
Cuba	894	0.0330	0.216
Czechoslovakia	406.1	0.189	0.259
Denmark	11,430	0.538	0.356
El Salvador	699	0.016	0.321
Finland	7,013	0.425	0.458
France	9,484	0.329	0.328
Federal Republic of Germany	11,558	0.374	0.337
Greece	3,376	0.248	0.294
Honduras	497	0.00552	0.0468
Indonesia	219	0.00241	0.0379
Ireland	3,466	0.161	0.295
Jordan	1,080	0.0240	0.240
Kenya	376	0.00968	0.0346
Korea	1,240	0.0534	0.271
Malaysia	1,264	0.0294	0.114
Morocco	729	0.0111	0.0794
Netherlands	10,184	0.419	0.295
New Zealand	4,821	0.552	0.874
Panama	1,191	0.0858	0.153
Paraguay	712	0.0144	0.0623
Peru	410	0.0239	0.123
Singapore	3,468	0.195	0.176
Spain	3,569	0.257	0.251
Sweden	11,178	0.716	0.394
Switzerland	15,192	0.654	0.337
Taiwan	6,775	0.0994	0.0880
Tanzania	208	0.00454	0.00184
Thailand	486	0.00820	0.122
Turkey	576	0.0320	0.0988
United Kingdom	6,276	0.415	0.717
United States	9,754	0.743	2.064
Venezuela	3,053	0.0645	0.384

Source: Calculated from data given in *The Encyclopedia Brittanica Yearbook*, 1980.

Table A3.33. Residential Consumption of Electricity in the United States

Year	Kilowatt-hours per year per customer
1912	264
1917	268
1920	339
1925	396
1930	547
1935	677
1940	952
1945	1229
1950	1845
1955	2773
1960	3854
1965	4933
1970	7066
1971	7300
1972	7700
1973	8100
1974	7900
1975	8200
1976	8400
1977	8700
1978	8800
1979	8800

Sources: Historical Statistics of the United States, U.S. Department of Commerce, Washington, DC; *Statistical Abstract of the United States,* U.S. Bureau of Census, Washington, DC.

Table A3.34. Installed Horsepower in Motor Vehicles (Automobiles, Trucks, Buses, and Motorcycles) and Total Number of Motor Vehicles in the United States

Year	Total horsepower (thousands)	Number of vehicles
1900	100	8,000
1909	7,714	312,000
1919	230,432	7,576,888
1920	280,900	9,239,161
1929	1,424,980	26,704,825
1930	1,420,568	26,749,853
1939	2,400,000	31,009,927
1940	2,511,312	32,453,233
1950	4,403,617	49,161,691
1952	5,361,386	53,265,406
1955	6,632,121	62,688,700
1960	10,366,880	73,868,565
1961	10,972,210	75,958,200
1962	11,930,000	79,173,300
1963	12,713,712	82,713,700
1965	14,306,300	90,357,000
1967	16,152,371	96,930,900
1969	18,075,000	105,096,600
1970	19,325,000	108,407,300
1971	20,732,000	113,000,000
1972	21,736,000	118,800,000
1973	23,029,000	125,700,000
1974	23,224,000	129,900,000
1975	23,752,000	132,949,000
1976	24,339,000	138,500,000
1977	25,025,000	142,400,000
1978	25,892,000	148,778,000

Sources: Historical Statistics of the United States, U.S. Department of Commerce, Washington, DC; *Statistical Abstract of the United States,* U.S. Bureau of the Census, Washington, DC.

Table A3.35. Highest Electrical Transmission Voltages Used in North America

Year introduced	Voltage (kV)
1890	11
1896	33
1904	69
1907	110
1910	140
1912	150
1922	220
1934	287
1953	345
1964	500
1966	735
1969	765

Source: Edison Electric Institute.

Table A3.36. Aircraft Speeds (Flown Under Other Than Official Conditions)

Date	Aircraft	Speed (mph)
27 August 1939	Heinkel He 178	180
27 August 1940	Caproni-Campini CC-2	176
5 April 1941	Heinkel He 280	400
15 May 1941	Gloster E28/39	Over 362
27 July 1942	Messerschmit Me 262A-1a	540
1 October 1942	Bell XP-59	404
July 1943	Republic XP-47H	490[a]
December 1943	Arado Ar-234B	461
8 January 1944	Lockheed XP-80	500
July 1944	Gloster Meteor F.1	410
5 August 1944	Republic XP-47J	504[a]
September 1944	Lockheed P-80 (with Rolls-Royce Nene Engine)	580
December 1944	Heinkel He 162	465
17 July 1945	Gloster Meteor F.4	590
20 August 1969	Modified F7F	478[a]

[a] Propeller-driven aircraft.

Table A3.37. Official Aircraft Speed Records

Date	Country	Speed (mph)
Nov. 12, 1906	France	25.66
Oct. 26, 1907	France	32.75
May 20, 1909	France	34.06
Aug. 23, 1909	France	43.38
Aug. 24, 1909	France	46.18
Aug. 28, 1909	France	47.84
Apr. 23, 1910	France	48.21
July 10, 1910	France	66.18
Oct. 29, 1910	United States	68.20
Apr. 12, 1911	France	69.47
May 11, 1911	France	74.42
June 12, 1911	France	77.67
June 16, 1911	France	80.91
June 21, 1911	France	82.73
Jan. 13, 1912	France	90.20
Feb. 22, 1912	France	100.22
Feb. 29, 1912	France	100.94
Mar. 1, 1912	France	103.66
Mar. 2, 1912	France	104.33
July 13, 1912	France	106.12
Sept. 9, 1912	United States	108.18
June 17, 1913	France	111.73
Sept. 27, 1913	France	119.24
Sept. 29, 1913	France	126.67
Feb. 7, 1920	France	171.04
Feb. 28, 1920	France	176.14
Oct. 9, 1920	France	181.86
Oct. 10, 1920	France	184.36
Oct. 20, 1920	France	187.98
Nov. 4, 1920	France	192.01
Dec. 12, 1920	France	194.52
Sept. 26, 1921	France	205.22
Sept. 21, 1922	France	211.90
Oct. 13, 1922	United States	222.97
Feb. 15, 1923	France	233.01
Mar. 29, 1923	United States	236.59
Nov. 2, 1923	United States	259.15
Nov. 4, 1923	France	266.58
Dec. 11, 1924	France	278.48
Sept. 5, 1932	United States	294.38
Sept. 4, 1933	United States	304.98
Dec. 25, 1934	France	314.319
Sept. 13, 1935	United States	352.388
Nov. 11, 1937	Germany	379.626
Mar. 30, 1939	Germany	463.917
Apr. 26, 1939	Germany	469.220

(continued)

Table A3.37. (*continued*)

Date	Country	Speed (mph)
Nov. 7, 1945	Great Britain	606.255
Sept. 7, 1946	Great Britain	615.778
June 19, 1947	United States	623.738
Aug. 20, 1947	United States	640.663
Aug. 25, 1947	United States	650.796
Sept. 15, 1948	United States	670.981
Nov. 19, 1952	United States	698.505
July 16, 1953	United States	715.745
Sept. 7, 1953	Great Britain	727.624
Sept. 25, 1953	Great Britain	735.702
Oct. 3, 1953	United States	752.943
Oct. 29, 1953	United States	755.149
Aug. 20, 1955	United States	822.266
Mar. 10, 1956	Great Britain	1132.136
Dec. 12, 1957	United States	1207.6
May 16, 1958	United States	1404.09
Oct. 31, 1959	U.S.S.R.	1483.85
Dec. 15, 1959	United States	1525.965
Nov. 22, 1961	United States	1606.48
July 7, 1962	U.S.S.R.	1665.9
May 1, 1965	United States	2070.102

Sources: Aeronautics and Astronautics: 1915–1960, National Aeronautics and Space Administration, 1961, and various issues of Jane's *All the World's Aircraft.*

Table A3.38. Accuracy of Astronomical Angular Measure

Year	Astronomer	Accuracy (arc measure)
127 BC	Hipparchus	20′
1430 AD	Ulugh Beg	10′
1580	Tycho Brahe	4′
1725	Flamsteed	10″
1800	Piazzi	1.5″
1838	Bessel	0.3″
1870	Auwers	0.1″
1900	Newcomb	0.03″
1935	Van Maanen	0.003″

Source: H. T. Pledge, *Science Since 1500* (New York: Philosophical Library), 1947.

Table A3.39. Accuracy of Measurements[a] of Mass

Year	Accuracy (grain)
1550	0.1
1644	0.05
1653	0.03
1825 (steel knife edge; agate bearing)	0.01
1870 (vacuum weighing)	0.001
1930	0.00015

Source: H. T. Pledge, *Science Since 1500* (New York: Philosophical Library), 1947.

[a] Measurements made using beam balance.

Table A3.40. Efficiency of Incandescent Lights

Year	Lumens per watt
1959	17
1940	16
1932	14
1916	12
1915	10
1908	8
1907	5
1906	4
1888	2

Source: G.E. Bulletin TP-110.

Table A3.41. Lumens from 40-W Flourescent Light

Year	Lumens
1967	3350
1963	3300
1962	3000
1959	2900
1958	2750
1954.5	2600
1954	2550
1953	2500
1952	2450
1949	2400
1946	2300
1941	2100
1939	1850
1938	1400

Source: G.E. Bulletin TP-111.

Table A3.42. Growth of the Cable Television (CATV) Industry 1952–1978

Year	Households with TV (thousands)	Households with CATV (thousands)	Households with CATV (%)	TV stations	CATV systems
1952	15,300	14		108	70
1953	20,400	30		198	150
1954	26,000	65		402	300
1955	30,700	150	0.5	458	400
1956	34,900	300	0.9	496	450
1957	38,900	350	0.9	519	500
1958	41,424	450	1.1	556	525
1959	43,950	550	1.3	566	560
1960	45,750	650	1.4	579	640
1961	47,200	725	1.5	553	700
1962	48,855	850	1.7	571	800
1963	50,300	950	1.9	581	1,000
1964	51,600	1,085	2.1	582	1,200
1965	52,700	1,275	2.4	588	1,325
1966	53,850	1,575	2.9	613	1,570
1967	55,130	2,100	3.8	626	1,770
1968	56,670	2,800	4.4	642	2,000
1969	58,250	3,600	6.1	673	2,260
1970	59,550	4,500	7.7	686	2,490
1971	NA	5,300	8.8	688	2,639
1972	NA	6,000	9.7	690	2,841
1973	NA	7,300	11.3	692	2,991
1974	NA	8,700	12.7	694	3,158
1975	70,520	9,800	13.8	693	3,506
1976	71,460	10,800	14.8	701	3,681
1977	73,100	11,900	16.1	697	3,832
1978	74,700	13,000	17.1	708	3,997
1979	76,240	NA	NA	NA	NA

NA = not available.

Source: Statistical Abstract of the United States, U.S. Bureau of the Census, Washington, DC., various years.

Table A3.43. Magnitude of Engineering Projects

Project	Year of completion	Cost in current dollars
Santee Canal	1800	7.5×10^5
Lancaster Turnpike	1794	4.2×10^5
Middlesex Canal	1803	6×10^5
National Road	1818	1.4×10^6
Erie Canal	1825	9×10^6
London–Holyhead Highway	1828	3.5×10^6
Welland Canal	1829	7.7×10^6
Miami & Erie Canal	1832	1.6×10^7
Washington–Baltimore Telegraph	1844	3×10^4
Marseilles Canal	1847	8.7×10^6
B&O Canal	1850	1.5×10^7
Panama Railroad	1855	8×10^6
Union Navy Ironclad Boats	1862	9.1×10^5
Atlantic Cable	1866	1.2×10^7
Suez Canal	1869	7.72×10^7
Mont Cenis Tunnel	1871	1.5×10^7
Welland Canal	1877	2.2×10^7
Brooklyn Bridge	1883	1.5×10^7
Panama Canal (French)	1890[a]	2.2×10^8
Corinth Canal	1893	1.16×10^7
Manchester Canal	1894	7.43×10^7
Panama Canal (U.S.)	1914	2.83×10^8
Welland Canal	1932	1.3×10^8
Oakland Bay Bridge	1936	7.7×10^7
German V-2 Project	1944	4×10^7
Manhattan Project	1945	2.2×10^9
USAF Ballistic Missile Project	1958	3×10^9
Project Apollo	1970	2.4×10^{10}
Trans-Alaska Pipeline	1976	6×10^9
Interstate Highway System	1982	7.6×10^{10}
North Sea Oil	1987	4×10^{10}

[a] Year construction halted.

Table A3.44. U.S. Aircraft Production

Year	Production	Year	Production
1912	45	1945	48,912
1913	43	1946	36,418
1914	49	1947	17,739
1915	178	1948	9,839
1916	411	1949	6,137
1917	2,148	1950	6,200
1918	14,020	1951	7,532
1919	780	1952	10,640
1920	328	1953	13,112
1921	437	1954	11,478
1922	263	1955	11,484
1923	743	1956	12,408
1924	377	1957	11,943
1925	789	1958	10,938
1926	1,186	1959	11,076
1927	1,995	1960	10,237
1928	3,346	1961	9,054
1929	6,193	1962	9,308
1930	3,437	1963	10,125
1931	2,800	1964	12,492
1932	1,396	1965	15,349
1933	1,324	1966	19,886
1934	1,615	1967	19,141
1935	1,710	1968	19,414
1936	3,010	1969	16,481
1937	3,773	1970	10,943
1938	3,623	1971	10,390
1939	5,856	1972	12,693
1940	12,813	1973	16,081
1941	26,289	1974	16,436
1942	47,675	1975	16,620
1943	85,433	1976	17,605
1944	95,272	1977	19,077
		1978	19,960

Source: Aerospace Industries Association, *Aerospace Facts and Figures,* (various editions) (original data from U.S. Bureau of the Census, Washington, DC, "Current Industrial Reports," Series MQ37D)

Table A3.45. Characteristics of Reciprocating Aircraft Engines

Year	Engine	Horsepower	Weight (lb)	Displacement (in.^{3})
1902	Wright Bros.	12	179	410
1902	Manley	52	151	540
1912	Mercedes	80	312	443
1913	Gnome	100	270	783
1914	Rolls-Royce Eagle	350	880	1241
1914	Curtiss OX5	90	390	568
1916	Hispano-Suiza A	150	467	718
1917	Liberty	420	857	1650
1921	Bristol Jupiter	485	775	1253
1922	Curtiss D-12	435	680	1210
1926	Pratt & Whitney Wasp	600	685	1344
1929	Wright Cyclone 9	1525	1469	1823
1934	Rolls-Royce Merlin	1730	1450	1649
1936	Bristol Hercules	1980	2115	2360
1937	Pratt & Whitney Double Wasp	2050	2390	2800
1944	BMW	1700	1940	2550
1945	Pratt & Whitney Wasp Major	3250	3670	4363
1948	Wright Turbo-compound	3400	3675	3347

Source: Encyclopedia Brittanica, 1964 edition.

Table A3.46. Installed Hydroelectric Generating Capacity

Year	Capacity (thousands of kilowatts)
1978	71,000
1977	69,000
1976	68,000
1975	66,000
1974	64,000
1973	62,000
1972	56,000
1971	56,000
1970	55,751
1965	44,490
1960	33,180
1955	25,742
1950	18,674
1945	15,892
1940	12,304
1935	10,399
1930	9,650
1925	7,150
1920	4,804
1912	2,794
1902	1,140

Sources: Historical Statistics of the United States, U.S. Bureau of the Census, Washington, DC; *Statistical Abstract of the United States*, U.S. Bureau of the Census, Washington, DC.

Table A3.47. Inanimate Nonautomotive Horsepower (Thousands)

Year	Factories[a]	Mines	Railroads	Merchant ships, powered	Sailing vessels	Farms
1978	63,000	48,000	64,000	25,000		335,000
1977	62,000	47,000	62,000	23,000		328,000
1976	61,000	47,000	64,000	22,000		324,000
1975	60,000	47,000	62,000	22,000		318,000
1974	59,000	46,000	61,000	21,000		315,000
1973	58,000	46,000	57,000	20,000		308,000
1972	57,000	46,000	57,000	21,000		305,000
1971	56,000	45,000	56,000	21,000		300,000
1970	54,000	45,000	54,000	22,000	1	288,500
1965	48,400	40,300	43,838	24,015	2	269,822
1960	42,000	34,700	46,856	23,890	2	237.020
1955	35,579	30,768	60,304	24,155	5	207,742
1950	32,921	8,500[b]	110,969	23,423	11	57,533[b]
1940	21,768	7,332	92,361	9,408	26	57,472
1930	19,519	5,620	109,743	9,115	100	28,610
1920	19,422	5,146	80,182	6,508	169	21,443
1910	16,697	4,473	51,308	3,098	220	10,460
1900	10,309	2,919	24,501	1,663	251	4,009
1890	6,308	1,445	16,980	1,124	280	1,452
1880	3,664	715	8,592	741	314	668
1870	2,453	380	4,462	632	314	
1860	1,675	170	2,156	515	597	
1850	1,150	60	586	325	400	

Sources: Historical Statistics of the United States, U.S. Bureau of the Census, Washington, DC; *Statistical Abstract of the United States,* U.S. Bureau of the Census, Washington, DC.

[a] Includes electric motors

[b] Beginning 1955, not strictly comparable with earlier years.

Table A3.48. Mill Consumption of Man-made Versus Natural Fibers (Millions of Pounds)

Year	Natural	Man-made
1950	5340	1491
1951	5371	1471
1952	4957	1464
1953	4966	1502
1954	4526	1484
1955	4815	1851
1956	4825	1685
1957	4444	1745
1958	4207	1702
1959	4781	2065
1960	4657	1817
1961	4506	2064
1962	4629	2418
1963	4465	2788
1964	4615	3174
1965	4923	3512
1966	5016	3989
1967	4747	4244
1968	4489	5306
1969	4256	5550
1970	4102	5500
1971	4185	6345
1972	4091	7566
1973	3819	8665
1974	3412	7700
1975	3141	7416
1976	3543	8053
1977	3296	8889
1978	3162	9236
1979	3195[a]	9465[a]

Sources: Historical Statistics of the United States, U.S. Bureau of the Census, Washington, DC; *Statistical Abstract of the United States,* U.S. Bureau of the Census, Washington, DC.

[a] Preliminary.

Table A3.49. Locomotives in Service

Year	Steam	Electric	Diesel	Total[a]
1978		100	26,900	27,000
1977		100	27,100	27,300
1976		200	29,100	29,600
1975		200	29,700	30,200
1974		200	29,700	30,200
1973		200	29,400	29,900
1972		300	29,000	29,300
1971		255	28,831	29,185
1970		270	28,773	29,122
1965	89	365	29,552	30,001
1960	374	498	30,240	31,178
1955	6,266	639	26,563	33,533
1950	26,680	827	15,396	42,951
1945	41,018	885	4,301	46,253
1940	42,410	900	967	44,333
1935	48,477	884	130	49,541
1930	59,406	663	77	60,189
1925	67,713	379	1	68,098
1920	68,554	388		68,942
1918	67,563	373		67,936
1916	65,253	342		65,595

Sources: Historical Statistics of the United States, U.S. Bureau of the Census, Washington, DC; *Statistical Abstract of the United States,* U.S. Bureau of the Census, Washington, DC.

[a] Summation of steam, electric, and diesel does not equal total because of other types of locomotives in the total.

Table A3.50. Installed Steam-Generating Capacity

Year	Capacity (thousands of kilowatts)	Year	Capacity (thousands of kilowatts)
1978	400,000	1955	101,698
1977	388,000	1950	61,495
1976	368,000	1945	45,248
1975	353,000	1940	37,138
1974	338,000	1935	32,429
1973	321,000	1930	31,503
1972	294,000	1925	22,937
1971	285,000	1920	14,635
1970	298,803	1912	8,186
1965	205,423	1902	1,847
1960	149,161		

Sources: Historical Statistics of the United States, U.S. Bureau of the Census, Washington, DC; *Statistical Abstract of the United States,* U.S. Bureau of the Census, Washington, DC.

Table A3.51. Maximum Steam Turbine Capacity

Year	Capacity (thousands of kilowatts)
1960	500
1958	350
1957	275
1956	260
1955	217.26
1953	217
1929	208
1928	94
1925	60
1924	50
1915	45
1912	20
1906	12.5
1905	9
1904	5

Table A3.52. Maximum Hydroturbine Capacity

Year	Capacity (thousands of kilowatts)
1966	220
1963	204
1957	167
1945	108
1936	82.5
1932	77.5
1930	66.7
1929	45
1928	40.6
1926	40
1922	31.25
1921	30
1920	17.5
1919	13.5
1916	7.5

Appendix 4

Computer Programs

1. Introduction

This appendix contains three computer programs in BASIC: KSIM, GROWTH, and REGRESS. KSIM is a version of the simulation model developed by Julius Kane and described in Chapter 8. GROWTH can be used to fit either a Pearl or a Gompertz curve to a set of data. REGRESS performs the linear regression of one variable on another, with three options: the dependent variable is regressed on the independent variable, the logarithm of the dependent variable is regressed on the independent variable, or the logarithm of the dependent variable is regressed on the logarithm of the independent variable.

These programs could have been made interactive, requesting input values from the user; however, one of the features of BASIC is its ability to include data within a program in a DATA statement. These programs have been written to use this capability. Entering data into a program is quicker than responding to a set of prompts, and the data can be edited more readily; in addition, a program can be made self-documenting if the data statements are incorporated into it. Each program is intended to have the data incorporated into a series of DATA statements following the main body of the program.

2. KSIM

This program carries out the algorithm presented in Chapter 8. After reading in the initial values of the variables, the cross impacts, and the plotting symbols for the variables, the program plots the initial values. The program then goes through 1000 iterations of recomputing new values for the variables. During each iteration the positive and negative impacts are summed respectively as D and N. An exponent is formed as 1 plus the sum of the negative impacts multiplied by a time step, divided by 1 plus the sum of the positive impacts multiplied by a time step. The previous value of the variable is then raised to this exponent. This is done for each variable during each iteration. If a variable is smaller than 10^{-70}, it is set to 0 to avoid underflow problems. The values of the variables are printed out every tenth iteration. Thus 100 lines of print are produced for each simulation of 1000 iterations. DATA statements go between lines 2000 and 5999, at the end of the main body of the program. Subroutine 6000 initializes the array of plotting symbols to blank spaces; subroutine 7000 inserts the plotting symbols at the proper locations in the array; and subroutine 8000 prints the array of plotting symbols.

```
100  REM THIS PROGRAM IS DIMENSIONED FOR UP TO TEN VARIABLES. IF
     MORE
101  REM VARIABLES ARE DESIRED, THE ARRAYS IN THE FIRST "DIM"
     STATEMENT
102  REM MUST BE INCREASED APPROPRIATELY. NOTE THAT C MUST BE
103  REM DIMENSIONED C(N1,N1+1).
104  REM
105  DIM X(11),C(10,11),P$(10),N(10),D(10),E(10)
106  DIM L$(51)
107  LET S$=" "
108  LET a$="*"
109  REM
110  REM READ NUMBER OF VARIABLES
111  REM
112  READ N1
113  REM
114  REM READ INITIAL VALUES
115  REM
116  FOR I=1 TO N1
117  READ X(I)
118  NEXT I
119  REM
120  REM OUTSIDE WORLD HAS VALUE 1
121  REM
122  LET X(N1+1)=1
123  REM
124  REM READ CROSS IMPACTS
125  REM
126  FOR I=1 TO N1
127  FOR J=1 TO N1+1
128  READ C(I,J)
129  NEXT J
130  NEXT I
131  REM
132  REM READ PLOTTING SYMBOLS
133  REM
134  FOR I=1 TO N1
135  READ P$(I)
```

```
136  NEXT I
200  REM
201  REM PRINT HEADER FOR PLOT
202  REM
203  PRINT
204  PRINT
205  PRINT
206  PRINT
207  PRINT USING 251
208  PRINT USING 250
250  :      +....+....+....+....+....+....+....+....+....+....+
251  :      0    .1   .2   .3   .4   .5   .6   .7   .8   .9   1.0
300  REM
301  REM PRINT INITIAL VALUES
302  REM
303  GOSUB 6000
306  LET K=0
307  GOSUB 7000
308  GOSUB 8000
400  REM
401  REM COMPUTE SUCCESSIVE VALUES AND PLOT
402  REM
403  LET T=.001
404  FOR K=1 TO 1000
405  FOR I=1 TO N1
406  LET N(I)=D(I)=0
407  FOR J=1 TO N1+1
408  N(I)=N(I)+((ABS(C(I,J)))-C(I,J))*X(J)
409  D(I)=D(I)+((ABS(C(I,J)))+C(I,J))*X(J)
410  NEXT J
411  E(I)=(1+T*(.5)*N(I))/(1+T*(.5)*D(I))
412  NEXT I
413  FOR I=1 TO N1
414  X(I)=X(I) ↑ E(I)
415  IF X(I)>1E-70 THEN 417
416  X(I)=0
417  NEXT I
418  REM
419  REM PRINT EVERY TENTH TIME INCREMENT. IF PRINTING AT SOME
     OTHER
420  REM INCREMENT IS DESIRED, CHANGE THE DIVISOR OF K IN LINE 422
421  REM
422  IF K/10<>INT(K/10) THEN 426
423  GOSUB 6000
424  GOSUB 7000
425  GOSUB 8000
426  NEXT K
500  PRINT USING 250
501  PRINT USING 251
502  PRINT
503  PRINT
600  PRINT "THIS RUN IS COMPLETE. TO CHANGE INITIAL VALUES ALTER
     LINE"
601  PRINT "3001. TO CHANGE IMPACTS ON VARIABLE J ALTER LINE 4000+J"
```

```
1999  END
2000  REM NUMBER OF VARIABLES, N1
2001  DATA 4
3000  REM INITIAL VALUES, EACH OF N1 VALUES
3001  DATA .1,.3,.3,.2
4000  REM CROSS IMPACTS: N1 ROWS, N1+1 COLUMNS
4001  DATA 3,5,4,0,0
4002  DATA -2,-2,0,0,6
4003  DATA 3,0,0,7,0
4004  DATA 8,0,0,0,0
5000  REM PLOTTING SYMBOLS, ONE FOR EACH OF N1 VARIABLES
5001  DATA "S", "F", "P", "R"
5999  STOP
6000  FOR I=1 TO 51
6001  LET L$(I)=S$
6002  NEXT I
6003  RETURN
7000  FOR I=1 TO N1
7001  Z=INT(50*X(I)+1.5)
7002  IF L$(Z)=S$ THEN 7005
7003  L$(Z)=A$
7004  GOTO 7006
7005  L$(Z)=P$(I)
7006  NEXT I
7007  RETURN
8000  FOR I=1 TO 51
8001  IF I>1 THEN 8025
8010  IF K>99 THEN 8015
8011  PRINT "      "; K;":";
8012  GOTO 8025
8015  IF K>999 THEN 8020
8016  PRINT "      "; K;":";
8017  GOTO 8025
8020  PRINT "      "; K;":";
8025  PRINT L$(I);
8040  IF I<51 THEN 8050
8041  PRINT ":"
8050  NEXT I
8051  RETURN
```

3. Growth

This program applies a transformation to linearize a growth curve and then fits a straight line to the linearized data. The results of the regression are printed out as the parameters of the selected growth curve.

The choice between a Pearl and a Gompertz curve is made by inserting 1 or 2 in a DATA statement after line 900. The upper limit is placed in a DATA statement after line 950. The number of data points is placed in a DATA statement after line 1000. The data pairs, with the independent variable preceding the dependent variable, must be placed in DATA statements after line 2000. The T variate corresponding to the desired confidence interval and the number of degrees of freedom goes in a DATA statement after line 3000. The value should be obtained from Appendix 2. If the fitted curve is to be extrapolated or interpolated, the number of points for extrapolation/interpolation is placed in a DATA statement after line 4000. The values of the independent variable for which extrapolation/interpolation is desired are placed in a DATA statement after line 6000. If no extrapolation/interpolation is desired, the 0 in the DATA statement of line 6003 automatically terminates the program after the curve is fitted, and a table with the original data, the fitted value, and the confidence limits is printed.

The transformation selected for the input data is reversed before the fitted values are printed out. The output includes the two parameters of the growth curve, the correlation coefficient, the standard error of the slope of the linear regression, and a table of the data, the fitted values, and the confidence limits.

```
100  DIM Y(100),T(100)
101  DIM W(100)
102  DIM O(100)
103  DIM N(100)
150  DEF FNR(Z)=A+B*Z
151  DEF FNP(Z)=L1/(1+EXP(Z))
152  DEF FNG(Z)=L1*EXP(-EXP(Z))
153  DEF FNF(Z)=O3*SQR(1+1/N1+(N1*(Z-M2)**2)/S1)
200  READ M1
201  READ L1
202  READ N1
300  FOR I=1 TO N1
301  READ T(I)
302  READ Y(I)
303  NEXT I
304  ON M1 GOTO 325,350
305  PRINT "INCORRECT VALUE FOR OPTION. CORRECT LINE 901"
306  END
325  FOR I=1 TO N1
326  W(I)=LOG(L1/Y(I)-1)
327  NEXT I
328  GOTO 400
350  FOR I=1 TO N1
351  W(I)=LOG(LOG(L1/Y(I)))
352  NEXT I
353  GOTO 400
400  FOR I=1 TO N1
401  P=P+W(I)
402  Q=Q+W(I)**2
403  R=R+T(I)
404  S=S+T(I)**2
405  U=U+W(I)*T(I)
```

```
406  NEXT I
407  S1=N1*S-R**2
408  M2=R/N1
500  B=(N1*U-P*R)/S1
501  A=(P-B*R)/N1
502  V=B*SQR(S1/(N1*Q-P**2))
510  FOR I=1 TO N1
511  N(I)=FNR(T(I))
512  O(I)=W(I)-N(I)
513  NEXT I
520  FOR I=1 TO N1
521  O1=O1+O(I)**2
522  NEXT I
523  O2=O1/(N1-2)
524  O3=SQR(O2)
525  B1=(SQR(N1)*O3)/SQR(S1)
600  REM PRINT OUT RESULTS
601  READ T1
602  ON M1 GOTO 625,650
625  PRINT "A = ";EXP(A)
626  PRINT "B = ";-B
627  PRINT "CORRELATION COEFFICIENT ="; V
628  PRINT "STANDARD ERROR OF B =";B1
629  PRINT
630  PRINT
631  PRINT USING 700
632  FOR I=1 TO N1
633  Z1=FNP(FNR(T(I)))
634  Z2=FNP(FNR(T(I))+T1*FNP(T(I)))
635  Z3=FNP(FNR(T(I))-T1*FNF(T(I)))
636  PRINT USING 702,T(I),Y(I),Z1,Z3,Z2
637  NEXT I
638  READ J1
639  IF J1=0 THEN 649
640  FOR I=1 TO J1
641  READ J
642  Z1=FNP(FNR(J))
643  Z2=FNP(FNR(J)+T1*FNF(J))
644  Z3=FNP(FNR(J)-T1*FNF(J))
645  PRINT USING 703,J,Z1,Z3,Z2
646  NEXT I
649  END
650  PRINT "B = ";EXP(A)
651  PRINT "K = ";-B
652  PRINT "CORRELATION COEFFICIENT ="; V
653  PRINT "STANDARD ERROR OF K =";B1
654  PRINT
655  PRINT
656  PRINT USING 701
657  FOR I=1 TO N1
658  Z1=FNG(FNR(T(I)))
659  Z2=FNG(FNR(T(I))+T1*FNF(T(I)))
660  Z3=FNG(FNR(T(I))-T1*FNF(T(I)))
```

```
661  PRINT USING 702,T(I),Y(I),Z1,Z3,Z2
662  NEXT I
663  READ J1
664  IF J1=0 THEN 674
665  FOR I=1 TO J1
666  READ J
667  Z1=FNG(FNR(J))
668  Z2=FNG(FNR(J)+T1*FNF(J))
669  Z3=FNG(FNR(J)-T1*FNF(J))
670  PRINT USING 703,J,Z1,Z3,Z2
671  NEXT I
674  END
700  :        T           Y          PEARL      UPPER CONF  LOWER CONF
701  :        T           Y        GOMPERTZ     UPPER CONF  LOWER CONF
702  : #####.#### #####.#### #####.#### #####.#### #####.####
703  : #####.####             #####.#### #####.#### #####.####
 900 REM OPTION: 1 FOR PEARL CURVE; 2 FOR GOMPERTZ CURVE
 950 REM READ UPPER LIMIT, L1
1000 REM READ NUMBER OF DATA POINTS
2000 REM READ DATA PAIRS: T,Y
3000 REM READ T VARIATE FOR DESIRED CONFIDENCE INTERVAL, T1
4000 REM READ NUMBER OF EXTRAPOLATE/INTERPOLATE POINTS, J1
5000 REM READ EXTRAPOLATE/INTERPOLATE POINTS
6000 REM THE FOLLOWING DATA ENTRY AUTOMATICALLY TERMINATES THE
     PROGRAM
6001 REM AFTER FITTING THE CURVE AND PRINTING RESULTS
6002 REM IF NO EXTRAPOLATE/INTERPOLATE POINTS ARE TO BE READ.
6003 DATA 0
```

4. Regress

This program performs a linear regression of a dependent variable on a single independent variable. Prior to the regression the logarithm of the dependent variable or those of both the dependent and independent variables may be taken. The option chosen should be placed in a DATA statement after line 901. The number of data points to be read should be placed in a DATA statement after line 1000. The data pairs, with the independent variable preceding the dependent variable, should be placed in DATA statements after line 2000. Instructions for the *T* variate and the extrapolate/interpolate points are the same as for program GROWTH.

If a transformation of the input has been selected, this transformation is reversed before the fitted values are printed out. The output includes the two regression coefficients, the correlation coefficient, the standard error of the slope of the linear regression, and a table of the data, the fitted values, and the confidence limits.

```
100  DIM Y(100),T(100)
101  DIM X(100),W(100)
102  DIM O(100)
```

```
103  DIM N(100)
150  DEF FNR(Z) = A + B*Z
151  DEF FNF(Z) = O3*SQR(1 + 1/N1 + (N1*(Z - M2)**2)/S1)
200  READ M1
201  READ N1
300  FOR I = 1 TO N1
301  READ T(I)
302  READ Y(I)
303  NEXT I
304  ON M1 GOTO 325,350,375
307  PRINT "INCORRECT VALUE FOR OPTION. CORRECT LINE 902."
308  END
325  FOR I = 1 TO N1
326  W(I) = Y(I)
327  X(I) = T(I)
328  NEXT I
329  GOTO 400
350  FOR I = 1 TO N1
351  W(I) = LOG(Y(I))
352  X(I) = T(I)
353  NEXT I
354  GOTO 400
375  FOR I = 1 TO N1
376  W(I) = LOG(Y(I))
377  X(I) = LOG(T(I))
378  NEXT I
379  GOTO 400
400  FOR I = 1 TO N1
401  P = P + W(I)
402  Q = Q + W(I)**2
403  R = R + X(I)
404  S = S + X(I)**2
405  U = U + W(I)*X(I)
406  NEXT I
407  S1 = N1*S - R**2
408  M2 = R/N1
500  B = (N1*U - P*R)/S1
501  A = (P - B*R)/N1
502  V = B*SQR(S1/(N1*Q - P**2))
510  FOR I = 1 TO N1
511  N(I) = FNR(X(I))
512  O(I) = W(I) - N(I)
513  NEXT I
520  FOR I = 1 TO N1
521  O1 = O1 + O(I)**2
522  NEXT I
523  O2 = O1/(N1 - 2)
524  O3 = SQR(O2)
525  B1 = (SQR(N1)*O3)/SQR(S1)
600  REM PRINT OUT RESULTS
601  READ T1
602  ON M1 GOTO 625,650,675
625  PRINT "INTERCEPT = "; A
```

```
626  PRINT "SLOPE ="; B
627  PRINT "CORRELATION COEFFICIENT ="; V
628  PRINT "STANDARD ERROR OF SLOPE =";B1
629  PRINT
630  PRINT
631  PRINT
632  PRINT
633  PRINT USING 700
634  FOR I=1 TO N1
635  Z1=FNR(X(I))
636  Z2=Z1+T1*FNF(X(I))
637  Z3=Z1-T1*FNF(X(I))
638  PRINT USING 703,T(I),Y(I),Z1,Z2,Z3
639  NEXT I
640  READ K1
641  IF K1=0 THEN 649
642  FOR I=1 TO K1
643  READ K
644  Z1=FNR(K)
645  Z2=Z1+T1*FNF(K)
646  Z3=Z1-T1*FNF(K)
647  PRINT USING 704,K,Z1,Z2,Z3
648  NEXT I
649  END
650  PRINT "CONSTANT TERM =";EXP(A)
651  PRINT "GROWTH RATE ="; B
652  PRINT "CORRELATION COEFFICIENT ="; V
653  PRINT "STANDARD ERROR OF GROWTH RATE =";B1
654  PRINT
655  PRINT
656  PRINT
657  PRINT
658  PRINT USING 701
659  FOR I=1 TO N1
660  Z1=EXP(FNR(X(I)))
661  Z2=EXP(FNR(X(I))+T1*FNF(X(I)))
662  Z3=EXP(FNR(X(I))-T1*FNF(X(I)))
663  PRINT USING 703,T(I),Y(I),Z1,Z2,Z3
664  NEXT I
665  READ K1
666  IF K1=0 THEN 674
667  FOR I=1 TO K1
668  READ K
669  Z1=EXP(FNR(K))
670  Z2=EXP(FNR(K)+T1*FNF(K))
671  Z3=EXP(FNR(K)-T1*FNF(K))
672  PRINT USING 704,K,Z1,Z2,Z3
673  NEXT I
674  END
675  PRINT "CONSTANT =";EXP(A)
676  PRINT "POWER ="; B
677  PRINT "CORRELATION COEFFICIENT = "; V
678  PRINT "STANDARD ERROR OF POWER =";B1
```

```
679  PRINT
680  PRINT
681  PRINT
682  PRINT
683  PRINT USING 702
684  FOR I=1 TO N1
685  Z1=EXP(FNR(X(I)))
686  Z2=EXP(FNR(X(I))+T1*FNF(X(I)))
687  Z3=EXP(FNR(X(I))-T1*FNF(X(I)))
688  PRINT USING 703,T(I),Y(I),Z1,Z2,Z3
689  NEXT I
690  READ K1
691  IF K1=0 THEN 699
692  FOR I=1 TO K1
693  READ K
694  Z1=EXP(FNR(LOG(K)))
695  Z2=EXP(FNR(LOG(K))+T1*FNF(LOG(K)))
696  Z3=EXP(FNR(LOG(K))-T1*FNF(LOG(K)))
697  PRINT USING 704,K,Z1,Z2,Z3
698  NEXT I
699  END
700  :          T                Y              A + BT        UPPER CONF   LOWER CONF
701  :          T                Y             AEXP(BT)       UPPER CONF   LOWER CONF
702  :          T                Y              AT**B         UPPER CONF   LOWER CONF
703  :   #####.####  #####.####  #####.####  #####.####  #####.####
704  :   #####.####              #####.####  #####.####  #####.####
900  REM OPTION: 1 FOR LINEAR REGRESSION; 2 TO REGRESS LOG ON
     LINEAR;
901  REM 3 TO REGRESS LOG ON LOG
1000 REM READ NUMBER OF DATA POINTS, N1
2000 REM READ DATA PAIRS: T,Y
3000 REM READ T VARIATE FOR DESIRED CONFIDENCE INTERVAL, T1
4000 REM READ NUMBER OF EXTRAPOLATE/INTERPOLATE POINTS, K1
5000 REM READ EXTRAPOLATE/INTERPOLATE POINTS
6000 REM THE FOLLOWING DATA ENTRY AUTOMATICALLY TERMINATES THE
     PROGRAM
6001 REM AFTER FITTING THE CURVE AND PRINTING RESULTS
6002 REM IF NO EXTRAPOLATE/INTERPOLATE POINTS ARE TO BE READ.
6003 DATA 0
```

Index

A

Agriculture, 219
  mechanization of, 219
  migration, 220
  productivity, 221
Aircraft
  combat, speed, 71
  composite materials, 99
  parts movable in flight, 81
  transport, productivity, 78
  transport, speed, 101
Aircraft hazard dector, 88
American Woolen Corporation, 199
Analogy, dimensions of, 40
Anomalous data points, 275
Arnold, H. H., 228

B

Baldwin Locomotive Corporation, 199
Benz, Carl, 216
Beverage cans, 241
Binding energy, 131
Boeing Company, 199
Bright, James R., 9, 228
Bubble chambers, 190
Business firm
  impacts from technology, 199
  nature of its business, 193

C

Cable television, 63
Capital formation, U.S., 218
Central Leather Company, 199
Chen, Kuei-Lin, 127
Coates, Joseph F., 215
Codes of ethics for engineers, 234
Consensor, 22
Curtis-Wright Corporation, 199

D

Daimler, Gottlieb, 216
Dalkey, Norman, 22, 23, 25
Diesel locomotive, 235
Dodson, E. N., 97

E

Electric power
  production, 73
  transmission voltage, 191
Engels, Friedrich, 74
Exploratory forecasting, 159

F

Farm population, 219
Federal Communications Commission, 44, 63, 205
Fisher, John C., 60
Floyd, Acey L., 112
Flying wing, 228
Food and Drug Administration, 204
Ford, Henry, 216
Forecast (definition), 1
Forrester, Jay, 125
Functional capability, 3
Fusfeld, Alan R., 103

G
Gompertz, Benjamin, 58
Gordon, Theodore J., 93, 154, 221
Ground effect machines, 213

H
Hadcock, Richard N., 99
Heil, Oskar, 133
Heinlein, Robert A., 132
Helmer, Olaf, 154, 226
Highway debt, 217
Holton, G., 75
Horse population, 259

I
IBM, 231
Ikle, Fred, 238, 239
Innovation, stages of, 9
Interstate Commerce Commission, 205
Investment
  farms, 219
  railroads, 218

J
Jantsch, Erich, 227
Johnson, Lyndon B., 263

K
Kahn, Herman, 148
Kane, Julius, 121
KSIM, 121

L
Lamar, William E., 81
Laser, number of papers about, 112
Lenz, Ralph C., 5, 7, 127
Levassor, E. C., 216
Lilienfeld, J. E., 133
Linstone, Harold, 16, 165
Lockheed Aircraft Corporation, 199

M
Machnic, John, 241
Malthus, Thomas R., 74
Manhattan Project, 39
Mansfield, Edwin, 117
Marchetti, Cesare, 61
Martino, Joseph P., 121
Martino, Theresa, 54
Marx, Karl, 215
  interest rates, 43
Maser, number of papers about, 112
Mass defect, 137
Mass spectrometer, 137
McCormick's reaper, 43
Median, 298
Moderator (Delphi), 18
Monitoring, 134
Munson, Thomas R., 93

N
Nader, Ralph, 233
Neutron, discovery of, 132
Newcomen steam engine, 41
Normative forecasting, 159
Northeast curve, 273
Nuclear fission, 132

O
Occupational Safety and Health Administration, 204
O'Connor, Thomas J., 228

P
Patents
  petroleum-refining, 217
  railroad, 218
PATTERN, 163, 167
Pearl, Raymond, 57
PERT, 231
Petroleum-refining patents, 217
Polaris missile, 231
Precursor events, 134
Productivity, farm, 219
Pry, Robert, 60

Q
Qualitative trends, 81
Quartile, 299

R
Rand Corporation, 16
Rescher, Nicholas, 46

S
Salancik, J. R., 23, 24
Scenario, 148
Schoeffler, Sidney, 126, 238
Seamans, Robert, 76
Sears Roebuck, 199
Solar energy, 123
Stages of innovation, 9

T

Technical approach, 2
Technological breakthrough, 129
Technological forecast, definition of, 2
Technology, 2
  definition of, 1
Turoff, Murray, 16

V

von Karman, Theodore, 228
von Ohain, Hans, 141

W

Watt steam engine, 41
Whittle, Frank, 138
Wiener, Anthony, 148
Wilkinson lathe, 41

X

Xerox, 231

Z

Zwicky, Fritz, 163